Rosetta: The Remarkable Story
of Europe's Comet Explorer

Peter Bond

Rosetta: The Remarkable Story of Europe's Comet Explorer

 Springer

Published in association with
Praxis Publishing
Chichester, UK

Peter Bond
Cranleigh
Surrey
UK

SPRINGER-PRAXIS BOOKS IN SPACE EXPLORATION

Springer Praxis Books
Space Exploration
ISBN 978-3-030-60719-7   ISBN 978-3-030-60720-3   (eBook)
https://doi.org/10.1007/978-3-030-60720-3

Project Editor: David M. Harland

This Springer imprint is published by the registered company Springer Nature Switzerland AG
The registered company address is: Gewerbestrasse 11, 6330 Cham, Switzerland

Contents

This book is dedicated to the thousands of people who committed many years of their careers to the Rosetta mission. Through their efforts, the dream of a European comet chaser became a reality, revolutionizing our knowledge of these once-mysterious cosmic icebergs.

Foreword

Writing just before the Space Age dawned, Roland Barthes described the Citroen DS car as the modern equivalent of a medieval cathedral, conceived by passionate, unknown artists but seen with awe by everybody. I have always felt that the grand missions of space exploration are similar. Like the great cathedrals, they are built by teams of highly skilled people working together, but most of the names of those involved are completely unknown. There is no doubt in my mind that the passion and commitment in the science and engineering teams for these missions must surely reflect that which drove the medieval craftsmen in their skilled tasks.

Peter Bond reports here on the full history of the Rosetta space mission, one of the great steps to explore not only the Solar System as it is today, but also to reveal critical clues to help to decode how it formed. No doubt there were ecclesiastical dreamers behind the conception of a cathedral, and there must have been both sacred and secular authorities whose endorsement and finance had to be secured before it could be built.

It is just so in the grand schemes of space exploration. The Rosetta project involved much politics and lobbying to get the resources required to ensure that everything could come into place. Ultimately, as Barthes said about the cathedrals, it was the craftsmen whose skills and artistry finally delivered the dream, and it fell to the engineering and science teams working together to create the final achievement. Peter's book illustrates how an idea can grow, gather support, surmount obstacles, and eventually achieve a magnificent reality.

Rosetta was a European idea, and one where Europeans had to recognize that they had to be prepared to fall back upon their own resources. Although cooperation with the United States might bring the resources for even grander science, if the US was not ready to join in, Europe needed to go it alone. The European scientists and engineers would have to define what they could achieve with their own resources and, if necessary, accept, on their own, a host of new technical challenges.

It was not simple, and there was much argument and compromise on the way. Big problems needed addressing and resolving on the technical front. Rosetta produced technical advances such as developing solar panels that could operate five times farther from the Sun than Earth, and setting up a European deep space communications network that could monitor a craft far out in the Solar System – continuously, if necessary. Nonetheless, perhaps the most unnerving aspect was putting the spacecraft into hibernation and out of communication for just over two and a half years while it made its way out to rendezvous with its comet.

If the Rosetta project had its share of known challenges to deal with, it also had to face the unexpected. Perhaps the most dramatic event was the decision to delay the launch by a year due to a failure of the Ariane launcher in the month before the planned date in 2003. Having to store the spacecraft presented its challenges, but so did dealing with the cost of the delay, coupled with the fact that comets do not wait for late arrivals. Finding an alternative target became an urgent major task.

The new comet chosen, Churyumov-Gerasimenko, or 67P, turned out to be a very unexpected sight once Rosetta was close enough for imaging. Its resemblance to a 'cosmic duck' grabbed everyone's imagination, but also led to concern within the team as to how stable its internal structure was. However, rendezvous and insertion into orbit were accomplished, Philae was sent down to the surface, and, after a voyage around the Sun, the Rosetta spacecraft itself was deliberately dropped onto the surface at the mission's finale. At that point, I do not know how many people globally felt that a little part of them had been involved in the great adventure. What is clear is that everyone knew it was a great human achievement.

At various times, in the past 30 years, I had my own small part in the great adventure that was Rosetta. I relived many personal memories as I read this book. The Rosetta mission, as a true milestone in European space exploration, has found a very fine chronicler in Peter Bond.

David Southwood

After an academic career as a space scientist, including being Head of Imperial College London's Physics Department (1994-1997), David Southwood joined the European Space Agency in 1997. In 2001, he became ESA Science Director, retiring in 2011. He was president of the Royal Astronomical Society 2012-2014. He is currently a senior research investigator at Imperial College. He is a Fellow of the Royal Aeronautical Society, was awarded the NASA Distinguished Public Service Medal, and won the 2011 Sir Arthur C. Clarke award for space achievement. He is the past chairman of the Steering Board of the UK Space Agency and served on the Board 2011-2019. He received a CBE in the 2019 Queen's Birthday Honours for services to space science and industry in the United Kingdom and Europe.

Acknowledgments

I would like to express my sincere thanks and appreciation to everyone who kindly agreed to assist me in writing what I hope will become the definitive account of the remarkable Rosetta mission.

Most of the information about the early years of the mission's genesis and evolution came from documents and online status reports by the European Space Agency (ESA), NASA, and Daimler Chrysler Aerospace. The details of the mission itself were covered in great depth on the ESA science website and the website of the German Aerospace Agency (DLR), as well as the daily blogs written by Emily Baldwin, Claudia Mignone and Daniel Scuka.

The plethora of scientific results was summarized on the ESA science and exploration web pages and on the blog pages of the Planetary Society, edited by Emily Lakdawalla. Numerous papers detailing analysis of Rosetta's treasure trove of data were also made available in open source issues of leading scientific journals – Science, Nature, Monthly Notices of the Royal Astronomical Society, and Astronomy & Astrophysics.

The vast majority of the illustrations used in the book were provided by ESA, many of them originating with the German OSIRIS camera team. Holger Sierks, the OSIRIS principal investigator, kindly gave me his perspective on the debate regarding the public release of the high resolution images from this wonderful instrument.

I was pleased to be able to correspond with two people with whom I had worked at ESA, and who had played leading roles in the early years of the mission: John Ellwood, who was the project manager prior to Rosetta's launch, and Gerhard Schwehm, who was a leading light in the Giotto mission to Comet Halley and later became Rosetta project scientist. Both of them kindly read through early drafts of chapters and gave me their recollections of key moments in the mission's development.

The current project scientist, Matt Taylor, was also most helpful by reading the chapter about the science results and providing advice on other aspects of the book. Charlotte Götz, one of his colleagues, went out of her way to provide an updated plot of Rosetta's orbital distances after its arrival at Comet 67P/

Churyumov-Gerasimenko. Patrick Martin, the current Rosetta mission manager, kindly provided some historical information.

Sylvain Lodiot, who was deeply involved in Rosetta flight operations for most of the mission, not only read through several draft chapters but willingly gave his time to answer numerous requests for information and clarification.

Stephan Ulamec, the DLR project manager for the Philae lander, was an invaluable help by reading through the sections about the small craft which made history by landing on a comet.

David Southwood, the ESA Director of Science around the time that the Rosetta mission was launched, graciously wrote a Foreword which puts the wonderful achievements of Rosetta in perspective.

I also gratefully acknowledge assistance with the biographical section from Luigi Colangeli, Wlodek Kofman, Thurid Mannel, Stephan Ulamec, John Ellwood and Gerhard Schwehm. Those for Klim Churyumov and Svetlana Gerasimenko are largely based upon interviews that I conducted with them in 2014.

I am grateful to David M. Harland, who not only meticulously edited the manuscript but also improved the quality of several images, as well as to Clive Horwood of Praxis Publishing and Hannah Kaufman at Springer in New York, for their support throughout the development and completion of this endeavor.

Finally, I must thank my wife, Edna, for her support, forbearance and countless cups of coffee.

Preface

By the early 1980s, planetary exploration was dominated by the space superpowers, namely the United States and the Soviet Union. Eager to find a niche research area in which it could make a ground-breaking contribution, the European Space Agency (ESA) decided to focus on the smaller members of the Solar System, the comets and asteroids which represent 'building blocks' left over from the era of planet formation, some 4.5 billion years ago.

ESA's first sortie into in-situ comet research was as a member of an international effort to study Comet Halley, which was returning to the inner Solar System in 1986 after 76 years in the frigid depths of space. Inspired by this once-in-a-lifetime event, ESA, the Soviet Union, and Japan sent an armada of spacecraft (the ESA one being named Giotto) to study the famous intruder at close range. The resulting treasure trove of data transformed the field of cometary research, and provided new insights into the early stages of how the planets came into being.

Even before the accomplishment of this pioneering endeavor was confirmed, ESA and NASA scientists were coming together to discuss the next giant leap in the exploration of comets and asteroids. Their ambitious vision was a landing on the nucleus of a comet to retrieve pristine material and return it to laboratories on Earth for detailed analysis.

As we shall see in the following chapters, the scientists' dream encountered major obstacles, some of which proved to be insurmountable. However, even after the United States pulled out of the comet sample return venture, the ESA Member States decided to press ahead with their own remarkable comet chaser, soon named 'Rosetta'. Despite further obstacles and setbacks, their foresight and commitment produced a truly historic mission.

This is the story of that monumental mission – the people, the hardware and the science that culminated in the unprecedented, close range exploration of a tiny chunk of ice and dust as it swept through space, hundreds of millions of kilometers from Earth. Its scientific results are revolutionizing our understanding of the billions of small, icy objects that populate the Solar System.

Peter Bond
June 2020

1

Comets and Asteroids

When beggars die, there are no comets seen:
The heavens themselves blaze forth the death of princes.
(Shakespeare's Julius Caesar)

By the late-1980s, all of the planets of the Solar System had been visited by space-craft. However, in order to understand the formation and evolution of these worlds, including Earth, scientists were aware that they needed to study the small plane-tary 'building blocks' – comets and asteroids.

Inspired by the once-in-76-years return of Comet Halley, scientists from many nations began to propose new missions and instruments to explore these elusive chunks of rock and ice. In response to this demand, the European Space Agency (ESA) included a planetary cornerstone mission, subsequently named Rosetta, in its new, long-term Horizon 2000 science program.

Although the original plan to land on a comet's nucleus, retrieve samples of pristine material, and bring them back to Earth for analysis was eventually shelved, Rosetta survived as a mission to survey two main belt asteroids *en route* to a ren-dezvous with a periodic comet. After arrival, Rosetta would deploy a small lander on the nucleus and then fly alongside the comet to monitor changes in activity as it entered the inner Solar System and was warmed by the Sun.

This chapter is intended to put Rosetta's ambitious mission into context by describing what we knew of cosmic debris at the time that ESA's comet chaser began its 12-year adventure in March 2004.

© Springer Nature Switzerland AG 2020
P. Bond, *Rosetta: The Remarkable Story of Europe's Comet Explorer*,
Springer Praxis Books, https://doi.org/10.1007/978-3-030-60720-3_1

1.1   COSMIC DEBRIS

Earth is just one out of billions of planets that reside in an enormous spiral galaxy, the Milky Way. In one of the galaxy's spiral arms is an unremarkable star, the Sun, which lies at the center of our Solar System. It is accompanied by eight planets and a handful of dwarf planets, many of which have lesser companions orbiting around them. Less familiar are the swarms of cosmic debris that populate the seemingly empty spaces between the planets. Ranging in size from a few thousand kilometers across to mere specks of dust, these innumerable pieces of ice and rock represent the leftovers from the formation of the planets, some 4.5 billion years ago.

It is generally believed that the Solar System started with the collapse of an enormous cloud of interstellar gas. The trigger for this collapse could have been the passage of an externally generated shock wave from one or more exploding stars – supernovas – that occurred when giant stars in the cloud ran out of fuel and reached the end of their short lives.

Over millions of years, the original cloud may have broken up into smaller segments, each mixed with heavier elements from the dying stars, as well as the ubiquitous hydrogen and helium gas. Once a cloud reached a critical density, it overcame the forces associated with gas pressure and began to collapse under its own gravitational attraction.

The contracting cloud began to rotate, slowly at first, then faster and faster – rather like an ice skater who draws in her arms. Because material falling from above and below the plane of rotation collided at the mid-plane of the collapsing cloud, its motion was canceled out. The cloud began to flatten into a disk, with a bulge at the center where a protostar started to form. The disk could have been thicker at a greater distance from the evolving Sun, where the gas pressure was lower.

The solar nebula would almost certainly have been rotating slowly in the early stages, but as it contracted, conservation of angular momentum would have made it spin faster. This process naturally formed a spiral-shaped magnetic field that helped to generate polar jets and outflows associated with very young stars. Gravitational instability, turbulence, and tidal forces within the 'lumpy' disk may also have played a role in transferring much of the angular momentum to the outer regions of the forming disk.

The center of the protoplanetary disk was heated by the infall of material. The inner regions, where the cloud was most massive, became hot enough to vaporize dust and ionize gas. As contraction continued and the cloud became increasingly dense, the temperature at its core soared until nuclear fusion commenced. As a result, the emerging protostar started to emit copious amounts of ultraviolet

Fig. 1.1: Around 4.5 billion years ago, the infant Sun was surrounded by a rotating disk of dust and gas. Fledgling planets grew as the result of gravitational instabilities and turbulence within the disk, often followed by gigantic collisions. At the end of this process, smaller pieces of debris remained as rocky asteroids and meteorites, or icy comets. (NASA-JPL/Caltech/T. Pyle, SSC)

radiation. Radiation pressure drove away much of the nearby dust, causing the nebula to separate from its star.

The young star may have remained in this so-called T Tauri stage for perhaps 10 million years, after which most of the residual nebula had evaporated or been driven into interstellar space.[1] All that remained of the original cloud was a rarefied disk of dust grains, mainly rocky silicates and ice crystals.

Meanwhile, the seeds of the planets began to appear within the nebula. Rocky, less volatile material condensed in the warm, inner regions of the nebula, while icy grains condensed in the cold, outer regions.

---

[1] T Tauri is a variable star in the constellation of Taurus and is the prototype of the T Tauri stars.

Individual grains collided and stuck together, growing into centimeter-sized particles. These swirled around at different rates, partly due to turbulence and partly due to differences in the drag exerted by the gas. After several million years, these small accumulations of dust or ice grew into kilometer-sized planetesimals and gravitational attraction took over.

The Solar System now resembled a shooting gallery, with objects moving at high speed in a chaotic manner, giving rise to frequent collisions. Some high speed impacts were destructive, causing the objects to shatter, generating a lot of dust or meteoritic debris. Slower, less violent collisions enabled the planetesimals to grow via a snowballing process. Over time, the energy loss resulting from collisions meant that planetary construction became the dominant process.

Eventually, the system contained a relatively small number of large bodies or protoplanets. Over millions of years, these continued to mop up material from the remnants of the solar nebula and collided with each other, producing a small population of widely separated worlds that occupied fairly stable orbits and traveled in the same direction around the young central star.

The largest planets in the Solar System – Jupiter and Saturn – probably formed first. They presumably accumulated their huge gaseous envelopes of hydrogen and helium prior to the dispersal of the solar nebula.

The small, rocky planets formed in the warmer, inner regions of the Solar System, whereas the gaseous and icy giants originated in the outer reaches. Observations of young star systems show that the gas disks that form planets usually have lifetimes of only 1 to 10 million years, which means that the giant gas planets probably formed within this brief period. In contrast, the much smaller, rocky Earth probably took at least 30 million years to form, and may have needed as long as 100 million years.

Theorists believe that for a while the outer planets interacted in a chaotic way, due to mutual gravitational interactions. Jupiter and Saturn may well have migrated inward before reversing direction. Farther from the Sun, the ice giants Uranus and Neptune may also have swapped places.

Vast numbers of small, leftover pieces of rock and ice avoided being swept up during this planet-building process. Any pieces of debris approaching too close to the giant planets would have been deflected either inward, toward the Sun, or outward, into the frigid depths. Some would even have been ejected from the Solar System completely.

Much of the rocky debris was shepherded into the asteroid belt that lies between the orbits of Mars and Jupiter. The overwhelming gravitational influence of Jupiter prevented this material from coalescing into a single planet, so its largest inhabitant, dwarf planet Ceres, has a modest diameter of 965 km; much smaller than Earth's Moon.

Much of the icy debris was removed to a region we now know as the Edgeworth-Kuiper Belt, lying just beyond the orbit of Neptune, 30 to 100 times Earth's distance from the Sun.[2] As a convenient metric for the Solar System, Earth's average distance from the Sun of about 150 million km is known as 1 astronomical unit (AU). Since 1992, dozens of objects, each several hundred kilometers across, have been discovered in this outer belt, as well as many thousands of smaller objects. Dwarf planet Pluto is its largest known member.

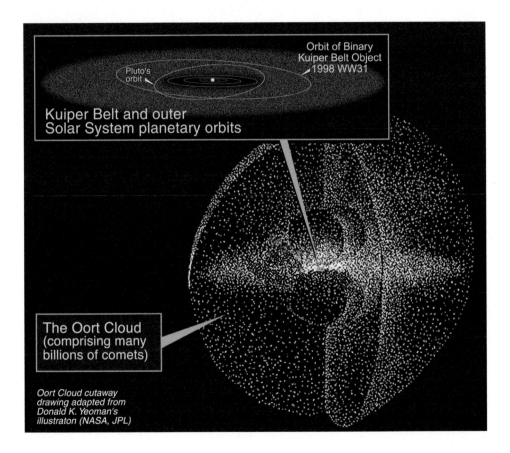

Fig. 1.2: The Oort Cloud is a spherical swarm of icy bodies 2,000 to 100,000 AU from the Sun. The diagram shows its presumed size and shape in relation to the Kuiper Belt and the region inside Pluto's orbit. (STScI/A. Field)

[2] It is named after two astronomers, Kenneth Edgeworth and Gerard Kuiper, who independently suggested the existence of a swarm of comets beyond the orbit of Neptune. The name is usually abbreviated to Kuiper Belt. Much further from the Sun is the Oort Cloud, whose existence was first proposed by Dutch astronomer Jan Oort.

Many billions of icy objects were also ejected even farther, to the so-called Oort Cloud, a vast spherical region that is believed to lie between 2,000 and 100,000 AU.

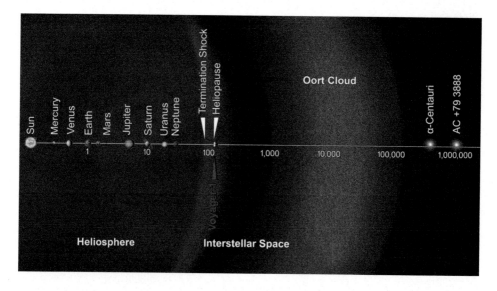

Fig. 1.3: The scale of the Solar System in units of AU, showing the planets, the Kuiper Belt, the Oort Cloud, and two nearby stars. (NASA)

By tracking the orbits of incoming comets, it is possible to determine where they came from. Comets that have fairly short period orbits – less than 200 years – originate in the Kuiper Belt. Those with much longer periods, often taking many thousands of years to orbit the Sun, come from the Oort Cloud. These were ejected into their extremely elliptical or parabolic orbits by gravitational interactions with the young gas giants. This process also scattered objects out of the ecliptic, the plane of Earth's orbit, producing a spherical distribution of the icy population.

Comets and asteroids (together with asteroid fragments known as meteorites) provide clues to the processes that led to the formation of the planets, some 4.5 billion years ago. But comets are the more useful objects for investigating the primordial Solar System. Whereas asteroids formed in the environment between the orbits of Mars and Jupiter, comets formed in the frigid regions much farther out and because their material is much less processed it is much closer to the pristine composition of the early Solar System.

1.2   LONG-HAIRED STARS

Comets are small, ice-rich objects which are most notable for sprouting long tails of gas and dust when their volatiles are vaporized in approaching the Sun. Every year, dozens of comets travel through the inner Solar System, passing close to the

Sun and then returning to whence they came. Most are not visible without the aid of binoculars or a telescope, but, occasionally, a very bright comet may blaze a trail across the night sky.

For thousands of years, these brilliant naked-eye comets have inspired awe and wonder – as anyone who saw the blue gas tail and yellowish dust tail of Comet Hale-Bopp in 1995 or the spiraling tails of Comet C/2006 P1 (McNaught) can testify.

Fig. 1.4: Comet Hale-Bopp, discovered by Alan Hale and Thomas Bopp on 23 July 1995, was one of the 'great comets' of the 20th century. As it approached the Sun from the Oort Cloud, it became extremely bright and active, developing a bluish ion tail some 8 degrees long and a yellowish dust tail 2 degrees long. The nucleus was estimated to be 35 to 40 km in diameter, which is huge compared with most comets that reach the inner Solar System. (ESO/Eckhard Slawik)

Many ancient civilizations saw these sudden apparitions as portents of death and disaster, and omens of social and political upheavals. Shrouded by luminous comas with tails streaming behind them, these 'long-haired stars' were assigned the name 'comets' by the ancient Greeks (from their word 'kome' meaning 'hair').

1.3   HALLEY AND PERIODIC COMETS

By the beginning of the 18th century, it was understood that comets were celestial objects that appeared without warning, illuminated the skies for several weeks or months as they moved closer to the Sun and then withdrew, presumably never to be seen again.

However, our understanding of the nature of comets was revolutionized by the British astronomer Edmond Halley (1656-1742). In 1705, when Halley began to calculate the orbits of 24 comets, he noticed that the path followed by a bright comet observed in 1682 was very similar to the orbits of other bright comets recorded in 1607 and 1531. He concluded the only reasonable explanation was that the same comet had reappeared over a period of 75-76 years. The slight variations in the timing of each return were attributed to small gravitational tugs on the comet by the giant planets.

Working forward in time, Halley predicted that the comet should return again in December 1758. Although he did not live to see the event, his theory was proved correct when the comet duly reappeared on schedule. The first periodic comet to be recognized was named 1P/Halley in his honor.[3]

Trawls through ancient records have revealed that this famous comet was recorded by the Chinese as long ago as 240 BC. It was later given a starring role in the Bayeux Tapestry – which told the story of the Norman Conquest of England in 1066 – and it may have inspired Giotto to include a comet in his 14th century painting, 'Adoration of the Magi'.

Since Edmond Halley's first successful prediction of a comet apparition, almost 400 periodic comets have been discovered and confirmed. They all follow recurring, elliptical orbits which last less than 200 years, but a large proportion of them have orbits that have been modified by close encounters with Jupiter, whose gravity dominates the Solar System.

Consequently, the farthest points of their orbits (aphelia) lie fairly close to the orbit of Jupiter, typically about 6 AU from the Sun. Each solar orbit takes about six years, although their paths are always being deflected by Jupiter and other planets. One of these Jupiter family comets is 67P/Churyumov-Gerasimenko, the target of Europe's Rosetta mission (see Chapter 6).

The shortest period belongs to Comet 2P/Encke, which races around the Sun every 3.3 years. Some 150 known comets, including Halley's, follow a more leisurely route, traveling beyond the orbit of Neptune prior to returning to the inner

---

[3] The letter P after the number denotes a periodic comet.

Fig. 1.5: An image of Comet 1P/Halley taken on 8 March 1986 by W. Liller, as part of the International Halley Watch. Note the large dust tail and ion tail. (NASA/W. Liller)

Solar System. Although these comets have also been perturbed by encounters with the giant planets, their orbits are more random and are often steeply inclined to the ecliptic. Many of these, including Halley, travel in a retrograde direction.[4]

The orbits of periodic comets have evolved greatly since they were first formed. Comets with orbits of less than 200 years are believed to have originated in the Kuiper Belt, the doughnut-shaped region which ranges from the orbit of Neptune out at least 50 AU. They were probably ejected to their present location billions of years ago by gravitational interactions with Uranus and Neptune. Since the first Kuiper Belt Object was discovered in 1992, many hundreds more have been found.

As mentioned, the census of comets is increased when newcomers arrive from the depths of space, far beyond the Kuiper Belt. These intruders from the Oort Cloud, such as Hale-Bopp, appear without warning, moving along parabolic paths at high speeds. After sweeping rapidly around the Sun, they head back out, where they will remain for thousands of years.[5]

---

[4] In terms of orbits, retrograde means 'backward' or clockwise when viewed from the north celestial pole.

[5] Occasionally, objects may enter our Solar System from interplanetary space. Traveling on hyperbolic paths, their velocities are so great that the Sun's gravity cannot capture them. Two of these have been discovered in recent years.

1.4 DIRTY SNOWBALLS?

Although comets had been studied by ground-based telescopes for more than three centuries, we had little idea what they were made of, or where they came from, until the introduction of photography and the spectroscope.

The problem was that it is impossible to observe a comet's tiny nucleus from Earth. Even for the largest comets, such as Hale-Bopp, this icy heart measures only about 35 km in diameter. Furthermore, as soon as one of the wandering chunks of ice was close enough to make detailed observation, it was obscured by a coma of gas and dust. However, the growth of a coma and gas and dust tails as the nucleus was warmed by the Sun led to the reasonable hypothesis that the nucleus was a mixture of volatile ices and rocky material.

The key breakthrough came with the introduction of spectroscopy – a method of analyzing the light from the coma and tail. As early as the 1860s, the presence of compounds of hydrogen (H) and carbon (C) was revealed. Nitrogen (N) was also a common constituent.

Over the next century, spectral analysis of cometary gas revealed neutral molecules of CH (methylene), CN (cyanogen), and C_2 (carbon) beyond the orbit of Mars. Inside the orbit of Mars, the spectra included ionized (i.e. electrically charged) molecules (CO^+, N_2^+ and OH^+), along with CH_2 and NH_2. As the comets passed inside Earth's orbit, spectral lines for metallic elements such as sodium, iron and nickel began to be detected.

The most popular theory about the nature of comets was put forward in 1950 and 1951 by the American astronomer Fred Whipple, who is widely regarded as the 'grandfather' of modern comet science. Aware that some periodic comets must have made thousands of orbits around the Sun, he realized that they would have broken apart if they had comprised only a large pile of sand mixed with hydrocarbons.

Whipple concluded that comets were like dirty snowballs – large chunks of water ice and dust mixed with ammonia, methane and carbon dioxide. As the snowball approached the Sun, its outer ices started to vaporize, releasing large amounts of dust and gas that, in turn, formed the characteristic tails. He assumed that water vapor released from sublimating water ice was the main propulsive force behind the jets of material seen to originate on comet nuclei, but later data indicated that it is solar heating of frozen carbon dioxide beneath the surface that powers the jets of material that erupt from comet nuclei.

By the mid-1980s, when the Rosetta mission was being proposed, it was known that cometary nuclei were often amongst the blackest objects in the Solar System, despite their bright comas and tails. This is because the nucleus is coated in dark organic (carbon-rich) material, and dust is apparently thoroughly mixed with the ices inside. Scientists began to regard comets more as 'icy dirtballs' than 'dirty snowballs'.

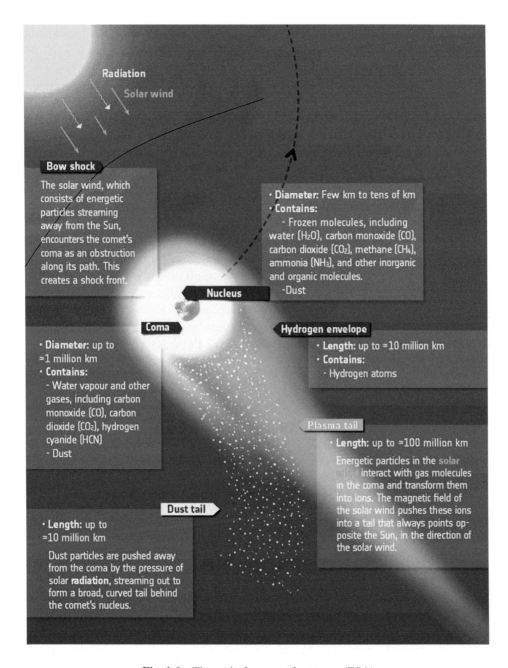

Radiation

Solar wind

Bow shock

The solar wind, which consists of energetic particles streaming away from the Sun, encounters the comet's coma as an obstruction along its path. This creates a shock front.

Nucleus

· **Diameter:** Few km to tens of km
· **Contains:**
  - Frozen molecules, including water (H_2O), carbon monoxide (CO), carbon dioxide (CO_2), methane (CH_4), ammonia (NH_3), and other inorganic and organic molecules.
  -Dust

Coma

· **Diameter:** up to ≈1 million km
· **Contains:**
  - Water vapour and other gases, including carbon monoxide (CO), carbon dioxide (CO_2), hydrogen cyanide (HCN)
  - Dust

Hydrogen envelope

· **Length:** up to ≈10 million km
· **Contains:**
  - Hydrogen atoms

Plasma tail

· **Length:** up to ≈100 million km

Energetic particles in the solar wind interact with gas molecules in the coma and transform them into ions. The magnetic field of the solar wind pushes these ions into a tail that always points opposite the Sun, in the direction of the solar wind.

Dust tail

· **Length:** up to ≈10 million km

Dust particles are pushed away from the coma by the pressure of solar **radiation**, streaming out to form a broad, curved tail behind the comet's nucleus.

Fig. 1.6: The main features of a comet. (ESA)

Each time a comet approaches the Sun, it loses some of its material and mass. During its peak activity, near the Sun, Comet Halley was losing about 20 tonnes of gas and 10 tonnes of dust every second from seven jets of vaporized ice erupting from its nucleus.

Over time, a nucleus is depleted until all of its ices have been vaporized, at which point it may become inactive, resembling a small rocky asteroid. Alternatively, the comet might fragment into a swarm of dust particles.

Measuring the density of a nucleus is not easy, even by monitoring the trajectory of a nearby spacecraft, but estimates for various comets indicate they are typically 0.3-0.5 g/cm^3, which is considerably less than the density of water. This is probably due to a largely icy composition in combination with a porous, fluffy texture, or perhaps to a 'rubble pile' structure containing large voids.

Despite their insubstantial nature, their high impact velocity enables comets to cause a lot of damage if they collide with another object. Craters created by ancient comet and asteroid impacts can still be seen on the Moon, Mercury, Earth, and many planetary satellites.

In the case of Earth, only the largest nuclei survive to strike the ground and excavate a large crater. Most break apart in the atmosphere and explode in an enormous airburst that sends out shock waves in all directions. One of the most famous examples occurred on 30 June 1908, when an object, most likely a comet, exploded above the Tunguska region of Siberia and the blast flattened trees for a radius of hundreds of kilometers. If such an event were to take place above a conurbation such as London, the entire city would be flattened.

Fig. 1.7: This photo taken in 1927 shows parallel trunks of trees that were flattened by the shock wave from the 'Tunguska Event'. Note how the branches have been stripped off the trees. (ESA)

The most spectacular example of a comet collision occurred in 1994 when some 20 fragments of Comet Shoemaker-Levy 9 plunged into Jupiter, leaving a string of dark 'bruises' where the icy chunks exploded in the atmosphere.

Comets (and asteroids) may also have provided much of the water which now forms Earth's oceans, and possibly even delivered the complex organic chemicals that gave rise to the first primitive life forms.

1.5 TRANSIENT TAILS

Comets spend most of their lives far from the Sun, when they are invisible to even the largest instruments. However, any comet that enters the inner Solar System develops a shroud of gas and dust known as the coma. The roughly spherical coma is fed by jets of material that erupt into space as the surface of the nucleus is warmed by solar radiation.

The coma is mainly composed of water vapor and carbon dioxide. Some comas display the greenish glow of cyanogen (CN) and carbon when illuminated by sunlight. Other compounds of carbon, hydrogen and nitrogen have been found. Ultraviolet images by spacecraft have also shown that the visible coma is surrounded by a huge, sparse cloud of hydrogen gas.

Fig. 1.8: Comets travel around the Sun in highly elliptical orbits, and when they venture into the inner Solar System the warmer environment causes volatiles in the nucleus to vaporize to produce a dense coma and tails of gas and dust. The tails always point away from the Sun. (Wikimedia https://en.wikipedia.org/wiki/Comet#/media/File:Cometorbit01.svg)

If the production of rates of dust and gas are sufficient, a comet can develop several tails. One is the yellowish dust tail. Usually broad, stubby and curved, these are formed when tiny dust particles in the coma are pushed away by solar radiation pressure, as photons of light impact the grains. Meanwhile, the gases released by vaporization of the nucleus are ionized by solar ultraviolet light. The ions are influenced by the magnetic field associated with the solar wind, a flow of electrically charged particles emanating from the Sun. The ions are swept out of the coma to produce a long, distinctive ion tail (also called a gas or plasma tail). Because the most common ion (carbon monoxide) scatters blue light better than red light, ion tails often appear blue to the human eye (see Figure 1.4).

Gusts in the solar wind can cause the ion tail to swing back and forth, sometimes developing temporary ropes, knots and streamers that can break away and then reform. These features are not seen in the dust tail. The ion tail is usually narrow and straight, often streaming away from the nucleus for many millions of kilometers. In 1998, analysis of data from the Ulysses probe indicated it had passed through the ion tail of Comet Hyakutake at the remarkable distance of 570 million km from the nucleus.

Fig. 1.9: Comet C/2006 P1 (McNaught) provided a spectacular sight close to the horizon in the southern hemisphere in January and February 2007. At least three jets of gas and small dust particles were seen to spiral away from the nucleus as it rotated, stretching over 13,000 km into space. The larger dust particles, which were ejected on the sunlit side of the nucleus, followed a different pattern. They produced a bright fan, which was then blown back by the pressure of sunlight. (ESO/Sebastian Deiries)

One of the most characteristic features of a comet's tail, is a shift in its alignment as the comet pursues its orbit. The solar wind sweeps past a comet at about 500 km/s, shaping the tails and making them point away from the Sun, particularly the ion tail. As a result, on the outward leg of its orbit, the solar wind causes the tails of a comet to point ahead of it, not trail behind it.

In extreme cases, comets have been observed to lose their tails temporarily when subjected to strong gusts in the solar wind. In 2007, NASA's Stereo spacecraft observed the collision of a coronal mass ejection (CME) – a huge cloud of magnetized gas ejected by the Sun – and the tail of Comet Encke, which was cut in two. This was triggered by a process known as magnetic reconnection, when the magnetic fields around the comet and the CME were spliced together.

When Earth passes through streams of material that are strewn along comets' orbits, the tiny particles burn up on entering the atmosphere, creating short luminous trails known as meteors or 'shooting stars'. More than twenty major meteor showers occur around the same time each year (see Table 1.1), with the shooting stars appearing to radiate from a point in the sky, like the spokes of a wheel.[6]

Table 1.1: Major Meteor Showers

Shower	Dates	ZHR*	Parent Comet
Quadrantids	Jan 1-6	100	96P Macholz 1?
Lyrids	Apr 19-25	10-15	C/1861 G1 Thatcher
Eta Aquarids	Apr 24-May 20	50	1P Halley
Delta Aquarids	Jul 15-Aug 20	20-25	96P Machholz 1?
Perseids	Jul 25-Aug 20	80	109P Swift-Tuttle
Orionids	Oct 15-Nov 2	30	1P Halley
Leonids	Nov 15-20	100	55P Tempel-Tuttle
Geminids	Dec 7-15	100	Asteroid 3200 Phaethon

*Approximate zenithal hourly rate

One of the best known showers is the Orionids, whose peak occurs in October. This stream of debris originated from Halley's Comet and the meteoroids penetrate the Earth's atmosphere at 237,000 km/h, which is faster than every other major annual shower apart from the Leonids in November. The Leonids are associated with dust from Comet 55P/Tempel-Tuttle. When that comet approaches the Sun, the Leonids can be spectacular. The displays from the apparitions in 1833 and 1966 produced over 100,000 meteors an hour.

---

[6] Sporadic meteors may also appear at any time and from any direction throughout the year.

Fig. 1.10: This drawing shows the famous Leonid meteor storm of 12 November 1833, when the skies over the United States were ablaze with shooting stars. It was, one eye-witness said, as if "a tempest of falling stars broke over the Earth… The sky was scored in every direction with shining trails and illuminated with majestic fireballs." (ESA)

Sometimes Earth passes directly through a comet's tail, as happened during the apparition of Comet Halley in 1910. Many people were alarmed, since it was known that the tail contained cyanide (CN), a highly poisonous gas. But despite the doom-laden prophecies, the event had no noticeable effects. This was hardly surprising, since the tails of comets are so insubstantial that it is possible to observe stars through them. It was rather like hurling a bowling ball into a cloud of cigarette smoke.

1.6   BREAKING UP IS EASY TO DO

As long as they remain in the frigid depths of the Solar System, where temperatures plummet to −230°C, comets can survive intact for billions of years. But when they are nudged toward the Sun their nuclei can become susceptible to disruption. The fragility of a nucleus, owing to its porous or fractured nature, was indicated by objects that split into several pieces and often failed to reappear at their next expected return.

Comet 73P/Schwassman-Wachmann 3 began to splinter in 1995, during one of its numerous ventures into the inner Solar System. Shortly after experiencing a major outburst of activity, it split into four nuclei. The comet subsequently disintegrated into dozens of fragments. Hubble Space Telescope images indicated a hierarchical destruction process was taking place as large pieces continued to break up. Dozens of 'mini-fragments' were observed trailing behind each main object.

Occasionally, close encounters with planets can have a similar effect. One famous example is Comet Shoemaker-Levy 9, which was pulled apart by the tidal forces of Jupiter's tremendous gravity in 1992, forming a chain of around 20 pieces that crashed into the planet in July 1994.

An even more catastrophic example was Comet LINEAR (C/1999 S4), which disintegrated on 25 July 2000 during its first passage through the inner Solar System. Having appeared completely normal, the comet rapidly evolved into a fuzzy, extended, and much fainter object. By early August, all that could be seen of it was a cloud of debris, with no sign of the nucleus or any active fragments larger than a few meters across. It seemed that rapid vaporization near perihelion had caused the small nucleus to run out of ice, and with nothing to bond the solid material together it started to fall apart, leaving behind a loose conglomerate of particles that dispersed into space.

Comet nuclei can also be broken up by rapid rotation, thermal stresses as they pass near the Sun, or explosive disruption caused by trapped gases under pressure suddenly breaking out.

One example was Comet 17P/Holmes, which exploded on two separate occasions, firstly in November 1892 and again then in October 2007 as it approached the asteroid belt. During its 2007 apparition it unexpectedly changed from magnitude 17 to 2.8, thereby becoming almost a million times brighter in only 42 hours

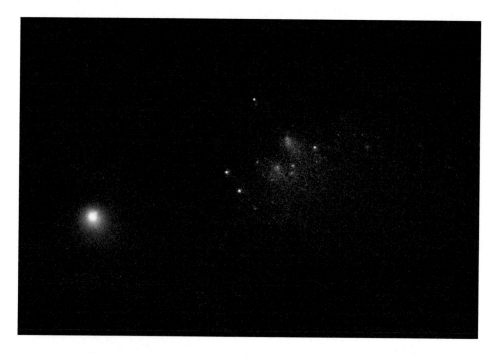

Fig. 1.11: This Hubble Space Telescope image, obtained on 27 January 2016, reveals Comet 332P/Ikeya-Murakami disintegrating as it approaches the Sun. The main nucleus (lower left) measures about 490 meters across. The debris, visible near the center of the image, comprises a cluster of about 25 pieces, roughly 20-60 meters wide. They form a 4,800 km long trail of fragments that is drifting away from the comet. At that time, the comet was 240 million km from the Sun, slightly beyond the orbit of Mars. The small comet seems to be breaking up as jets of material erupting from the nucleus speed up its rate of rotation. (NASA, ESA, and D. Jewitt/UCLA)

and enabling it to be observed by the naked eye. The gradually expanding coma eventually grew larger than the Sun before finally dissipating.

Infrared observations in November 2007 by the Spitzer Space Telescope revealed a lot of fine silicate dust. This was apparently created by a violent explosion that destroyed larger particles in the interior of the nucleus. Calculations indicated the energy of the blast to be equivalent to 24 kilotons of TNT, and the total mass of the ejected material to be about 10 million tonnes.

The presence of jets and a spherical cloud (and particularly two similar events more than a century apart) indicated a collision was an unlikely reason. Thermal stresses are also unlikely to have been the cause, as in both cases the comet was well past perihelion and heading away from the Sun. It appears the nucleus acted like a pressure cooker, with trapped gases suddenly erupting through weaknesses in the surface.

Some comets undergo violent eruptions far beyond the warmth of the Sun. Perhaps the most impressive was the discovery on 15 February 1991 that Halley's

Comet was surrounded by a coma at least 300,000 km in diameter. This indicated a tremendous outburst. At the time, the object was 14.3 AU from the Sun, so this renewed activity was explained as sublimation of a more volatile compound than water, perhaps carbon monoxide or carbon dioxide, which built up sufficient pressure beneath the dark crust to trigger a major outburst.

1.7   VERMIN OF THE SKIES

Throughout most of recorded history, scientists and astrologers thought there were six planets (including Earth) in the Solar System. No one dreamt that there might be other worlds beyond Saturn, but there were occasional suggestions of one or more objects lurking unseen between the planets. Johannes Kepler, the 17th century mathematician who made a major advance by formulating the laws of planetary motion, was one who believed that another world must exist between Mars and Jupiter.

Then, in 1772, the German astronomer Johann Bode drew attention to a simple arithmetic progression which matched the distances of the planets from the Sun fairly accurately.[7] This so-called Bode's Law involved the series 0, 3, 6, 12, 24, and so on. Each successive number after 3 was double the previous number. After adding 4 to each number and dividing by 10, the sequence gave a good approximation to the actual distances of the known planets in units of AU (see Table 1.2).

Table 1.2: The Titius-Bode Law

Planet	Bode's Distance	Actual Distance from Sun (AU)
Mercury	0.4	0.39
Venus	0.7	0.72
Earth	1.0	1.0
Mars	1.6	1.52
(Ceres)	2.8	2.76
Jupiter	5.2	5.20
Saturn	10.0	9.53
Uranus	19.6	19.19
Neptune	38.8	30.06

Nine years later, in 1781, William Herschel in England discovered the planet Uranus more or less where the Titius-Bode Law predicted it ought to be. This

[7] Bode failed to mention that this mathematical curiosity had first been pointed out six years earlier by Johann Titius!

'confirmation' of the arithmetic progression encouraged others to search for new worlds.[8]

In 1800, a group of six astronomers, headed by Johann Schröter, gathered in the German town of Lilienthal with the intention of beginning a systematic search for a planet whose orbit lay at Bode's empty location of 2.8 AU, between the orbits of Mars and Jupiter. Their plan was to search the ecliptic, because the orbital planes of all the known planets were closely aligned to this. They divided it into 24 sectors, each of which was to be carefully scanned by a different observer. However, before the 'celestial police' could start their grand survey, some exciting news arrived from Sicily.

On 1 January 1801, Giuseppe Piazzi, director of the Palermo observatory, had discovered a mysterious, star-like object wandering among the stars of the constellation of Taurus. Initially he presumed it to be a comet, but the absence of a gaseous coma soon prompted doubts.

Unfortunately, on 11 February, Piazzi became seriously ill, and by the time he had recovered, the mysterious object had disappeared in the Sun's glare. With only limited information to go on, frustrated astronomers all across Europe began a frantic search for the mysterious object.

The breakthrough came when a brilliant young mathematician, Carl Gauss, devised a way of predicting planetary positions from a limited set of observations. Using Gauss's calculations, Franz Xavier von Zach, a member of Schröter's team, relocated it on 31 December, just half a degree from where Gauss had predicted.[9]

To everyone's delight, the newcomer was in an almost circular, planet-like path between the orbits of Mars and Jupiter. Piazzi suggested the name Ceres, the Roman goddess who was the patron of Sicily, which was soon adopted. Its distance from the Sun of 2.77 AU was an almost precise match for the empty 2.8 AU slot in Bode's Law. The one concern was that at only 8th magnitude Ceres had to be very small, more of a minor planet than a fully-fledged member of the Sun's family.

Then, on 28 March 1802, Wilhelm Olbers (another member of the 'celestial police') found a second object at Bode's distance of 2.8 while searching for comets. It was in an eccentric and inclined orbit. Named Pallas, it appeared to be even smaller than Ceres.

It was clear that these minor planets were rather different from their larger neighbors. Having noticed that they lacked planet-like disks even through his largest telescope, William Herschel suggested that they be collectively called

---

[8] The relationship was not infallible. Neptune, which was only discovered in 1846, does *not* coincide with the predicted number. However, the dwarf planet Pluto orbits quite close to location 38.8.

[9] For comparison, the disk of the Moon is half a degree in diameter.

'asteroids' due to their star-like appearance. Olbers suggested they were remnants of a larger body that had exploded.

Further possible fragments were discovered by Karl Harding (an assistant of Schröter) in 1804 and by Olbers in 1807. Named Juno and Vesta, they were smaller than Pallas. The orbit of Vesta did not even approach those of the other three asteroids.

The scattered family seemed complete for nearly 40 years until a fifth minor planet (Astraea) was found in 1845, followed by three more in 1847. Since then, at least one asteroid has been discovered each year, with a marked increase in the discovery rate following the introduction of photographic techniques in the late 19th century.

Asteroids came to be regarded as small, unimportant rocky objects of little scientific interest. They appeared as unwanted trails on long photographic exposures, causing one astronomer to dub them "the vermin of the skies".

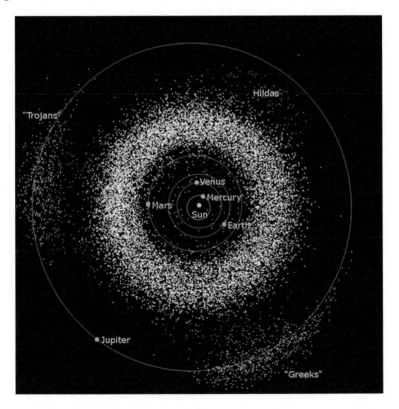

Fig. 1.12: The vast majority of the asteroids travel around the Sun in the main belt (white) in the gap between the orbits of Mars and Jupiter. The two 'clouds' of objects (green) which are centered 60 degrees ahead of and behind Jupiter in its 12 year orbit, are dubbed the Trojans and the Greeks. Also shown are the Hildas (orange) and some asteroids that approach (and in some cases cross) Earth's orbit. (Wikimedia https://en. wikipedia.org/wiki/Asteroid#/media/File:InnerSolarSystem-en.png)

Most of the asteroids proved to orbit the Sun between Mars and Jupiter, a region termed the main asteroid belt. In the early 20th century, two large groups of asteroids, collectively known as Trojans, were found to share Jupiter's orbit, on average leading it by 60 degrees and trailing it by 60 degrees.

Although most astronomers lost interest in the thousands of seemingly unimportant lumps of rock in the broad zone between the four terrestrial planets and their giant, gaseous cousins, the discovery of (433) Eros, the first of the so-called near-Earth asteroids, in 1898, caused quite a stir. As subsequent discoveries would establish, there are thousands of potentially hazardous objects capable of striking Earth (see The Impact Threat below).

1.8   ASTEROIDS

With a diameter of 970 km, Ceres, the biggest member of the asteroid belt, is large enough to have achieved an almost spherical shape. It accounts for about 25% of the overall mass of the belt. A handful of other asteroids are also nearly spherical, but the vast majority are small and irregular in shape.

Asteroids that have been visited by spacecraft or imaged using ground-based radar are often double-lobed. Those with two components in contact at a narrow waist were evidently formed when separate asteroids gently collided and became gravitationally bound.

However, many asteroids appear to be very weak internally, rather like clumps of rubble that are loosely held together by gravity, and they may be readily broken apart by more energetic collisions. Even an increase in their rotation rate may be sufficient to break them asunder.

Close binaries are quite common, with two objects circling their common center of gravity. A few triples have also been discovered. Many of the meteorites arriving on Earth are thought to have originated from the shattering of asteroids (see Meteorites below).

Asteroids have been classified according to the spectra of their reflected sunlight. The most common are the C-type of carbon-rich or carbonaceous asteroids. They are more numerous in the middle and outer regions of the main belt. They are very dark and black-brown in color.

The next most common group is the S-type of stony or silicaceous objects that are composed of metal-rich silicates. These predominate in the inner main belt, at solar distances of less than 2.4 AU. They reflect more light than the C-types.

The third most populous group is the M-type of metallic asteroids. These are believed to be derived from the nickel-iron cores of objects which were large enough to melt internally and then differentiate, with the denser metals sinking

Fig. 1.13: Asteroids vary considerably in composition, shape and size. Top left: The S-type (433) Eros measures 33 × 13 × 8 km. Top right: The C-type (253) Mathilde is 59 × 47 km, and is darker than asphalt. Images from the NEAR spacecraft by NASA-JPL. Center right: Diamond-shaped (162173) Ryugu is a C-type asteroid less than 1 km wide. Bottom left: The peanut-shaped S-type near-Earth asteroid (25143) Itokawa is 535 × 294 × 209 meters. Images from the Hayabusa project by JAXA.

toward the center. When such objects were broken apart by violent collisions, some of the resulting asteroids were fragments of the metal cores. Based on its reflectance spectrum, asteroid (21) Lutetia, one of the targets of the ESA Rosetta mission, was either a C-type or an M-type.

Many other types and subtypes are recognized. For example, the other asteroid target for the Rosetta mission, (2867) Steins, was classified as a rare E-type object, based on its high albedo (reflectivity) and its visual and near-infrared spectral characteristics.

1.9   METEORITES

Every day, about one hundred tons of interplanetary material drifts down to Earth's surface. Most numerous are the tiny dust particles that are released by comets as their ices vaporize in the solar neighborhood. However, larger pieces of rock also arrive from deep space and some of these survive to reach the surface in the form of meteorites.

Most meteorites come from the main asteroid belt, between Mars and Jupiter. The first clear evidence of this link between asteroids and meteorites came in 1959, when scientists were able to photograph an incoming meteorite with sufficient accuracy to calculate its orbit. Since then, many other meteorites have been traced back to the asteroid belt.

In the 1980s, scientists recognized that a few rare meteorites found on Earth have been ejected by impacts on the Moon and Mars. These provide invaluable information about the chemical composition of the Solar System, past and present, and they help scientists to understand the composition and nature of the interstellar cloud in which our star and planets were born.

The most common meteorites are stony, some are metallic, some are mixtures of stone and metal, and a few fragile specimens are rich in carbon. The type of meteorite is related to the type and size of the asteroid that was its parent.

The iron-rich meteorites are derived from the molten cores of large asteroids that were at least several tens of kilometers across. The heat came from radioactive decay of elements and large impacts with other objects.

The core formed where denser material sank toward the center, whereas silicate rock rose to create a less dense, rocky crust. Material of intermediate buoyancy produced the mantle. This process is known as differentiation.

Iron meteorites are made of varying proportions of iron and nickel alloys. Their existence is evidence of the catastrophic destruction of large, differentiated M-type asteroids by collisions.

Stony-iron meteorites are thought to have originated at the core-mantle boundary inside large asteroids. Stony meteorites were formed either in undifferentiated

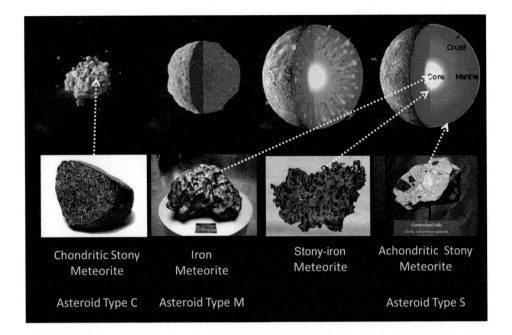

Fig. 1.14: When an asteroid or a protoplanet accretes sufficient material, it begins to become roughly spherical in shape. Heat from radioactive decay of elements and numerous collisions causes melting. In a process known as differentiation, the densest, metal-rich material sinks to the core, and the lightest rock floats to the surface. The type of meteorite derived from one of these asteroids depends on the part of the asteroid where it originated. (After the Smithsonian Museum of Natural History)

objects or at the surface of a differentiated asteroid, where constant impact bombardment caused the rocks to be mixed. Primitive stony meteorites are called chondrites because they contain spherical grains known as chondrules, which consist of silicates and are typically about 1 mm in diameter.

1.10 THE IMPACT THREAT

As the numbers of known near-Earth objects steadily grew, awareness of the potential threat spread beyond the scientific community and, in 1994, the U.S. Congress gave NASA the task of discovering, within a decade, 90% of all near-Earth objects (NEOs) larger than 1 km in size. The impact of such an object could cause global-scale devastation sufficient to send humanity the way of the dinosaurs.

Fig. 1.15: On 28 February 2009, Peter Jenniskens, an astronomer at NASA Ames Research Center and the SETI Institute in California, found two pieces of the 2008 TC3 asteroid, an SUV-sized object that broke apart over the Nubian Desert of northern Sudan in 2008. This was the first time scientists had the opportunity to study meteorites definitively linked to a particular asteroid that exploded on entering the atmosphere. (NASA/ SETI/P. Jenniskens)

To follow this up, Congress issued a new challenge in 2005 by instructing NASA to discover 90% of NEOs with diameters larger than 140 meters by 2020. Their estimated average impact rate is about 10,000 years, which equates to a 1 in 100 chance every 100 years.

Although considerably less destructive than a 1 km impactor, such an object would be capable of destroying a large city, laying waste to a country-sized area, or producing a tidal wave that would inundate low lying coastal areas.

Although they are much more numerous, NEOs in this more modest size range are not easy to detect or to characterize, because most of them are spotted

Fig. 1.16: Meteor Crater (also called Barringer Crater) in Arizona, is one of the most recent impacts on Earth. It was created about 50,000 years ago when an iron meteorite excavated a hollow about 1.2 km wide and 180 meters deep. Surrounding the basin is a wall of material 30-45 meters high where the target rock was uplifted and, in some cases, overturned. The 30 meter wide meteorite probably weighed about 100,000 tons and struck the surface at a speed of around 12 km/s. The energy released was equivalent to about 2.5 megatons of TNT. (D. Roddy/USGS, Lunar and Planetary Institute)

only when they pass close to Earth. To date, only about 40% of the smaller members of this group have been found, together with around 80% of the larger members.

Meanwhile, every month small chunks of rock fly within a few Earth-Moon distances of our planet, with the possibility that, at some time in the future, they will collide with Earth. At the time of writing (in 2020) 2,018 of these potentially hazardous objects had been detected.

The detection of as many NEOs as possible, and the determination of their orbital paths, sizes, shapes and compositions will be essential if we are to avoid a catastrophic impact, but there is not yet an accepted strategy for preventing such an event.

As dramatized in Hollywood blockbusters, the most obvious solution would appear to be to launch a missile carrying a nuclear warhead to break up a large asteroid. Unfortunately, this is likely to create a swarm of smaller objects that would continue on a similar course and cause multiple impacts across the world, delivering a similar total energy release. It would be rather like the fragments of Comet Shoemaker-Levy 9 colliding with Jupiter.

The less drastic alternative of deflecting an object from its collision course has been widely studied. In most cases, a change in orbital speed of several centimeters per second would be sufficient. Deflection methods generally call for either a sudden, fairly large lateral force, or a slow, relatively gentle, but prolonged thrust.

Once again, nuclear devices have been suggested because they can supply around a million times more energy per unit of mass than conventional explosives. Impacts by non-explosive projectiles have also been proposed.

A more subtle approach is the so-called 'gravity tractor', whereby the minuscule gravitational pull provided by a nearby thrusting spacecraft would very slowly accelerate the asteroid in the spacecraft's direction. Although the acceleration imparted to the asteroid would be very small, this method would be able to alter an asteroid's orbit slightly. Calculations show that a 20 ton gravity tractor could deflect a 200 meter asteroid after a year of such 'towing'.

Another innovative proposal involves placing an electromagnetic 'mass driver' on the surface of a NEO. Material excavated on the asteroid's surface would be fired into space, creating an equal and opposite reaction that adjusts the orbit. A similar concept involves anchoring either ion engines or a solar sail on an asteroid, in order to apply a weak propulsive force over a long period of time.

The need for the difficult task of anchoring such devices can be overcome if a solar sail or a mirror can be placed in orbit. By focusing incoming solar radiation onto the NEO, some of the surface would be strongly heated and vaporized, with the resultant jet of material adjusting its trajectory.

Whichever method is chosen, it should be borne in mind that a deflection mission could take years to develop and another 5 years to reach its target. Furthermore, the propulsive procedure itself may require many years to have the desired effect. Hence the threat must be recognized well in advance.

1.11   WHY STUDY COSMIC DEBRIS?

As the innumerable impact craters on our Moon demonstrate, Earth has been bombarded by comets, asteroids and meteorites for billions of years. Although the storm has abated in more recent times, there are close encounters on a weekly basis. At least once per century, an object large enough to cause widespread damage, and possibly loss of life, enters the atmosphere and explodes, sending shock waves to the surface.

Most recently, on 15 February 2013, a 10-20 meter wide object exploded in the air above the city of Chelyabinsk in Russia. The blast injured about 1,500 people and damaged more than 7,000 buildings, collapsing roofs and breaking thousands of windows.

As mentioned above, there is little that we can do to avoid the fate of the dinosaurs unless we use our technology to mitigate the impact risk. One essential prerequisite is to catalogue all of the most threatening objects, so that we can prepare well in advance for their arrival from the depths of space and hopefully have time to take preventative measures.

Understanding their trajectories is necessary but not sufficient. We must also understand their physical properties: composition, density, size, and so on.

Comets and asteroids represent a diverse menagerie of ancient debris, ranging in size from less than one kilometer to hundreds of kilometers. Some are mainly composed of solid metal, most are stony, while others are friable and fragile. Some are comparable to fluffy snowballs, some are loose rubble piles, and others are dense, solid chunks of rock. Some are shaped like peanuts with two lobes separated by a waist, some are close binaries, a few are triple systems orbiting their common center of gravity.

Although ground-based observations will always play a role, the only way to achieve a true understanding of these objects is to visit them with spacecraft. To date, most of these missions have involved relatively short fly-bys that provided snapshots of the object over a period of a few days.

Now, however, a new era of intense examination is under way, with the first asteroid sample return missions such as Japan's Hayabusa series and NASA's OSIRIS-Rex. NASA and ESA are also planning a joint planetary defense mission that will examine a technique for changing the trajectory of an asteroid using a so-called kinetic impactor. The Deep Impact mission fired a projectile into a comet to investigate its composition. The NASA Stardust mission collected tiny samples of comet dust and returned them to Earth for study. (See Appendix 1 for a list of missions.)

The following chapters tell the story of the mother of all missions to explore cosmic debris, the ESA Rosetta 'comet chaser' that flew alongside the nucleus of a comet for two years and sent a lander down to its craggy surface.

REFERENCES

Exploring the Solar System (2nd edition), Peter Bond, Wiley-Blackwell, 2020
Minor Planet Center: https://minorplanetcenter.net/
Comets (NASA): https://www.nasa.gov/comets
Comets (NSSDC): https://nssdc.gsfc.nasa.gov/planetary/planets/cometpage.html
Comets: https://en.wikipedia.org/wiki/Comet
Asteroids (NSSDC): https://nssdc.gsfc.nasa.gov/planetary/planets/asteroidpage.html
Asteroids: https://en.wikipedia.org/wiki/Asteroid
Eight billion asteroids in the Oort cloud, Andrew Shannon et al: https://arxiv.org/pdf/1410.7403.pdf

List of Potentially Hazardous Asteroids (PHAs): https://www.minorplanetcenter.org/iau/lists/
   PHAs.html
Center for Near Earth Object Studies: https://cneos.jpl.nasa.gov/
Asteroid Watch (NASA-JPL): https://www.jpl.nasa.gov/asteroidwatch/
Asteroid Radar Research: https://echo.jpl.nasa.gov/
Meteoroids: https://en.wikipedia.org/wiki/Meteoroid
Meteorites: https://en.wikipedia.org/wiki/Meteorite

2

Beginnings

2.1 EARLY PLANS

During the 1970s, scientists on both sides of the Atlantic began to consider a new generation of deep space missions that would advance our knowledge of the primitive planetary building blocks known as comets.

In 1974, ESA made its first preliminary studies of a possible mission to Comet Encke. The proposal was put aside because the brief fly-by was considered to be too expensive for the limited scientific return. However, a future comet mission continued to generate interest, and, after a European workshop on potential comet missions, held in April 1978, the ESA Solar System Working Group received proposals for participation in a future NASA rendezvous mission and a multi-comet ESA fly-by mission.

After further review and consultation, ESA accepted an invitation to consider participation in a comet mission being studied by NASA that would involve making a fast fly-by of Comet Halley at the end of 1985 and a rendezvous with Comet Tempel 2 during its 1988 apparition.

ESA's contribution to the project was envisaged to be a probe that would be released from the spacecraft about 15 days before the encounter with Halley and targeted at the nucleus.

NASA's share of the project was to be provision of the primary spacecraft and a solar electric propulsion system (SEPS) – a new, highly efficient type of low thrust, continuous propulsion that was considered to be a necessary element for a comet rendezvous mission. Following the announcement in January 1980 that the initial funding for SEPS was not in the budget request, it was agreed that the cooperative mission was no longer feasible.

© Springer Nature Switzerland AG 2020

P. Bond, *Rosetta: The Remarkable Story of Europe's Comet Explorer*,
Springer Praxis Books, https://doi.org/10.1007/978-3-030-60720-3_2

Not to be thwarted, ESA scientists began to consider Europe-only alternatives. In view of the limited time remaining before the arrival of Halley's Comet in the inner Solar System, ESA decided not to consider entirely new spacecraft or mission concepts. Instead, the focus turned to a low cost (maximum $80 million) project that would be based on the existing GEOS spacecraft design and would be launched by either a European Ariane or an American Delta rocket.

The mission, now known as Giotto, was accepted by the Science Program Committee as the Agency's next scientific mission during the meeting of 8-9 July 1980. U.S. scientists were to be allowed to participate as co-investigators on the European-led experiments. The scientific payload was decided by February 1981.

It was agreed that the ideal target for Giotto would be a fairly active comet that followed a well-known trajectory. Compared with most short period comets that have shed much of their volatile material during frequent passages through the inner Solar System, Halley seemed the ideal candidate. Not only was it very bright and active, but it could be expected to generate large dust and ion tails. Although it followed a retrograde or 'backward' path, the comet's orbit was well characterized and only a modest launch velocity would be required to reach the nucleus. Furthermore, the historic nature of this occasional visitor from deep space elevated it above all other potential targets.

2.2   THE FIRST COMET ENCOUNTERS

The first return of Halley's Comet since the dawn of the Space Age prompted a great deal of excitement world-wide. A ground-based observation campaign, known as the International Halley Watch, was initiated by NASA. Comprising both professional and amateur observers, it was a tremendous success and inspired an interest in cometary science in people from many nations and walks of life.

In 1981, four space agencies – Intercosmos, ISAS, NASA and ESA – established the Inter-Agency Consultative Group (IACG) for Space Science, whose objectives were to maximize opportunities for multilateral scientific coordination among approved space science missions in areas of mutual interest. Until 1986, its primary focus was coordinating the planned space missions to Halley's Comet.

During late 1984 and 1985, five spacecraft – two from the Soviet Union, one from Europe and two from Japan – were launched on a historic endeavor to investigate the intruder, which was once again heading sunward after 76 years spent in the dark, frigid reaches of the outer Solar System.

Two Soviet probes, Vega 1 and Vega 2, were the first to set off for Halley, on trajectories that took advantage of fly-bys of Venus.[1] Although they used a proven

---

[1] In the Russian language there is no 'H' so their full name was Venera-Gallei, meaning Venus-Halley.

design, the spacecraft were modified to enable them to undertake a dual mission involving deploying small balloons into the dense atmosphere of Venus before making fly-bys of the famous comet. In a rare example of cooperation with non-communist countries, it was agreed that they would act as pathfinders for ESA's Giotto by helping to guide the European craft to an even closer encounter with the nucleus. The Vegas were launched by powerful Proton rockets on 15 and 20 December 1984. With a launch mass of 4,920 kg, they were by far the largest and heaviest of the international comet flotilla.

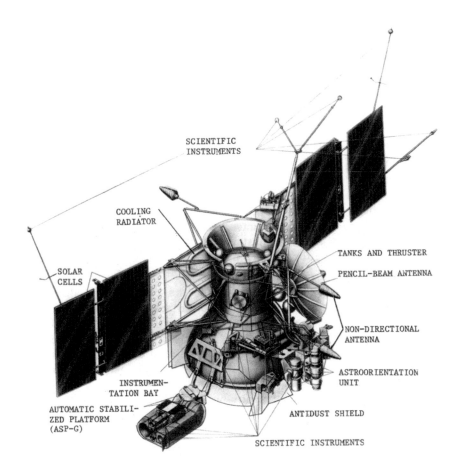

Fig. 2.1: When making a fly-by of Venus on the way to Comet Halley, each Vega space-craft released a spherical capsule that delivered a lander and an inflatable balloon. This illustration shows the configuration on reaching the comet. (ESA)

The identical spacecraft both carried a descent sphere that enclosed a Venus lander similar to those delivered to the surface in the late 1970s and early 1980s, and a small, inflatable balloon to study the atmosphere. Each sphere was separated from the main spacecraft bus as it passed close to the planet.

The parent spacecraft continued on their way to rendezvous with Comet Halley. Vega 1 made its closest approach on 6 March 1986 at a range of 8,890 km from the nucleus. This was the first ever close encounter with a comet. Vega 2 flew even closer, passing within 8,030 km of the nucleus on 9 March 1986.

Both spacecraft imaged the nucleus, showing an elongated, irregular body that was emitting bright jets of gas and dust. Other experiments studied the dust-rich coma that surrounded the nucleus, as well as the nearby plasma, gas, energetic particles and magnetic field.

Japan's contribution to the Halley flotilla was two small, drum-shaped probes called Sakigake (Pioneer) and Suisei (Comet), launched on 8 January 1985 and 19 August respectively. They were that nation's first deep space missions.

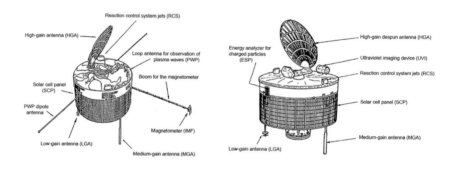

Fig. 2.2: The two Japanese spacecraft sent to study Comet Halley were almost identical, apart from variations in their scientific payloads: Sakigake (left) and Suisei (right). (ISAS-JAXA)

The 139.5 kg Suisei, which had an ultraviolet imaging system and a solar wind experiment, approached to within 151,000 km of the sunward side of Halley's Comet on 8 March 1986. Its task was to observe how the comet interacted with the solar wind, the stream of charged particles that flow away from the Sun. Other results included measurements of the comet's rotation period and variations in its water discharge rate.

The 138 kg, spin-stabilized Sakigake approached to within 7 million km of the comet on 11 March. It did not have an imager, but had a boom-mounted magnetometer to measure interplanetary magnetic fields, and instruments to study the solar wind and plasma waves.

Meanwhile, NASA, which did not directly participate in this once-in-a-lifetime international endeavor, had devised an innovative scheme to send a small spacecraft to a different periodic comet.

Launched in 1978 and originally designated ISEE-3 (International Sun-Earth Explorer 3), this spacecraft was diverted, after four years of service, on a path to pass through the tail of Comet 21P/Giacobini-Zinner and to study Halley's Comet from long range. After a complex series of thruster firings and five lunar fly-bys for gravitational assistance, the repurposed satellite was renamed the International Cometary Explorer (ICE).

Although ICE did not carry a camera, it achieved the first ever comet encounter when it flew within 7,800 km of the nucleus of Giacobini-Zinner on 11 September 1985. It was able to make unique measurements of particles, fields and waves while passing through the plasma (ion) tail. In late March 1986, ICE passed between the Sun and Halley's Comet, adding to the data from other spacecraft. Its minimum distance from the comet's nucleus was 28 million km.

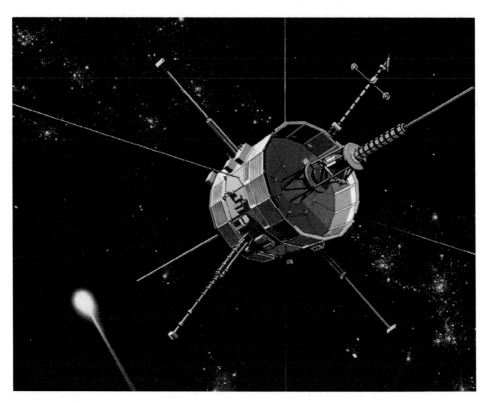

Fig. 2.3: The International Cometary Explorer (ICE), which was launched as International Sun-Earth Explorer 3 (ISEE-3), is depicted approaching Comet Giacobini-Zinner. (NASA)

2.3   GIOTTO

The flagship of the Halley flotilla was the first deep space mission ever flown by Europe. This was named in honor of Giotto di Bondone, an Italian artist who used the 1301 appearance of Halley's Comet to represent the Star of Bethlehem in his fresco 'Adoration of the Magi' in the Scravegni Chapel in Padua.

The mission was originally envisaged as a small probe that would make a fly-by of Halley's Comet in November 1985, while a much larger NASA vehicle would continue on to make a rendezvous with Comet Tempel 2. But when funding for a solar electric propulsion system (SEPS) was omitted from the NASA budget, ESA was obliged to consider flying its own solo mission.

Since the development program was tied to the arrival date of the comet, Giotto had to be redesigned in only five years, which was a short time scale for such a complex mission.

Built by British Aerospace in Bristol, UK, the cylindrical spacecraft was a modified version of the spin-stabilized GEOS Earth-orbiting research satellites that were launched in the late 1970s. The most significant modifications were the replacement of the central radio mast by an antenna tripod and the addition of a buffer, or shield, to protect the spacecraft from being battered by high speed dust particles during the encounter with the active Comet Halley. The drum-shaped bus was also larger, and capable of carrying more solar cells for greater power production.

Giotto was fairly modest in size. Its launch mass was 960 kg and its main body was a short cylinder with a diameter of 1.85 meters and a height of about 1.1 meters. It held three interior platforms: the top platform (30 cm thick), the main platform (40 cm), and the experiment platform (30 cm) – each being a disk on which various subsystems and science experiments were mounted. On top of the main body was a tripod which surrounded a 1.5 meter diameter high-gain dish antenna, giving the spacecraft a total height of 2.85 meters. The main rocket motor was in the center of the cylinder, with its nozzle protruding from the base.

Giotto's objective was to image the comet's nucleus at extremely close range – something akin to flying through an army firing range. The key to a successful mission was to ensure that Giotto survived long enough to snap its pictures of the nucleus when the spacecraft and the comet were heading towards each other at a combined speed of 245,000 km/h, which was equivalent to crossing the Atlantic Ocean in 11 minutes! At this speed, a dust particle with a mass of 0.1 g would be able to penetrate 8 cm of solid aluminum. Because it was out of the question to equip Giotto with a 600 kg aluminum shield, engineers used a more lightweight sandwich design that had been first proposed back in 1947 by American astronomer Fred Whipple.

Fig. 2.4: The spin-stabilized Giotto spacecraft was based on the Earth-orbiting GEOS research satellites, and was Europe's first deep space mission. (ESA)

The spacecraft's dust shield consisted of two protective layers, spaced 23 cm apart. At the front was a sheet of aluminum (1 mm thick) which would vaporize all but the largest of the incoming dust particles. A 12 mm thick sheet of Kevlar behind it would absorb any plasma and debris that pierced the front barrier. Together these two sheets could withstand impacts from particles up to 1 g in mass and traveling 50 times faster than a bullet.

Electrical power came from 5,032 silicon cells that were attached to its cylindrical exterior. These were to provide 190 watts of power during the comet encounter. Four silver-cadmium batteries were carried as back-up and also to sustain the spacecraft when it was in shadow.

The spacecraft was to rotate around the drum's axis at 15 revolutions per minute. During the encounter, it would fly with its dust shield and spin axis pointing towards the nucleus and its dish antenna pointed at the Earth for continuous communications.

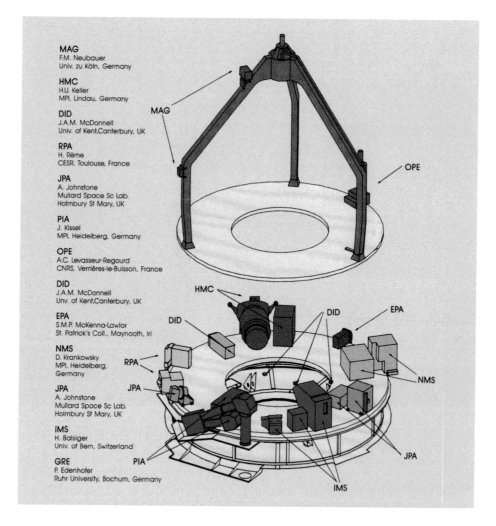

MAG
F.M. Neubauer
Univ. zu Köln, Germany

HMC
H.U. Keller
MPI, Lindau, Germany

DID
J.A.M. McDonnell
Univ. of Kent,Canterbury, UK

RPA
H. Rème
CESR, Toulouse, France

JPA
A. Johnstone
Mullard Space Sc Lab.
Holmbury St Mary, UK

PIA
J. Kissel
MPI, Heidelberg, Germany

OPE
A.C. Levasseur-Regourd
CNRS, Verrières-le-Buisson, France

DID
J.A.M. McDonnell
Unv. of Kent,Canterbury, UK

EPA
S.M.P. McKenna-Lawlor
St. Patrick's Coll., Maynooth, Irl

NMS
D. Krankowsky
MPI, Heidelberg, Germany

JPA
A. Johnstone
Mullard Space Sc Lab.
Holmbury St Mary, UK

IMS
H. Balsiger
Univ. of Bern, Switzerland

GRE
P. Edenhofer
Ruhr University, Bochum, Germany

Fig. 2.5: Giotto had 10 experiments. These included a multicolor camera that would scan the comet as the spacecraft rotated. It also had spectrometers to determine the composition of the comet; dust impact detectors to count the number of impacts and assess their size; plasma and particle analyzers, together with a magnetometer, for plasma studies; and an optical probe to study the dust and gas environment. (ESA)

Giotto carried 10 experiments. Its most high profile instrument was a multi-color camera that would scan the comet as the spacecraft rotated. Other instruments included spectrometers to determine the comet's composition; dust impact detectors to count the number of impacts and assess their size; an optical probe to investigate the dust and gas environment; plasma and particle analyzers, and a magnetometer to investigate the plasma environment.

The spacecraft was launched by a European Ariane 1 vehicle on 2 July 1985. By utilizing the data supplied by the Soviet Vegas, it was possible to localize the position of Halley's nucleus to within 75 km, and this enabled Giotto to make a close sweep past the nucleus on 13 March 1986. During the fly-by, it took 2,112 images, providing an unprecedented view of the nucleus of a comet.

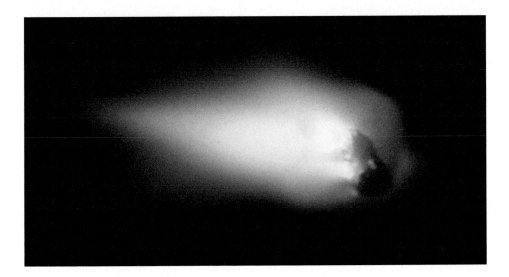

Fig. 2.6: Giotto obtained this composite image, the first detailed view of a comet's nucleus, on 13 March 1986 from about 18,000 km. The dark nucleus measured about 8 × 8 × 15 km. The warmer, sunlit side (left) issued bright jets of gas and dust. The rugged surface revealed sizeable valleys and depressions. A 'mountain' on the dark side was illuminated by sunlight. The rotation axis is approximately horizontal in this view. (ESA/MPI)

The multicolor camera images revealed a potato-shaped nucleus 15 km in length and 7-10 km in width. There were active regions on its sunlit side erupting jets of gas and dust through fractures in the surprisingly dark crust that was insulating the underlying ice from solar radiation.

As expected, Giotto was peppered by tiny, smoke-sized particles, particularly during its final minutes of approach, with 230 strikes occurring in only a few seconds whilst passing through a plume from a jet. Less than 8 seconds prior to

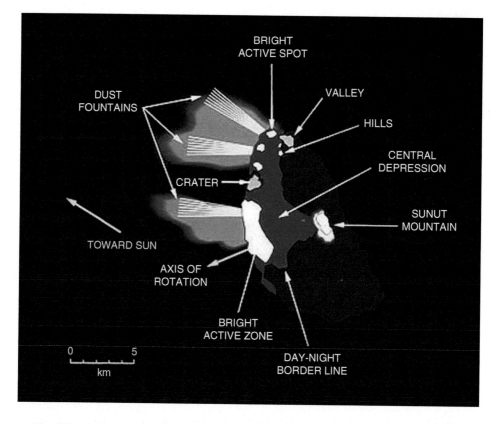

Fig. 2.7: A sketch of the main features revealed in Giotto's detailed images of the nucleus of Halley's Comet. (ESA)

closest approach, a single large (1 g) particle struck with such force that it made the spacecraft tumble. Contact with Earth was temporarily lost. Anxious Giotto team members – and the live TV audience – feared the worst, but then, to everyone's amazement, occasional bursts of data began to arrive at ground control. Over the next 32 minutes, the spacecraft's thrusters stabilized its motion and contact was fully restored. By then it had passed within 596 km of the nucleus and was heading back into interplanetary space.

Detailed analysis of the data confirmed that the comet had formed 4.5 billion years ago from ices condensing onto grains of interstellar dust. Since then, it has remained almost unaltered in the cold, outer Solar System, despite its occasional returns to the warmer inner regions.

The data indicated that the traditional idea of a comet as a 'dirty snowball' was misleading; it might be better described as an 'icy dirtball'.

The key results from Giotto were:

- Water accounted for about 80% by volume of all of the material thrown out by the comet. There were also substantial amounts of carbon monoxide (10 %), carbon dioxide (2.5 %), methane and ammonia. Traces of other hydrocarbons, iron and sodium were also found.
- Halley's nucleus was one of the blackest objects ever observed, with an albedo (reflectivity) of 4-6%. The very dark surface suggested a thick covering of dust.
- The surface of the nucleus was very irregular, with hills and depressions.
- With a density as low as 0.3 g/cc (one third the density of water) the nucleus has a fluffy porous texture.
- Only about 15% of the surface was really active, but the seven jets identified were together ejecting material at a rate of 3 tonnes per second. They imparted a strange, wobbling rotation that seems to be stable over centuries or even millennia.
- Most of the dust was no larger than specks of cigarette smoke. The largest detected was 40 milligrams (which was the upper limit of the dust detectors), but the particle that knocked the spacecraft out of alignment was independently estimated to have a mass approaching 1 g.
- Two classes of dust particles were found. One was dominated by the light elements carbon, hydrogen, oxygen and nitrogen, collectively named 'CHON'. The other was rich in the mineral-forming elements sodium, magnesium, silicon, iron and calcium.
- All of the light elements (except nitrogen) occurred in the same relative abundance as for the Sun, indicating that Comet Halley consists of the most primitive (i.e. original, unprocessed) material known in the Solar System.
- The plasma and ion mass spectrometers revealed the nucleus to be covered in a layer of organic (carbon-rich) material.
- Giotto was the first spacecraft to be deliberately placed in hibernation – the first time 1986-1990 and then again 1990-1992.
- Giotto was the first spacecraft to exploit an Earth gravity assist from an interplanetary trajectory.

Giotto's Second Comet Encounter

Although Giotto was damaged during the Halley encounter, most of its instruments remained operational. To the scientists' delight, the mission was extended to perform an unprecedented near encounter with a second comet, 26P/Grigg-Skjellerup.

In 1992, after a lengthy hibernation, Giotto was reawakened for the closest ever fly-by of a cometary nucleus. It passed within 200 km of its target on 10 July. This time, the spacecraft approached at an angle of 68 degrees instead of head-on, meaning its shield could afford little protection. However, the closing rate was much slower (14 km/s), with the comet approaching Giotto from below and behind.

Although the camera was no longer functional and there was very little electrical power, the eight instruments that were working returned high quality data about the ion tail of the comet and the interactions of the coma with the solar wind. The first evidence of cometary ions was detected at a distance of 600,000 km from the nucleus.

Since Grigg-Skjellerup was a much smaller and less active comet than Halley, it was expected to emit considerably less dust than its larger cousin, and this proved to be the case. Most of its ingredients have boiled away during numerous visits to the inner Solar System, so it was only capable of generating a small coma of gas and dust. The first indication of dust from the coma was detected at a range of 17,000 km, with a marked increase close to the nucleus.

Only three measurable individual impacts were detected, all soon after closest approach, and the spacecraft emerged from the encounter unscathed. The largest, nicknamed 'Big Mac' in honor of Professor Tony McDonnell of the University of Kent, UK, weighed in at about 100 micrograms. Other collisions with dust particles almost certainly took place, but the onboard instruments were unable to detect them directly. In particular, one suspected impact probably occurred three to four seconds prior to Big Mac. This event caused the spacecraft's velocity to change by 1 mm/s and imparted a slight wobble on its spin axis.

The overall picture from Giotto's two encounters was that the nucleus of a comet is more like a frozen mud ball that is rich in dust as well as ices, rather than the expected dirty snowball.

REFERENCES

The ESA Mission to Comet Halley, R. Reinhard, ESA, 1980: https://ntrs.nasa.gov/archive/nasa/casi.ntrs.nasa.gov/19820006152.pdf

Space Missions to Halley's Comet, R. Reinhard and B. Battrick, ESA SP-1066, March 1986

Suisei: https://nssdc.gsfc.nasa.gov/nmc/spacecraft/display.action?id=1985-073A

Sakigake: https://nssdc.gsfc.nasa.gov/nmc/spacecraft/display.action?id=1985-001A

International Cometary Explorer (ICE): https://en.wikipedia.org/wiki/International_Cometary_Explorer

Encounter '86, ESA BR-27, November 1986

Giotto (ESA): https://sci.esa.int/web/giotto

Giotto: https://nssdc.gsfc.nasa.gov/nmc/spacecraft/display.action?id=1985-056A

Halley's Comet (Wikipedia): https://en.wikipedia.org/wiki/Halley%27s_Comet

The Giotto Extended Mission, R. Reinhard. Paper in ESA SP-278, pp 523-529, September 1987

Giotto's Heritage: The Past and Future of Comet Exploration, 10 April 2001: https://solarsystem.nasa.gov/news/187/giottos-heritage-the-past-and-future-of-comet-exploration/

Symposium on the Diversity and Similarity of Comets, ESA SP-278, September 1987

Giotto to the Comets, Nigel Calder, Presswork, 1992

Giotto's Second Encounter, ESA BR-93, 1993

The Giotto Extended Mission (GEM) – High Risk, High Payoff, ESA Bulletin no. 73, February 1993, p. 32-41

Close Encounter with a Comet, Astronomy, November 1993, p. 42-47

3

The Birth of Rosetta

3.1 DEFINING THE FUTURE

Even before the exciting new data came flooding back from the pioneering missions to various comets (see Chapter 2), scientists on both sides of the Atlantic were starting to consider their next steps. This process of crystal-ball gazing came at a time when knowledge of the origin, physical properties and composition of these 'cosmic icebergs' was distinctly sketchy. Innumerable comets had been studied by ground-based observatories, but, as yet, none had been visited by spacecraft and inspected at close quarters.

In 1985, ESA was putting together its long-term science program, named 'Horizon 2000'. There was still a year to go before Giotto would encounter Comet Halley but the planetary science community was already looking beyond that, and the Solar System Working Group recommended a Comet Nucleus Sample Return (CNSR) mission as a cornerstone of the new program. This was seen as the next logical step in improving our knowledge of these familiar, yet mysterious, objects. The proposed mission would involve sending an advanced spacecraft to land on a comet's nucleus, collect material, and return it to Earth for laboratory analysis.

In September 1985, a joint Science Definition Team (SDT) was created by Reimar Lüst, the European Space Agency's Director General, and Geoffrey Briggs, Director of Solar System Exploration at NASA, to identify the scientific goals for such a mission. It comprised thirteen European experts, many of whom were involved in the Giotto mission, and seven American researchers (see Table 3.1). They worked in parallel with another joint ESA-NASA panel, the Primitive Bodies Science Steering Group, whose task was to investigate possible missions to primitive bodies, namely asteroids and comets.

© Springer Nature Switzerland AG 2020
P. Bond, *Rosetta: The Remarkable Story of Europe's Comet Explorer*,
Springer Praxis Books, https://doi.org/10.1007/978-3-030-60720-3_3

Table 3.1: The Members of the CNSR Science Definition Team

European delegation:

E. Grün (Co-Chairman)	Max-Planck-Institut fur Kernphysik, Heidelberg, West Germany
F. Begemann	Max-Planck-Institut fur Chemie, Mainz, West Germany
P. Eberhardt	Physikalisches Institut, Universitat Bern, Switzerland
A. Coradini	Istituto Astrofisica Spaziale – CNR, Rome, Italy
M. C. Festou	Institut d'Astrophysique, Paris, France
Y. Langevin	University Paris-Sud, Paris, France
J. A. M. McDonnell	University of Kent, Canterbury, UK
C. T. Pillinger	Open University, Milton Keynes, UK
G. Schwehm	ESA/ESTEC, Noordwijk, Netherlands
D. Stöffler	Universitat Munster, Munster, West Germany
H. Wänke	Max-Planck-Institut fur Chemie, Mainz, West Germany
R. M. West	European Southern Observatory, Garching, West Germany

U.S. Delegation:

T. J. Ahrens (Co-Chairman)	California Institute of Technology, Pasadena, California
H. Campins	Planetary Science Institute, Tucson, Arizona
D. E. Brownlee	University of Washington, Seattle, Washington
S. Chang	NASA Ames Research Center, Moffett Field, California
A. W. Harris	Jet Propulsion Laboratory, Pasadena, California
G. J. Wasserburg	California Institute of Technology, Pasadena, California
J. A. Wood	Harvard-Smithsonian Center for Astrophysics, Cambridge, Massachusetts

The SDT held five meetings between 1985 and 1987, and then released a report. Meanwhile, NASA's Solar System Exploration Committee's Augmented Program rated such a mission as having high scientific merit.

3.2   HORIZON 2000

1985 also saw a major innovation in Europe's approach to space science, when ESA Member States approved Horizon 2000 as a long-term program of scientific research. This ambitious program was to ensure that Europe continued to play a key role in space science over the next 15 years, and beyond. It was also the starting point for a ground-breaking space mission that would soon become known as Rosetta.

- The primary objectives of Horizon 2000 were:
- To contribute to the advancement of fundamental scientific knowledge.
- To establish Europe as a major participant in the worldwide development of space science.
- To offer a balanced distribution of opportunities for frontline research to the European scientific community.
- To provide major technological challenges for innovative industrial development.

This new mandatory program would necessitate an increased financial commitment from the ESA Member States during the period of implementation.

Specifically, the realization of the Horizon 2000 plan would be critically dependent on an annual increase in expenditure of 5% until at least 1994, which was an unprecedented requirement at the time. The 5% growth rate for 1985-1989 was unanimously accepted at the ESA Council Meeting at Ministerial Level in 1985. However, the abstention by the UK at another vote at the next meeting in The Hague in 1987 effectively blocked the increase for the time being.

The Horizon 2000 missions would be separate from, but complementary to, the space science programs of ESA Member States. The countries that made the largest financial contributions to the Agency's science program would receive the largest industrial contracts – a procedure known as *juste retour* (fair return). The scientific payloads would be developed and provided by the Member States, under the leadership of principal investigators from the lead countries.

Although it was seen as a key means of promoting European space science, the program also allowed for the possibility of cooperation with agencies outside Europe, such as the United States and the Soviet Union, particularly for projects which were prohibitively expensive and required the development of new, advanced technologies.

Horizon 2000 was to include a balanced sequence of large and medium/small projects in all of the traditional science disciplines. In particular, it would include four major Cornerstone missions that would be developed over the next 15 years.

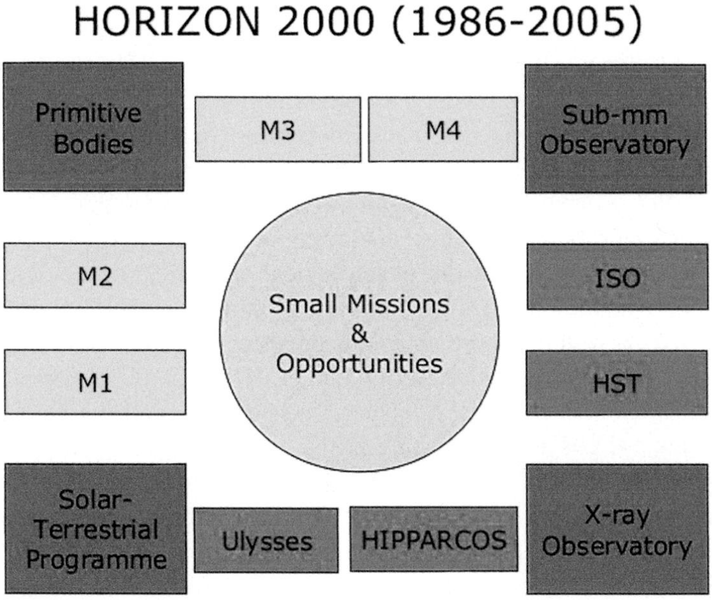

Fig. 3.1: ESA's Horizon 2000 long-term science program included four major Cornerstone space missions. The Primitive Bodies Cornerstone became the Rosetta mission. (ESA/Alvaro-Gimenez)

Cornerstone 1 was the Solar-Terrestrial Science Program (STSP). It comprised two medium-sized missions that were to investigate the complex interactions between the Sun and Earth. The Solar and Heliospheric Observatory (SOHO) was to be placed in a stable orbit between the Sun and Earth, a location known as the L1 Lagrangian point, from which it could observe continuously our nearest star, its corona and the solar wind.

The second STSP mission, named Cluster, involved the deployment of four identical, drum-shaped spacecraft into elliptical Earth orbits, to study the response of the near-Earth magnetic and particle environment to variations in solar activity.

The STSP was to be a co-operative venture by ESA and NASA, with its investigations led by scientists from both sides of the Atlantic.

The second Cornerstone to be authorized was an astrophysical observatory that would be a follow-on to Europe's successful Exosat mission. It would be devoted to the investigation of extremely hot, energetic objects that emit much of their energy in X-rays. Examples of such cosmic X-ray sources include supernova remnants and active galaxies. Originally called the High-Throughput X-Ray Spectroscopy Mission, it would be flown as the X-ray Multimirror Mission (XMM).

Bringing up the rear were the two remaining Cornerstones. One of these was to be a mission devoted to submillimeter astronomy. By providing the first space-based observations of the Universe at submillimeter wavelengths, the observatory would break new ground in the study of interstellar dust clouds and the formation of stars and planets. It was initially known as the Far Infra-Red Spectroscopy Mission, and then the Far Infrared and Submillimeter Telescope (FIRST), but in 2000 it was named the Herschel Space Observatory in honor of the famous discoverer of the planet Uranus and also the existence of infrared light.

The fourth Cornerstone was to be an international venture that was identified as a "mission to primordial bodies, including return of pristine material". This Cornerstone would build on the technological and scientific knowledge gained from the fly-by missions to Comet Halley, in particular the Giotto mission (see Chapter 2). This would require an advanced spacecraft that could return samples of material from either an asteroid or a comet. However, this ambitious next step was soon refined to become the first space mission designed to collect a sample of a comet's icy nucleus and return it to Earth.

This program introduced an unprecedented degree of complexity and sophistication to Solar System exploration, including long-term reconnaissance of a comet, soft landing on a comet's nucleus, drilling into its surface, collecting a sample, and safely returning it to Earth for study using the most advanced laboratory techniques.

To ensure that the new technologies and techniques required for a successful implementation of this pioneering enterprise would be available, ESA initiated a special Preparatory Program in 1986, with an anticipated completion date of 1991. This would define the mission scenario and system design, with the participation of European industry and the international scientific community. NASA would be involved as the likely provider of support in areas such as the carrier spacecraft, the launcher and deep space communications.

Following the formation of a joint ESA-NASA Science Definition Team in late 1985 (see Defining the Future above), the University of Kent at Canterbury, UK, held a workshop to inform the wider scientific community of this work. After the completion of major industrial studies, there were follow-up workshops in Granada (1990) and in Cagliari (1991).

Meanwhile, ESA's Technology Research Program studied most of the enabling technologies required for a sample return mission, confirming the feasibility of the ambitious project, now known as Rosetta.

3.3 WHY ROSETTA?

The ESA comet chaser was named after the Rosetta Stone, a famous, ancient slab of basalt (a dark volcanic rock) that is on display in the British Museum in London. It was unearthed in 1799 by French soldiers near the town of Rashid (in English, Rosetta) on the Nile Delta. Two years later, after the surrender of Napoleon's army in Egypt, the 762 kg slab was handed over to the British.

The carved inscription on the stone was unique because, for the first time, it included the same text in different languages – ancient Greek, which was readily understood, and ancient Egyptian. Furthermore, the latter was present in two forms – the common Demotic script and the pictorial hieroglyphs that were famous from their presence in the tombs of the pharaohs of ancient Egypt.

By comparing the inscriptions, scholars were able to painstakingly decipher the meaning of the mysterious hieroglyphs. Most of the pioneering work was done by an English physician and physicist, Thomas Young, and by French scholar Jean François Champollion. As a result of their breakthroughs, it became possible, for the first time, to piece together the language and literature of a long-lost civilization that dominated the Nile valley a thousand years.

Almost 200 years later, European scientists were seeking an appropriate name for the newly proposed Cornerstone mission to explore a comet in unprecedented detail. At the time, this international venture was simply the 'Comet Nucleus Sample Return' mission, but the Solar System Working Group was eager to come up with a more memorable name.

Eberhard Grün at the Max-Planck-Institute for Nuclear Physics in Heidelberg, Germany, was the chairman of the joint ESA-NASA Science Definition Team created to specify the science goals (see Defining the Future above), and he suggested the name 'Rosetta'.

Fig. 3.2: The Rosetta Stone, now located in the British Museum, London, has three different carved inscriptions showing the same text in ancient Greek and two different forms of ancient Egyptian. This permitted scholars such as Thomas Young and Jean François Champollion to reveal the meaning of the mysterious hieroglyphics. (Wikimedia https://en.wikipedia.org/wiki/Rosetta_Stone#/media/File:Rosetta_Stone.JPG)

As Grün later explained:

Meteorites are debris from (mainly) asteroids that make it to the surface of Earth, and their study helped us piece together the chemical composition of the Solar System today. But if we want to study the primordial Solar System, comets are the objects to explore. While asteroids formed between Mars and Jupiter, comets formed much farther away from the Sun, so comet material has been much less processed and is much closer to the pristine composition of the Solar System.

With a space mission to visit a comet and collect samples from its nucleus, we could gather brand new information to decipher the history of our Solar System. And it was then that it occurred to me that this is the same story as the Rosetta Stone!

When this idea first crossed my mind, towards the end of 1986, I went to the library at the University of Heidelberg to learn more about the Rosetta Stone and how it revolutionized the study of Ancient Egypt. There was clearly a parallel with comets and their role to interpret the history of the Solar System, and besides, the name 'Rosetta' was much more powerful than what we had before. I phoned Gerhard Schwehm, the mission study scientist at ESA, and he liked the idea. The entire Science Definition Team thought it was a great name and also Roger Bonnet, ESA's Science Director at the time, so it stuck pretty quickly.

By the end of 1987, we referred to the mission as 'Rosetta – Comet Nucleus Sample Return' in the first study report from the Science Definition Team.

As Grün would later observe during the mission, "Every new day with Rosetta at the comet, we are gathering more clues on how a comet works and how to decipher the Solar System hieroglyphs and piece together our cosmic origins."

3.4   COMET RENDEZVOUS OR SAMPLE RETURN?

Although the return of pristine material from a comet's nucleus was ESA's preference as the planetary Cornerstone in its Horizon 2000 program, it was recognized that such an ambitious enterprise would be difficult to achieve.

An acceptable alternative would be a mission that returned to Earth samples of dust and gas from a cometary 'atmosphere'. Accordingly, in response to ESA's Call for Mission Proposals in July 1985, a group of scientists led by J.A.M. (Tony) McDonnell of the University of Kent at Canterbury put forward a plan to develop a mission named CAESAR (Comet Atmosphere and Earth Sample Return) and a detailed Assessment Study was undertaken between July and November 1986.

CAESAR would carry nine scientific experiments to investigate a comet's nucleus and its surroundings. However, the most notable objective was to be the first mission to return to Earth samples of comet dust and gas. Unlike dust particles of extraterrestrial origin that are collected as they penetrate our planet's atmosphere, these samples of volatile and non-volatile materials would come from a known source and have a well-defined history.

As it passed through the comet's coma, the spacecraft would collect this pristine, unprocessed material and later return it to Earth for analysis in advanced laboratories. Protected by a nose cone made of ablative material, the return capsule would parachute to a fairly soft landing at a designated site.

It was anticipated that the mission would, for the first time, enable scientists to look back to the earliest stages in the history of our Solar System, and thus obtain important data on the composition of the solar nebula, including the composition of individual interstellar grains and the processes that resulted in the condensation of material in the solar nebula.

The Assessment Study recognized that the scientific objectives of the mission could be more easily fulfilled if the fly-by velocity was very small. This would mean that the number of dust motes obtained, and the percentage of those that would survive the collection process intact, would increase as the impact hazard to the spacecraft diminished. The optimum case would be a "zero velocity fly-by", meaning a rendezvous in which the spacecraft flew alongside the comet.

Although the CAESAR team was unable to find an international partner that was willing and able to conduct a slow fly-by, they did suggest that a "truly exciting" scientific mission could be achieved if CAESAR could be carried out as part of NASA's CRAF (Comet Rendezvous and Asteroid Fly-by) mission (see Section 3.5 below).

The CAESAR team envisaged their small craft piggybacking on the much larger CRAF as it rendezvoused with and flew alongside Comet Temple 2. This would enable a sizeable sample of material to be collected when the comet became most active around its closest approach to the Sun, known as perihelion. Later, CAESAR could return to Earth with its precious payload of perhaps 10-100 g of cometary dust.

Although CAESAR never got off the ground, these preliminary studies would prove valuable when the Rosetta comet rendezvous mission became the Agency's priority a few years later. Interestingly, the baseline comet for the CAESAR proposal, 67P/Churyumov-Gerasimenko, was eventually visited by Rosetta.

3.5   CRAF – COMET RENDEZVOUS AND ASTEROID FLY-BY

In the late 1980s, as ESA was refining its plan to develop the international Rosetta Comet Nucleus Sample Return mission, NASA was preparing a complementary mission, known as CRAF (Comet Rendezvous and Asteroid Fly-by), which it intended to design in parallel with a mission to the outer Solar System (the Cassini Saturn orbiter) in order to make cost savings by having commonality of systems.

The motivation for CRAF was to address the major gaps in our knowledge of the origin and nature of small Solar System bodies. Specifically, it was to conduct a close fly-by of a main belt asteroid and then rendezvous with a short period comet (Kopff or Tempel 2) out near the orbit of Jupiter. It would fly alongside the nucleus for up to three years to study its changing surface activity at varying distances from the Sun.

Unlike Rosetta, CRAF was not intended to collect and bring back samples from the nucleus itself. Instead, its priority was to remotely study the nucleus and its environment, including onboard analysis of captured dust particles. It would also fire an instrumented penetrator into the nucleus.

The principal objectives were to:

- Determine the composition and character of a cometary nucleus, and characterize changes that occur as functions of time and orbital position.
- Characterize the comet's atmosphere and ionosphere and study the development of the coma as a function of time and orbital position.
- Determine comet tail formation processes and characterize the interaction of comets with the solar wind and radiation.
- Characterize the physical and geological structure of an asteroid.
- Determine the major mineralogical phases and their distribution on the surface of an asteroid.

Fig. 3.3: NASA's Comet Rendezvous and Asteroid Fly-by (CRAF) mission was to be based on a new, nuclear-powered, Mariner Mark II spacecraft. It would make an asteroid fly-by *en route* to Comet Tempel 2, then fire an instrumented penetrator (left) into the nucleus and fly alongside the comet for three years, examining the formation of its coma and tails. (NASA)

Selected in 1986, the scientific payload for CRAF included:

- Cameras to photograph the nucleus, the coma and tail, and changes that occurred as the comet moved around its orbit. The images would also help to determine the size and structure of the nucleus, the location of its poles, its rotation rate and geological make-up.
- A surface penetrator would be fired into the nucleus and travel perhaps 1 meter into the icy crust. Its instrumentation would measure the abundances of up to 20 chemical elements. A gamma ray spectrometer would measure the elemental composition of both ice and non-volatile material; an accelerometer would measure the strength and structure of the surface; thermometers would measure the temperatures beneath the surface; a calorimeter would detect phase changes as an ice sample was heated and vaporized; and a gas chromatograph would determine types and amounts of gaseous molecules released from the ice sample. The data would be radioed to the spacecraft and relayed to Earth.
- Various mass spectrometers to study the composition of gases released by the nucleus and cloud of plasma (ionized gas) surrounding the nucleus.
- A visual and infrared mapping spectrometer to study the chemical composition of the coma and the surface of the nucleus as they changed over time.
- Dust counters, collectors and analyzers to capture samples of the comet's dust and to study them on board. This would help scientists to determine the chemical elements that make up the dust and ice. At the same, the mass, size, shape and composition of individual dust grains would be measured.
- A magnetometer and a plasma wave analyzer to measure interactions between the coma and electrically charged particles of the solar wind. The magnetometer would also measure any intrinsic magnetic field associated with the comet.

CRAF and Cassini were to be based on a new spacecraft platform, called Mariner Mark II, which was intended for flights to the outer planets and primitive bodies such as comets. The same platform was proposed during design studies for ESA's Rosetta Comet Nucleus Sample Return spacecraft.

Mariner Mark II comprised a central, 10-sided bus to hold most of the electronics; a large propulsion subsystem (fuel tanks, rocket engine and structure) beneath the bus; a high-gain antenna and radio feeds on top of the bus; two nuclear-powered radioisotope thermoelectric generators on a boom behind the bus; and booms on which to mount experiments. There was considerable international involvement, most notably for the spacecraft's chemical propulsion system, whose development was entrusted to the German Federal Ministry for Science and Technology (BMFT). While the spacecraft flew with its high-gain antenna pointing at Earth, two instrument-laden platforms would ensure its telescopes and other sensors trained on the target.

CRAF was initially penciled in for launch by a Titan IV/Centaur-G rocket in February 1993. On its way to Tempel 2, it would gain gravitational boosts using fly-bys of Venus and Earth. After an asteroid fly-by in January 1995, the spacecraft would reach the comet in November 1996, just inside Jupiter's orbit. It would fly alongside the nucleus for more than three years as it approached the Sun and reached perihelion. The warming of the icy nucleus would lead to the growth of a coma and tail, increased jet activity and expulsion of dust and gases. As the comet became more active, the spacecraft would recede to a safe distance of several thousand kilometers.

The flight plan was modified as time went by. When the mission was approved as a new start in NASA's Fiscal Year 1990 budget, it was scheduled for launch to Comet Kopff in February 1996, with arrival at its target in January 2003 and the nominal end of mission in June 2005, after the comet had passed perihelion. However, the overall flight plan remained very similar to the original.

Unfortunately, the Mariner Mark II program ran into financial and technological headwinds. In particular, CRAF was squeezed on two fronts. The projected cost of a key instrument, the comet nucleus penetrator, was increased from an initial estimate of $22 million to a projected $120 million. The soaring cost obliged NASA to cancel the instrument in 1990, undercutting a major part of the scientific justification for the mission. Furthermore, the Scanning Electron Microscope and Particle Analyzer (SEMPA) instrument was also eliminated.

With the proposed Freedom space station seen as NASA's top priority by Congress, funding for the Agency's space science program was put under extreme pressure. In autumn 1991 the launch date was put back from February 1996 to April 1997, a decision that would spread out the mission's cost but delay the comet rendezvous by three years, from 2003 to 2006. Despite these setbacks, a 1992 National Research Council report stated that "the CRAF mission had great scientific merit even without the penetrator experiment."

However, the final stumbling block was the continuing congressional budget restrictions on mission expenditures. In the 1992 NASA budget, the money allocated to CRAF and Cassini was slashed by a whopping 36%. After the comet rendezvous program was deleted from the President's 1993 budget request, NASA abandoned CRAF and redesigned Cassini to make it cheaper.

3.6   ROSETTA COMET NUCLEUS SAMPLE RETURN

As already mentioned, Rosetta was originally envisaged to be the first mission to land on a comet, drill into its nucleus, and return a sample of material to Earth for analysis. Alongside NASA's CRAF, the ESA-NASA Comet Nucleus Sample Return (CNSR) would complete an ambitious double-header to reveal the secrets of comets.

Analysis of the cometary samples would be a huge advance over the knowledge gained from ground-based observations and fly-bys, yielding insights into the chemical, mineralogical and physical properties of comet material. It was expected that the Rosetta CNSR mission would revolutionize scientists' ideas about comets and their role in the formation of stellar nebulae – the birthplaces of stars and planetary systems.

In order to achieve these objectives, there were to be four phases of science investigations in the vicinity of the comet:

- Target acquisition, characterization of the nucleus, precise orbit determination and definition of gas and dust emission patterns.
- Coma transit, assessment of nucleus activity, mapping of active areas of dust and gas jets and evaluation of hazards to the spacecraft, determination of the rotational state of the nucleus.
- Landing/sampling site selection, and definition of the approach strategy.
- Sample acquisition and surface characterization.

Rosetta CNSR would carry a suite of remote sensing and in-situ experiments to investigate the nucleus and the coma (see Table 3.2).

An imaging system and radar sounder would support spacecraft navigation and landing, in addition to carrying out detailed mapping of the nucleus. A thermal infrared radiometer was to provide information on the distribution of surface temperatures on the nucleus to support the selection of the landing site. A neutral mass spectrometer and dust monitor would study the cometary environment. Other instruments would measure the temperature in the borehole and the temperatures of the samples, and high-resolution images of the sampling would be provided by a stereoscopic camera.

Table 3.2: The Rosetta CNSR Model Payload

Site Selection	Imaging system
	Infrared/thermal mapper
	Radar altimeter/sounder
	Laser altimeter
Environment Monitoring	Neutral/ion mass spectrometer
	Dust counter
Surface Drilling and Sampling	In-situ stereo imaging system
	Sample thermal logger
	Temperature profiler in drill hole
	Thermal logger for surface temperature
	Borehole stratigraphic (layer) recorder

There was also the possibility of depositing a surface science package on the nucleus which would remain active after the completion of the sample mission, to monitor the changes there as the comet traveled a considerable fraction of its orbit.

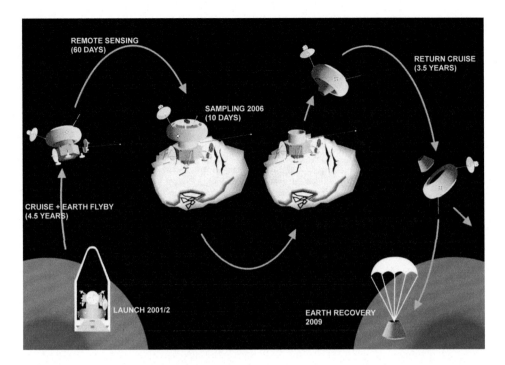

Fig. 3.4: An early artist's impression of the key stages in the Rosetta CNSR mission. From left to right: the spacecraft is launched from Earth; after a comet rendezvous, it makes a soft landing on the nucleus; the landing module remains behind as the return spacecraft lifts off with the Earth-Return Capsule; the capsule separates from the spacecraft and parachutes to Earth for recovery and analysis of its precious samples. (ESA)

The spacecraft was to comprise three modules: the Cruise module, the Lander, and the Earth-Return Capsule. The entire spacecraft would touch down on the nucleus of a short period comet such as 103P/Hartley. The Lander would remain on the nucleus after the sampling was finished and the sample had been stowed in the Earth-Return Capsule. After returning to the vicinity of Earth, the spacecraft would release the capsule on a trajectory that would enable its recovery.

Such a complex plan was beyond the financial and technological resources of ESA alone, so major U.S. involvement was deemed essential to the success of the mission.

As defined in 1991, the mission would be launched on a Titan/Centaur provided by NASA, and the spacecraft would be derived from the Mariner Mark II that was being developed by NASA. The Lander and Earth-Return Capsule would be supplied by ESA, thus allowing that Agency to focus on the cometary science. In flight, the mission would be primarily controlled by NASA's Deep Space Network of ground stations.

Fig. 3.5: An artist's impression of the Rosetta CNSR spacecraft on the surface of a comet. In the center is the Earth-Return Capsule (the red-brown conical module), with the drilling unit beneath it. The boom on the right carries the High-Precision Scan Platform, with its suite of scientific instruments. (ESA)

According to the 1991 mission definition document, the overall dry mass of Rosetta CNSR would be 2,447 kg, of which the main spacecraft, referred to as the Cruiser, would account for 1,523 kg. The Lander would account for 474 kg, with the Earth-Return Capsule weighing in at 297 kg and the launcher adapter at 153 kg. In addition, the spacecraft would carry 3,611 kg of fluids, primarily propellants for its main engine.

The three-axis stabilized Cruiser would be carried on top of the Lander module, and was to provide attitude control, navigation, propulsion, power generation, telecommunications, and the overall mission management. The science and engineering instruments that required high pointing accuracies, notably the cameras, laser range finder and radar altimeter, would be on a High Precision Scan Platform that could be moved in elevation and azimuth with respect to the main body.

The propulsion subsystem included a 400 Newton bi-propellant main engine for large orbital maneuvers that would use around 3,500 kg of monomethylhydrazine and nitrogen tetroxide. There was a mono-propellant system with 24 × 10 Newton thrusters for attitude control and small trajectory changes. These thrusters would also be used for the initial landing operations, and for lift-off after the sample was safely aboard. These maneuvers would use about 70 kg of hydrazine. The final stages of the descent to the comet's surface would be performed using 24 × 20 Newton cold gas thrusters with a supply of 39 kg of nitrogen.

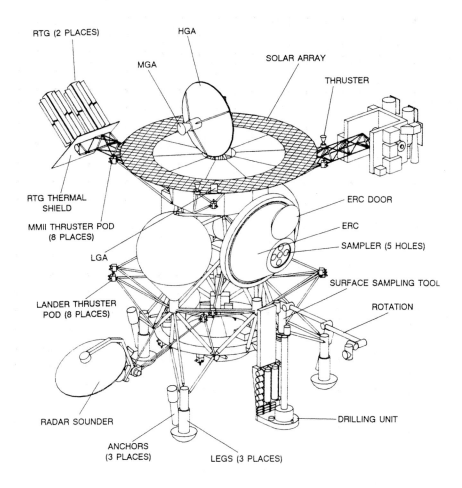

Fig. 3.6: The main features of the Rosetta CNSR spacecraft. The Lander is below the Cruiser and the Earth-Return Capsule is mounted on the side of the Cruise module. (ESA)

One technology that was not available in Europe was the pair of radioisotope thermoelectric generators (RTG) on a deployable boom. These would provide electrical power from the heat generated by the radioactive decay of plutonium-238, in the form of plutonium dioxide. The large difference in temperature between this hot fuel and the cold environment of space was applied across solid-state metallic junctions known as thermocouples to generate an electrical current without moving parts. If the mission were to proceed without RTGs, it would have to rely upon solar power, and, at the distance from the Sun at which the comet rendezvous was to occur, the solar arrays would have to be extremely large. Solar power became a requirement after NASA reduced its role in the Rosetta mission in 1993.

On top of the Cruiser was a ring-shaped Sun shield with an annular solar array on its exterior to provide 350 W/h after the RTGs were jettisoned prior to Earth-return – a safety precaution to eliminate the possibility of radioactive contamination of our planet's environment.

In the center of the sunshield was a steerable, 1.47 meter diameter high-gain antenna that was based on a design used by the Viking Orbiters that were sent to Mars. The boom-and-gimbal design ensured a clear field of view to Earth for data downlinking during comet approach and operations. The primary system operated in X-band, but a medium-gain antenna and a pair of low-gain antennas were available for back-up, including emergency commands and low-rate telemetry.

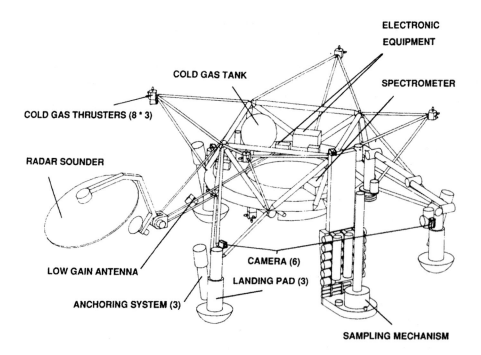

Fig. 3.7: The main features of the Rosetta CNSR Lander module. The three-legged module was to be anchored securely to the surface of the nucleus before drilling operations began. The module formed the lower part of the spacecraft that landed on a comet, and acted as a launch platform for the Cruiser once the surface sample had been collected and loaded into the Earth-Return Capsule. (ESA)

The Lander module – a tubular truss construction – would be attached beneath the Cruiser by four explosive bolts. It formed the basic load-carrying structure for the Cruise module during launch from Earth, and was attached at its base to the launch vehicle's adapter. It would carry the remaining science instruments, and have three legs with foot pads and anchors to secure it to the nucleus in the comet's low gravity. It also carried the surface sample acquisition unit. It would enable the entire Rosetta CNSR spacecraft to approach the nucleus and touch down.

For the last few hundred meters of the descent, control would be switched from the hydrazine thrusters to a set of cold gas thrusters to avoid contamination of the sample area.

The fixed legs were equally spaced around the periphery of the Lander for optimum stability, and devices were incorporated into each leg to reduce the shock of touchdown impact.

The anchoring system that would stabilize the spacecraft during the touchdown and sampling phases comprised a pyrotechnic (explosive) device on the pad of each leg. The design of the anchor had to accommodate a broad range of possible soil materials, hardness and strength. A telescoping system involving two aluminum or titanium tubes was proposed. In very soft soil, both tubes would be deployed. The minimum desired penetration in the hardest of soils was 50 cm. If both telescopic tubes were fully extended the maximum depth would be 1.2 meters.

In the event of the anchoring system failing, a back-up system comprising two 70 Newton bi-propellant thrusters mounted on the edge of the Cruiser's Sun shield would push the vehicle against the comet's surface and also provide the necessary resistance to the drilling forces and rotational torques.

3.7 SURFACE SAMPLING

The Sample Acquisition System (SAS) was to consist of a core sampler, a surface sampler, and a handling arm. The SAS and the Earth-Return Capsule were designed to collect and store five sample tubes, consisting of three core tubes, one volatile sample and one surface sample tube.

Surface sampling of non-volatile material (i.e. organic and inorganic compounds) would be the first task, using a dedicated tool that would be placed in selected locations by a robot arm. The device consisted of a rotary shovel head at the bottom and an attached sample container tube. Fluffy or coarse material with a mass of 1-5 kg could be collected to a depth of 10 cm, possibly containing solid fragments up to 5 cm in diameter.

The core sampler would consist of a rotary table equipped for independent axial and rotary articulation, a set of two outer core tubes that had a diamond-tipped drill head at the bottom, and six inner core sample containers.

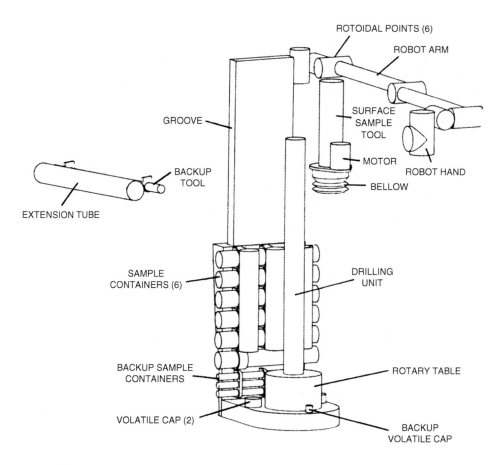

Fig. 3.8: The CNSR comet core sampler consisted of a vertical slide bar, a rotary table, two outer coring tubes with a drill head at the bottom, and six inner core sample containers. The drilling operations were to begin when the rotary table was lowered to the surface. The unit could extract material from a depth of 1-3 meters. (ESA)

The rotary table would be lowered to the surface by a vertical slide bar. Drilling would take place with a rotation rate in the range 10-100 rpm, and depth progression would be actively controlled in terms of the allowable temperature increase on the sample and the resistance of the surface material.

One coring tube would be installed on the rotary table prior to launch from Earth, to permit coring operations to start as soon as the spacecraft was safely anchored to the nucleus. The tube was 1.6 meters long, and could reach a depth of 1-1.3 meters, depending on the surface roughness beneath the rotary table. At that depth, the first volatile sample would be collected and retrieved. With extra core tubes, it might be possible to achieve a total sampling depth of 2.3-2.6 meters.

The core samples, having an overall mass of up to 10 kg, would form a continuous record of subsurface layering, although some compression of the core could be applied in a controlled manner. In addition, a small (10-100 g) sample of volatile material would be loaded into a sealed container capable of preserving it intact during the return to Earth.

The surface samples would be picked up by a manipulator arm and stored inside the ERC for delivery to Earth. An onboard camera would enable ground control to watch operations, and intervene if necessary.

The capsule would be a spinning, unguided, ballistic aeroshell with a sphere- or cone-shaped forward heat shield and a blunt rear end. Its baseline dimensions were 1.8 meters in diameter and just over 1 meter in height. At its heart was the container capable of holding five sample compartments, each with a length of 650 mm and a diameter of 130 mm. The passive thermal control regime involved the use of multiple layers of insulation, with low-power heaters for the battery, transmitter, parachute, and other mechanisms.

The samples would all be sealed independently. Their conditions would have to be carefully controlled during the return journey, with the core sample maintained below $-110°C$ for the entire return trek, until recovery. In addition, the samples would have to avoid extreme levels of acceleration or vibration.

Sometime after the largely autonomous sampling operations were completed, the Cruiser and its piggybacking ERC would lift off, using the Lander as a launch platform. On approaching Earth after a 2 year cruise, the RTG power pack would be jettisoned in order to comply with safety regulations. Several days later, the ERC would be released for a ballistic entry into the atmosphere. After a parachute assisted descent, it would splash into the Pacific Ocean. It was to be recovered by helicopter within 30 minutes. Then the priceless comet samples would be delivered to the sample receiving laboratory.

3.8 CANCELLATION

This imaginative and ambitious endeavor never came to fruition. By late 1991, it was clear that the CRAF mission was not the only NASA space science project in dire jeopardy. The Rosetta CNSR program was threatened by the rising cost of the Mariner Mark II spacecraft and other hardware, combined with financial cutbacks for NASA and difficulties associated with program planning and implementation timing.

ESA officials wisely began to initiate parallel studies for a smaller, all-European mission to investigate the primitive bodies of the Solar System. This revamped Rosetta mission would have to adhere to the budgetary constraints of a Horizon 2000 Cornerstone mission and rely on technologies and launch capabilities available in Europe.

The new studies looked at two mission alternatives:

- Multiple asteroid fly-bys *en route* to a near-Earth asteroid rendezvous and a possible landing on the asteroid at the end of the mission.

- A comet rendezvous with payload operations starting when both the spacecraft and the comet were within 2.5 AU of the Sun.

When NASA said in 1993 that it could no longer be a major participant in the Rosetta CNSR mission, ESA pursued a cheaper and less technologically challenging European project in the form of a comet rendezvous which was remarkably similar to the canceled CRAF proposal.

REFERENCES

Space Science – Horizon 2000, ESA SP-1070, 1 December 1984

Horizon 2000 Plus, ESA SP-1180, August 1995

Naming *Rosetta – An interview with Eberhard Grün*: https://rosetta.jpl.nasa.gov/news/naming-rosetta-%E2%80%93-interview-eberhard-gr%C3%BCn

CAESAR: A Comet Atmosphere Encounter and Sample Return, ESA SCI986)3, November 1986

The Comet Nucleus Sample Return Mission, ESA SP-249, December 1986

NASA Selects Science Investigations For CRAF, NASA-JPL press release, 28 October 1986 https://www.jpl.nasa.gov/news/news.php?feature=5905

The Comet Rendezvous Asteroid Flyby Mission, M. Neugebauer. Paper in ESA SP-278, pp 517-522, September 1987

Rosetta: A mission to sample the nucleus of a comet, John A. Wood. Paper in ESA SP-278, pp 531-537, September 1987a

ESA Report to the 27th COSPAR Meeting, SP-1098, July 1988

ESA's Report to the 28th COSPAR Meeting, ESA SP-1124, June 1990

ESA's Report to the 29th COSPAR Meeting, ESA SP-1154, September 1992

ESA's Report to the 30th COSPAR Meeting, ESA SP-1169, July 1994

Comet Rendezvous Asteroid Flyby https://en.wikipedia.org/wiki/Comet_Rendezvous_Asteroid_Flyby

Retrieving Samples From Comet Nuclei, Ernst Stuhlinger et al. Paper in ESA SP-278, pp 539-546, September 1987

Rosetta: The Comet Nucleus Sample Return, Report of the Joint ESA/NASA Science Definition Team, SCI(87)3, December 1987b

Report on ESA's Scientific Satellites, ESA SP-1110, May 1989

Rosetta/CNSR: A Comet-Nucleus Sample-Return Mission, ESA SP-1125, June 1991

Rosetta Comet Rendezvous Mission, SCI(93)7, September 1993

Outward to the Beginning: The CRAF and Cassini Missions of the Mariner Mark II Program, NASA, ESA and BMFT, 1988 https://archive.org/details/NASA_NTRS_Archive_19890001550

NASA Flight Project Data Book, Office of Space Science and Applications, 1989

NASA Flight Project Data Book, Office of Space Science and Applications, 1991

Assessment of Planned Scientific Content of the CRAF Mission, National Academies Press, 1987

On the Scientific Viability of a Restructured CRAF Science Payload, National Academies Press, 1990

Assessment of Solar System Exploration Programs, National Academies Press, 1991

On the CRAF/Cassini Mission, National Academies Press, 1992

Lessons Learned: GAO Report on the CRAF/Cassini Missions, 4 February 1994: https://www.aip.org/fyi/1994/lessons-learned-gao-report-crafcassini-missions

4

Creating A Comet Chaser

4.1 COMET RENDEZVOUS

In May 1985, following discussions by the European members of the ESA-NASA Primitive Bodies Science Study Group, a Comet Nucleus Sample Return mission was recommended by the ESA Solar System Working Group as a Planetary Cornerstone mission for the Horizon 2000 long-term science program (see Chapter 3).

This ambitious project, envisaged as a joint enterprise between ESA and NASA, was adopted as the mission with the highest potential for scientific return, since it would include the in-situ study of a comet and the return to Earth of samples from the nucleus.

Unfortunately, although the Rosetta Comet Nucleus Sample Return mission achieved a lot of scientific support, and was considered to be technologically feasible, various programmatic financial and scheduling difficulties put the implementation of the project in doubt.

By the end of 1991, ESA initiated a parallel effort to design a mission that could be achieved using European technology alone, and that could be accommodated by the budget for one of its Cornerstone missions.

The mission scenario that gained full support from the European scientific community was a comet rendezvous mission, with very similar objectives to those of CRAF, that would deliver much of the desired science about the structure, morphology, density and composition of the nucleus (see Chapter 3). Key technological constraints were the use of solar power instead of radioisotope thermoelectric generators, the use of a European Ariane launch vehicle, and the use of ground stations belonging to ESA Member States.

© Springer Nature Switzerland AG 2020

P. Bond, *Rosetta: The Remarkable Story of Europe's Comet Explorer*,

Springer Praxis Books, https://doi.org/10.1007/978-3-030-60720-3_4

Investigations established that improved solar array technology capable of operating at low temperatures and low solar flux, along with the enhanced capabilities of an Ariane 5 in which the ignition of the upper stage was delayed, would enable a European mission to be achieved within the technological and financial limits.

At its 68th meeting on 4-5 November 1993, ESA's Science Program Committee endorsed the recommendation of the Space Science Advisory Committee to implement Rosetta as a comet rendezvous. Upon formally becoming the third Cornerstone of the Horizon 2000 program, it was targeted for launch in 2003.

Rosetta project scientist, Gerhard Schwehm recalled:

> We had to compete with Herschel for the number three spot in the long-term program. We actually had to present the two missions at ESA Headquarters to the advisory bodies.

> For Rosetta it was Horst Uwe Keller who presented the science case. We had well prepared him, but I have never seen him so nervous. Basically we had to demonstrate that both missions were feasible, technologically and financially. The science had, of course, to be excellent, worth a Cornerstone. I believe in the end the higher risk for Herschel, which required developing a cryogenic telescope, was in our favor.

> Martin Hechler, leading the Rosetta mission analysis team at ESOC [the European Space Operations Center in Darmstadt], and Yves Langevin did a fantastic job to work out mission scenarios to find a mission that would bring sufficient payload mass to a comet. It was really an intense, but extremely interesting time in the Science Definition Team.

Although the revamped mission was deemed to be an ESA-led project based upon European technology and launch capability, the opportunity for other agencies to join and augment the science was left open, and international partners quickly indicated their interest in taking part.

The new baseline mission was a rendezvous with a short period comet and at least one (probably two) fly-bys of asteroids. Gravity assist maneuvers at Earth and Mars or Venus would be necessary to acquire the energy required to reach the comet at its aphelion (when it would be farthest from the Sun). The spacecraft would then fly alongside the comet during its penetration of the inner Solar System and through its perihelion passage (closest approach to the Sun). As far as possible, the plan retained the objectives of the original Comet Nucleus Sample Return mission by emphasizing the in-situ investigations of the comet's coma and the structure of the nucleus, as well as the evolution of the comet as the heliocentric distance varied. This would enable the mission to study the composition of the comet, as well as the physical and chemical processes that evolve with distance from the Sun. At least one Surface Science Station would be deployed onto the comet's surface to conduct in-situ studies of the nucleus.

"As we cannot bring the cometary material into our terrestrial laboratories, we will take our laboratories to the comet," said Dr. Roger Bonnet, ESA's Director of Science.

4.2 THE BASELINE MISSION

By 1995, the baseline Rosetta mission had been defined with 46P/Wirtanen as the primary comet candidate. An Ariane 5 would send the spacecraft on its way in January 2003, after which Rosetta would receive gravity boosts from one fly-by of Mars and two of Earth. There would also be fly-by opportunities with asteroids (3840) Mimistrobell and (2530) Shipka on the way to the comet. Rosetta would arrive in August 2011 and remain in its vicinity through perihelion in October 2013. To enhance the scientific return, there was a possibility that one or two surface science packages would be deployed onto the nucleus.

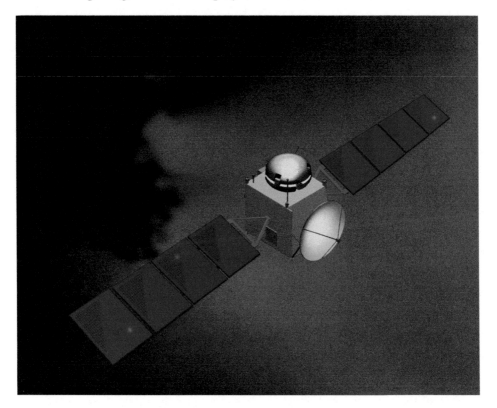

Fig. 4.1: A 1993 artist's impression of the Rosetta comet rendezvous spacecraft. Note the large solar arrays needed to provide electrical power at great distances from the Sun. The surface science package on the top of the craft had yet to be defined or selected. (ESA)

A number of consortia funded by national space agencies would supply the various scientific payloads. ESA would be responsible for the overall spacecraft and mission design; spacecraft procurement; orbiter payload integration; systems testing; spacecraft operations; acquisition, distribution and archiving of data; and the Rosetta Science Operations Center.

4.3   SURFACE LANDERS: CHAMPOLLION AND ROLAND

Although Rosetta would no longer be able to touch down on a comet or to return a sample to Earth, the revised mission scenario envisaged from the beginning some kind of surface probe or surface science package (SSP) that would piggy-back on the orbiter during the long trek to the comet.

One or two landers would be deployed after the nucleus had been mapped and examined in detail (the so-called observation phase) and a suitable landing site had been identified. They would be designed to carry out in-situ analysis of the surface over a period of at least several tens of hours, and possibly several months. It was envisaged that the probes/packages would be supplied by collaborations of national space agencies and/or major research institutes.

A request for SSP proposals was issued, and in November 1994 two were selected for further study. By this time, ESA had decided that two 45 kg packages could be accommodated on the orbiter and the intention was to deliver both of them to the nucleus.

One of these SSPs was the Champollion lander, proposed jointly by NASA-JPL in the USA and the French space agency CNES. The other SSP was RoLand (Rosetta Lander), proposed by an international consortium of European institutes, led by the German Aerospace Center (DLR) and establishments of the Max-Planck-Institute.

RoLand

RoLand was envisaged as a long term surface station capable of operating on the comet for a year or so, carrying out detailed studies and monitoring changes in the nature of the surface material that might enable scientists to learn more about the primordial matter from which it was derived.

The lander was originally drum-shaped (see Fig. 4.3), but by autumn 1995 it had become a pentagon. A scientific payload of at least 12 kg would allow investigation of the chemical and physical properties, and the mineralogy, of the surface material, the internal structure of the nucleus, and the processes driving its activity.

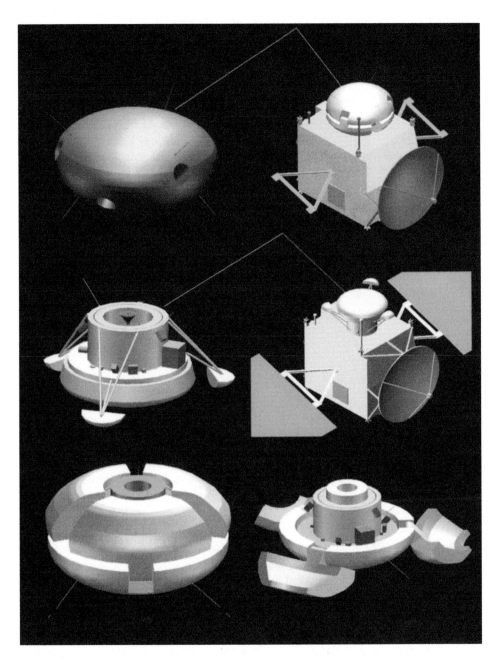

Fig. 4.2: Possible configurations of the probe to land on the nucleus of a comet, published in 1994 after the Rosetta comet rendezvous was accepted as ESA's baseline mission. (ESA)

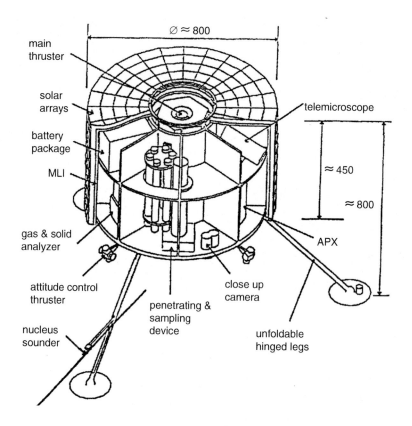

Fig. 4.3: The initial design of RoLand in autumn 1994. (RoLand consortium)

The mission scenario for RoLand foresaw separation from the main spacecraft at a distance of 3 AU from the Sun and 2 km above the comet's surface. The descent would be three-axis stabilized by cold gas thrusters, and the surface impact would occur at less than 1 m/s.

The lander would be covered with low-temperature solar cells that would provide about 4 W of power for the scientific instruments. Special consideration was made to provide optimized thermal insulation, since the length of the nights on the comet was unknown. RoLand would operate interactively for several months, including drilling activities and, towards the end of the mission, a degree of mobility.

The model payload of RoLand (see Table 4.1) was an alpha-proton-X-ray spectrometer, a gamma ray spectrometer, a gas and solid analyzer, a drill and sampling device, cameras (for panoramic viewing and microscopic observation), a seismometer, a magnetometer, several temperature sensors, and a microwave instrument that was part of a combined lander-orbiter instrument to investigate the interior of the nucleus.

Table 4.1:   Model Payload for RoLand

Abbreviation	Full Name	Principal Investigator
APXS	Alpha Proton X-ray Spectrometer	R. Rieder, MPI Chemistry, Mainz, Germany
COSAC	Cometary Surface and Subsurface Sampler and Evolved Gas Analyzer	H. Rosenbauer, MPI Ae, Germany
Sample Acquisition		ASI, Italy
MODULUS	Evolved gas Analyzer	Colin Pillinger, Open University, UK
ROLIS	RoLand Imaging System	S. Mottola, DLR, Germany
SESAME	Surface Electrical, Seismic and Acoustic Monitoring Experiment	D. Mühlmann, DLR, Cologne, Germany. H. Laakso, FMI, Finland. I. Apathy, KFKI, Hungary.
MUPUS	Multi-Purpose Sensor for Surface and Subsurface Science	T. Spohn, Univ. of Münster, Germany
ROMAP	RoLand Magnetometer and Plasma Monitor	U. Auster, DLR Berlin, Germany. I. Apathy, KFKI, Hungary.
CONSERT	Comet Nucleus Sounding	W. Kofman, CEPHAG, Grenoble, France

Champollion

Champollion was named after Jean-François Champollion, a French Egyptologist known for his role in translating the Rosetta Stone in 1824.

Following reconnaissance and gravity field mapping by the Rosetta orbiter, Champollion would be released and fall passively onto the surface of the comet from a height of several kilometers. During the descent, a flywheel would stabilize its attitude. Depending on the size of the comet, the lander would impact the nucleus at 1-4 m/s.

Most of the kinetic energy of the touchdown would be absorbed by landing pads of crushable material, such as aluminum honeycomb. A rebound would be prevented by solid propellant, hold-down thrusters that would fire for a few seconds at impact. Three 2.5 meter long spikes would then deploy to anchor the lander to the surface and damp out any forces that might be generated during sample acquisition.

Samples taken from depths of up to a meter would be transferred to science instruments for analysis. The results would be relayed back to Earth via the Rosetta orbiter as it continued its prime mission of remote sensing. Apart from detectors in science instruments, no radioactive materials would be flown. Thermal control was to be provided by electric resistance heaters.

The model payload for Champollion (see Table 4.2) was intended to determine the elemental, molecular, mineralogical, and isotopic compositions to a depth of a

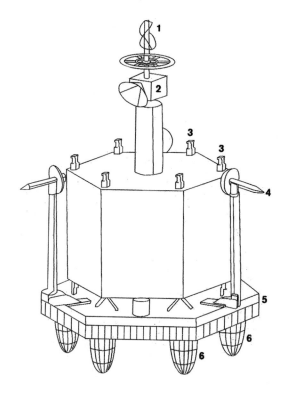

Fig. 4.4: An early 1995 illustration of the main features of the proposed Champollion lander. (1) low-gain patch antenna; (2) panoramic camera; (3) hold-down thrusters; (4) anchoring spike; (5) baseplate; (6) crushable honeycomb and feet. The battery-powered surface science package would operate for up to 84 hours. The three anchoring spikes are illustrated prior to deployment. The surface sample acquisition and transfer mechanism is not shown. (ESA)

Table 4.2: Model Payload for Champollion

Abbreviation	Full Name	Principal Investigator
CHAMPAGNE	Champollion Analysis of Gamma-Ray Emission from Comet Nucleus Elements	Claude d'Uston, Center d'Etude Spatiale des Rayonnements, Toulouse, France
CHARGE	Chemical Analysis of Released Gas Experiment	Paul Mahaffy, NASA Goddard Space Flight Center, Greenbelt, Maryland
CIRCLE	Champollion Infrared and Camera Lander Experiment	Roger Yelle, Boston University, Boston, Massachusetts
CPPP	Champollion Physical Properties Probe	Thomas J. Ahrens, CALTECH, Pasadena, California
ISIS	In-Situ Imaging and Spectrometry	Jean-Pierre Bibring, Institut d'Astrophysique, Orsay, France
CONSERT	Comet Nucleus Sounding	W. Kofman, CEPHAG, Grenoble, France

meter below the surface of the nucleus. The physical structure of the nucleus would be determined by measurements of the near-surface strength, density, texture, porosity, ice phases, and thermal properties. It was to obtain images at resolutions ranging from 1 meter down to 5 microns.

The lander would be powered solely by primary batteries, restricting its nominal lifetime on the surface of the comet to 84 hours. This would be sufficient for it to repeat measurement sequences, if necessary, and to obtain subsurface samples from at least three different depths.

Unfortunately, in September 1996, the cash-strapped NASA had to withdraw its support for Champollion, forcing its cancellation.[1] In response, the French half of the team opted to join the RoLand consortium. The expanded team then set out to develop a single, somewhat larger lander capable of addressing most of the scientific requirements. This was simply given the name Rosetta lander.[2]

Due to this late change in the membership of the RoLand consortium and the associated re-definition of its mission, the development of the SSP took longer to mature than the orbiter's payload. However, a "crash action" was initiated in order to define the interfaces sufficiently to enable the industrial Phase B to begin. The Rosetta lander became the subject of a specific design review by an independent team that confirmed the validity of the proposed design and praised the "excellent" engineering solutions in many critical areas.

4.4 STAGES OF DEVELOPMENT

The first step on the long road to realization of ESA's comet rendezvous mission came in early 1997, when the evaluation of the industrial offers for the role of prime contractor for the mission was completed.

ESA awarded a Phase B (System Design Phase) contract for the Rosetta spacecraft to the German company Dornier Satellitensysteme GmbH, with work to start in March 1997.

During this phase, the preliminary design of the spacecraft was elaborated and more closely defined. As with any major space mission, all of the proposed elements of Rosetta's flight to a comet underwent close examination at the end of Phase B. The extensive series of reviews scrutinized the design and development

---

[1] Champollion survived for a while in another iteration, as part of a comet lander/sample return technology demonstration associated with NASA's Deep Space 4 mission. A drill on the lander was to obtain a sample which would be taken to the spacecraft, where it would either be analyzed or returned to Earth. In 1999 the mission was canceled.

[2] The name 'Roland' was dropped because the eponymous Frankish hero wasn't favored by the German-led lander team.

status of the spacecraft, the payload of instruments, the ground segment, and the proposed launch vehicle.

In parallel, the industrial team that would build the spacecraft was assembled in two stages: first the major subcontractors were selected, then the equipment suppliers.

Once the overall system design was frozen, procurement specifications could be issued to the subcontractors. Matra Marconi Space UK and Matra Marconi Space France were assigned responsibility for the platform and avionics, respectively. Alenia Spazio of Italy was made responsible for the spacecraft's assembly, integration and testing.

The project passed its first major milestone on 19 December 1997, with the completion of the System Requirements Review. Nine months later, Phase B was completed with a successful System Design Review.

Meanwhile, at its meeting on 21 February 1996, ESA's Science Program Committee (SPC) endorsed the payload for the Rosetta orbiter. After a one-year science verification phase, final confirmation of the selected payload by the ESA Member States took place at the meeting of the SPC in February 1997.

Ground segment definition was also making progress, with ESA opting to augment its ground station network with a new deep space terminal in the southern hemisphere equipped with a 32 meter diameter dish antenna. This station would also serve other ESA missions, of course. After an Invitation-to-Tender was issued, the industrial procurement began in the second half of 1997. And finally, in 1999, a site was selected at New Norcia, about 130 km north of Perth in Australia. As the ESA Director's Annual Report for 1998 pointed out, "This concludes a rather complex selection process for the industrial contractors, which has at times seen very heavy competition within European industry, but which has ultimately enabled the given geographical return targets to be closely met."

As mentioned in Chapter 3, the principle of geographical return, often known as *juste retour* (fair return), required ESA to allocate industrial contracts to companies in its Member States based upon their financial contributions to the Agency's programs.

4.5   AN INTERNATIONAL ENTERPRISE

Rosetta's industrial team involved more than 50 contractors from 14 European countries and the United States. However, because Germany was the major financial contributor to ESA, followed by France, Italy and the UK, the *juste retour* policy automatically decided which countries would be handed the main industrial contracts.

Fig. 4.5: Rosetta was an international endeavor involving more than 50 contractors from 14 European countries, the United States, Canada and Australia. This map includes companies involved in building the orbiter and the ground station at New Norcia in Australia. The prime spacecraft contractor was Astrium Germany, and the major subcontractors were Astrium UK for the spacecraft platform, Astrium France for the spacecraft avionics and Alenia Spazio for spacecraft assembly, integration and verification. (ESA/AOES Medialab)

Dornier Satellitensysteme GmbH (a subsidiary of Daimler-Chrysler Aerospace AG/DASA) was selected as the prime contractor, having overall responsibility for the development and manufacture of the Rosetta orbiter.

Three major subcontractors were granted large shares of the work: Matra Marconi Space of Bristol and Stevenage (UK) was the lead contractor for the design and development of the spacecraft's platform, Matra Marconi Space of Toulouse (France) would design and develop the avionics, and Alenia Aerospazio of Turin (Italy) would assemble, integrate and verify the spacecraft.

After various corporate reorganizations resulted in the merger of DASA and Matra Marconi to create the giant Astrium company, Astrium Germany took over the role of Rosetta prime contractor. The companies having responsibility for the main subcontracts became Astrium UK (spacecraft platform), Astrium France (spacecraft avionics), and Alenia Spazio of Italy (assembly, integration and verification) (see Fig. 4.5).

ESA was responsible for funding the spacecraft platform, the launch services and operations. The Agency was also responsible for the overall management of the mission and coordination of the procurement and integration of the payload of scientific payloads and the lander, which were to be built by national research centers and universities under the direction of principal investigators with funding from ESA Member States and NASA.

The total cost to ESA of the Rosetta mission at its launch in 2004 was roughly €800 million, drawn from national subscriptions to the Agency's science program.

The lander was built by a European consortium led by the German Aerospace Center (DLR). Other members of the consortium were ESA, CNES, and institutes from Austria, Finland, France, Hungary, Ireland, Italy and the UK.

The overall mission cost was close to €1.4 billion, of which the lander cost €220 million (in 2014 economic terms). The costs covered development and construction of the spacecraft and its instruments, including the lander, along with launch and flight operations. Also included were the expenses arising from delaying the launch from 2003 to 2004 (see Chapter 6).

To put these figures in perspective, ESA noted that the overall cost was barely half the price of a modern submarine, or comparable to three Airbus 380 jumbo jets, and covered a period of 20 years, from the start of the project in 1996 until its conclusion in September 2016.

About 2,000 people from industry, ESA and scientific institutions were involved in Rosetta's development. The Agency noted, "It is difficult to establish exactly how many new jobs were created, but Rosetta has certainly helped contribute to the development of the space sector, both from the industrial and the scientific point of view."

4.6   ALL SYSTEMS GO

On 26 November 1998, the ESA Industrial Policy Committee (IPC) gave its approval for the start of the main development phase (Phase C/D) for Rosetta. This decision was confirmed by the Agency in January 1999. With the scheduled launch only four years away, there was little leeway to deal with any major problems that might arise.

However, all seemed to proceed fairly smoothly. The structural and thermal model (STM) of the spacecraft was assembled at the Finavitec plant in Halli, Finland, where the main section completed a series of static load tests. As hydraulic jacks pulled on the spacecraft's panels, a network of gauges at strategic locations measured the stresses imposed on the structure.

As Rosetta's systems engineering manager, Jan van Casteren, explained, "These tests tell us two main things. First we want to prove that nothing will break. For example, the Rosetta lander will be directly coupled to the shear panels, so it will be shaken during launch and pull on the side of the spacecraft. These stresses can be simulated during the static load tests by pulling on the lander side of the STM with a force of 2.7 tonnes. An even bigger load of 11.3 tonnes is applied to the lower fuel tank area of the central cylinder, which will eventually be filled with oxidizer. We can also see if the structure behaves as expected under such loads by comparing our measurements with predictions from computer models. If they don't match, we know something is wrong."

The STM was required to iron out any major problems in the design and assembly of the final operational spacecraft. Some 2.8 meters high and more than 2 meters wide, it was practically identical in size and shape to the final flight model.

It was also a good representation of Rosetta's thermal and mechanical design, because it was fitted with the same thermal blankets that would protect the flight model against the extreme temperatures of deep space. The scientific instruments, although not functional, were high fidelity replicas of the flight hardware, and a model of the lander which had already passed through vibration and thermal-vacuum testing was also attached.

However, there were some significant differences. For example, the STM carried only one solar array (the other was a dummy), the electronics boxes were empty, and only one of the attitude thrusters was real. Although the fuel tanks could be pressurized, they were filled with water instead of propellant for vibration testing.

After delivery to the Alenia Aerospazio plant in Turin during August 1999, the STM was separated into two sections: the payload module that would carry the scientific experiments, and the service module that would accommodate the spacecraft's main subsystems. Over the next several months, engineers worked in two shifts to carry out the assembly, integration and test programs on the individual modules, after which they were reintegrated in late October.

The first stage of the assembly process was installation of the propulsion system. This meant adding two propellant tanks and an intricate network of titanium pipes to the service module. Several kilometers of electrical cabling was then installed in the heart of each module. Next was the installation of 'dummy' scientific payloads on the payload module. The mechanical and thermal representations of the major subsystems, including the power, data handling and attitude control systems, were added in October. Finally, the lander STM, the solar arrays and the booms were added.

The STM was transported to the ESTEC test facility in the Netherlands on 4 December 1999, several days ahead of schedule, where it was subjected to a series of environmental tests over the next six months. This started with acoustic and vibration tests to find out if the spacecraft could survive the noise and shaking of an Ariane 5 launch. Then there were thermal-vacuum tests to verify its performance in the harsh temperatures and vacuum of outer space.

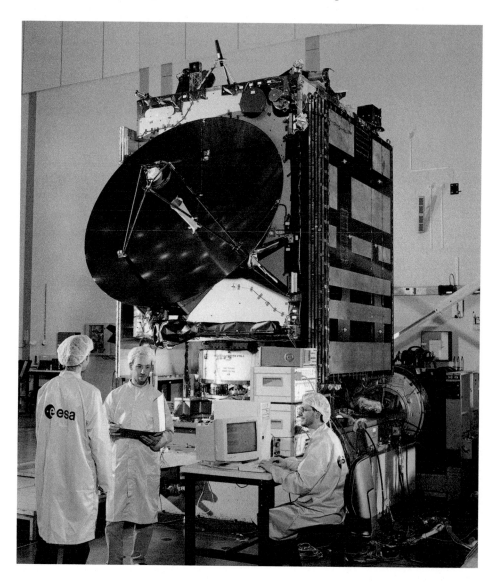

Fig. 4.6: The Rosetta structural and thermal model (STM) during vibration testing at ESTEC in January 2000. (ESA/A. Van Der Geest)

On 14 February 2000, a 14 meter long solar array was deployed. The checkout began when six 'thermal knives' were used to melt through the attachments that held the array to the side of the spacecraft. A special jig was used to compensate for the pull of gravity, while springs on the giant panel caused it to open out. After 3 minutes 47 seconds, the array was fully extended, enabling engineers to check its alignment and condition.

On 16 April 2000, after more than two weeks of being roasted and then frozen, the STM passed a series of tests in the Large Space Simulator at ESTEC. The largest thermal-vacuum chamber in Europe, this facility was a sealed cylinder 10 meters across and 16 meters high.

In the absence of the spacecraft's solar arrays, electricity was provided to different sections by about 100 external power sources that were attached to the STM by several kilometers of cable. A total of 450 temperature sensors placed all over the spacecraft were used to collect the most accurate information possible on its thermal behavior.

After all of the air was pumped out of the test chamber on 30 March, its black walls were flushed with liquid nitrogen. This caused their temperature to plummet to $-180°C$. The Sun was represented by an array of xenon arc lamps. Their output was directed and collimated by mirrors onto the STM, to simulate the variable exposure to sunlight and shadow that Rosetta would experience 1 AU from the Sun. In this 'hot case', all of the spacecraft's louvers were kept fully open to expel as much excess heat as possible.

The lamps were switched off a week later to simulate conditions beyond 5 AU (aphelion of the intended orbit), where the energy in sunlight is just 4% of that at 1 AU. In this case, the louvers automatically closed in order to prevent the heat escaping from the radiators, thus conserving as much heat as possible on the powered-down spacecraft. In addition to the flight model louvers procured from the U.S., there was a qualification louver entirely designed and built in Europe. Both types worked satisfactorily.

After completing the test program at ESTEC, the STM was placed in a protective container and loaded onto a lorry trailer for the five day trans-European trek back to Alenia Aerospazio in Turin, where it arrived on 19 May 2000. (Meanwhile, the lander STM was returned to the Max-Planck-Institut in Lindau.)

Stripped of its solar array, antenna and lander, the one tonne STM structure was refurbished for use as the engineering and qualification model (EQM), to demonstrate that the electrical and other systems would function correctly in deep space. All of the EQM instruments for the orbiter, as well as the EQM lander, arrived between May and December. The functional tests were completed in March 2001.

Major technological advances were also made in the development of the X-band transmitter, star tracker and software – new items which were critical for the success of the mission. All units passed their engineering model tests.

Meanwhile, working in parallel, engineers successfully assembled and tested the flight model (FM) hardware and software. The FM structure was transported from the Finavitec plant on 13-17 January 2001. Illuminated by flashing lights, the lorry (sometimes accompanied by a police escort) was ferried across to northern Germany prior to undertaking the 1,000 km trek through Germany and Austria and finally over the Brenner Pass. At the Alenia Spazio plant in Turin, the 250 kg structure was moved to a giant clean room, where the payload module (supporting all scientific instruments) and the service module (supporting all units needed to control the whole spacecraft) were separated.

Over the coming weeks, engineers integrated the orbiter's propellant tanks, attached their thermal blankets, installed the complex plumbing for the reaction control (thruster) system, and integrated the cabling harnesses. The platform's subsystems were attached in the spring, followed later in the year by all of the scientific payloads, with the exception of the lander.

All of the instruments underwent extensive scientific calibration between their assembly/test programs and delivery to the main spacecraft. By the end of the year, the experimenters were preparing flight operations procedures for the commissioning phase of the mission and future scientific operations.

In November 2001, the fully integrated orbiter FM was transported from Alenia to ESTEC in the Netherlands for an environmental program that included thermal-vacuum and mechanical testing. When the FM lander was delivered, it was mated to the orbiter.

4.7   PROBLEM SOLVING

As is the case with any major project, not everything went to plan. Late deliveries of certain units, and problems with the integration, meant that Alenia engineers often had to work three shifts per day on the FM and two shifts on the EQM. As ESA wryly noted, "The industrial contractors showed much dedication and flexibility in maintaining the very challenging schedule."

As early as December 2000, ESA's Rosetta project manager, John Ellwood, was informing the Science Working Team, "No major technological risks are outstanding, but we must all continue to make every effort to meet our deadlines. We are 'Go' for launch on 12 January 2003, but Comet Wirtanen will not wait!"

Meanwhile, delays in the development of the lander continued to raise alarm bells. As early as October 1999, experts making an independent review expressed concern about the "rather limited manpower" in the institutes that were participating in the lander consortium to design and build such a complex spacecraft.

By January 2000, it was clear that the lander's development was not going according to plan, so ESA stepped in to offer assistance to ensure the hardware development would be brought back on track. Regular meetings were arranged

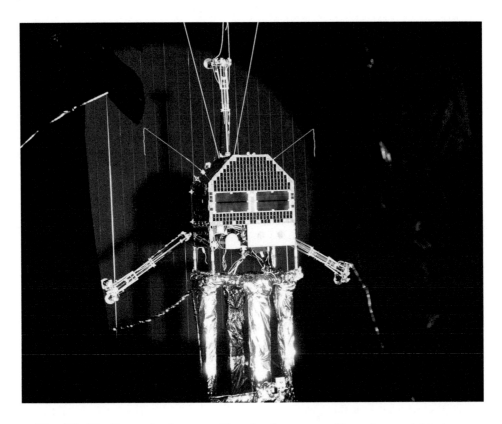

Fig. 4.7: The Rosetta lander was subjected to its own specific environmental test program at the Ottobrun (Munich) facilities of Industrieanlagen-Betriebsgesellschaft (IABG). Here, the spacecraft is shown from above (with respect to its launch attitude) during tests in a thermal-vacuum chamber, an 'oven' from which the air was extracted and the extreme temperatures of deep space simulated. The three legs of the landing gear are extended. The bright columns in the lower half of the frame are a support stand. The photograph was taken on 2 November 2001. (IABG, Ottobrun)

between the two groups in order to review the development program and develop a detailed course of action. The recommendations were presented to the Lander Steering Committee on 4 April.

"To start, we decided to create a mini project team within the Rosetta team to deal with this," explained Rosetta's newly appointed project manager, John Ellwood. "Originally there had been just one project member following the lander development, but now I recruited Philippe Kletzkine, a manager from the recently launched XMM team, and he had two engineers (an electrical/software and a mechanical engineer) working for him. We identified several areas where we could help. In particular, the lander was mechanically complex and so we provided some mechanical engineering support from our prime contractor to address these various landing mechanisms. We were also concerned with the amount of testing that was foreseen, and so offered to perform some environmental testing in ESTEC – alongside the main orbiter testing. By getting more involved, the lander consortium could also use the vast experience available at ESTEC."

The control system at ESA's Space Operations Center (ESOC), Darmstadt, Germany, was another cause for major concern. According to Paolo Ferri, head of mission operations, the first deliveries of the mission control system were full of problems, the system simulator arrived late, and the new interfaces with the ground stations progressed very slowly. Writing in an article published in 2016, Ferri noted, "When we first had the opportunity to connect our ground systems to the spacecraft, which was being integrated in Torino (Turin), we also found out that the onboard software did not behave as we had specified. The year 2001 was one of hard work and disappointment, delays, testing and retesting with slow progress and no major improvements."

Concerns over the tight schedule, the performance of the ground systems and ability to train the flight control team with a realistic and stable simulator caused the project manager, John Ellwood, to separate the development and testing of the spacecraft's onboard software from the integration and test campaign of the spacecraft flight model. At the same time he allowed ESOC to participate more directly in the onboard software test campaign on the spacecraft's engineering model.

A number of test sessions enabled the joint teams of ESA and industry to identify and resolve the main problems in the onboard software. At the same time, intensified testing with the spacecraft allowed them to improve the ground software and to increase the confidence in the ground systems interfacing with the spacecraft. Several long test sessions were conducted round the clock, with two shifts in ESOC and Turin.

Meanwhile, the simulation campaign at ESOC had started. Twice weekly the mission control team came together and used a software simulator to go through all operations, scenarios and flight procedures, including possible failures.

Paolo Ferri admitted, "To add to the stressful situation, our simulations uncovered new bugs in the spacecraft onboard software, which were not detected during system tests, and due to the limited time remaining could not be changed before flight. Many discussions took place at project level to evaluate the degree of risk we were taking by launching the spacecraft with known important software problems. The avionics team in industry immediately started to work on another version of the onboard software, to eliminate the problems, but this would only be ready for upload to the spacecraft in flight in July 2003. Rosetta would have to fly for a few months with the current software. A risk, certainly, but, given that it would be corrected in a short time compared to the long mission – and that during this period Rosetta would still be relatively close to Earth – it was considered acceptable."

4.8   FINAL TESTING AT ESTEC

During 2002, the FM spacecraft had to endure an intensive checkout, starting with a lengthy thermal-vacuum test that simulated the extreme temperatures to which it would be exposed in deep space.

One of the most dangerous activities undertaken by the ground team was the acrobatic high wire act of Alenia Spazio engineer Natalino Zampirolo. Before the thermal-vacuum checks could begin, the intrepid Italian was suspended from an electric hoist so that he could 'fly' alongside the spacecraft some 5 meters above the floor of the giant chamber. He gingerly removed the 'red tag' items, including protective covers and arming plugs on the explosive connectors that had been fitted as a safety precaution during normal work on the spacecraft. By the time he completed these hazardous activities and retreated to safety, Rosetta had been 'armed' and was ready to start its thermal trial.

In the Large Space Simulator, the orbiter, lander, and the full complement of 20 instruments were alternately baked and frozen to demonstrate that the insulation and louvers and radiators could restrict the spacecraft's internal temperature to between 40°C and –10°C.

The program continued with full mechanical, electromagnetic and functional testing, in order to confirm that the spacecraft and its payload would operate correctly during all phases of the mission.

In April 2002, once the high-gain antenna and solar arrays were mounted on the orbiter, the FM was subjected to a series of vibration tests in an acoustic-vibration chamber in which a blast of sound that peaked at 135 decibels was directed at the spacecraft by a huge amplifier. Anyone entering the chamber would have died almost instantly.

Another simulation to replicate the extreme conditions to be experienced during an Ariane 5 launch involved a rock-and-roll ride on a giant shaker. Attached to a movable table, the three tonne spacecraft was severely shaken, first horizontally, then vertically, over a wide range of frequencies. Hundreds of accelerometers monitored its response during each of the 3 minute endurance tests, and the results showed that Rosetta was ready for its rough ride to orbit.

Once the engineers at ESTEC were satisfied that the spacecraft's electronics had survived the stresses, it was time to find out whether the solar arrays, antennas and instrument booms had been adversely affected.

The solar wings, stretching one and a half times the length of a tennis court, were folded and held alongside the spacecraft's main structure by the six Kevlar cables that would bind them for launch. The cables were cut in sequence by 'thermal knives' which heated them to several hundred degrees Celsius. After the sixth cable was cut, the array gradually opened out until it reached its full length of 14 meters. In the absence of weightlessness, this was made possible by a special deployment rig in which the weight of the arrays was counterbalanced by a mass-compensation device incorporating dozens of springs. Once the solar arrays were successfully deployed, it was time to check the other movable parts on the orbiter.

First came a partial deployment of the 2.2 meter diameter high-gain antenna. This began with the firing of three pyrotechnic charges (explosives) to release the dish from its stowed launch position. Next was the firing of more charges to deploy the two booms, each 2 meters long, that would carry equipment to investigate the magnetic and particle environment around the comet. The fifth and final test was the explosive release of a wire antenna for the CONSERT radio experiment. Supported by five helium balloons, the H-shaped antenna gently unfolded as planned.

Fig. 4.8: Cocooned within protective plastic foil, the Rosetta spacecraft, with its high-gain antenna deployed, undergoes electromagnetic compatibility (EMC) testing in the Compact Payload Test Range at ESTEC in August 2002. The test chamber was lined with cones that absorbed radio signals and prevented reflections. The walls of the chamber formed a steel 'Faraday cage', impenetrable to electromagnetic signals from outside. In this radiation-free environment, engineers could study radio signals and electrical noise from the spacecraft's systems and determine whether there was any electromagnetic interference between them. (ESA/Anneke Le Floc'h)

After Rosetta was transferred to another test chamber, known as the Compact Test Range, the hectic schedule continued in June 2002 with a comprehensive electromagnetic compatibility (EMC) check. The interior of this chamber was lined with cones which absorbed radio waves and prevented reflections. The walls formed a steel 'Faraday cage' that was impenetrable to external radio signals. Engineers analyzed the radio signals and electrical noise coming from the spacecraft, to determine whether there was any electromagnetic interference between its systems.

Initially, the protective plastic foil in which Rosetta was cocooned was wrapped too tightly, causing the spacecraft's temperature to rise. Once this had been loosened, the test was able to proceed.

"For some of this time we were measuring the energy emitted by the spacecraft's high-gain antenna, and this is hazardous, so the chamber was completely closed and everyone had to remain outside it whilst the measurements were made," recalled Flemming Pedersen, a senior assembly, integration and verification engineer for Rosetta. "That would be like exposing the engineers to the radiation from thousands of mobile phones simultaneously."

The first series of tests studied behavior during 'launch mode' to ensure that the spacecraft's systems would not interfere with communications between the Ariane 5 and ground control. Other EMC tests simulated different periods of activity, from the quiet hibernation period of no instruments operating to times of hectic scientific operations.

"We could switch on each instrument individually and measure the electromagnetic waves coming from it," explained Bodo Gramkow, the principal payload engineer. "The rest of the instruments were put into listening mode to see if any of them detected any disturbance. On other occasions we switched on all of the instruments, including the lander, to see whether we got any unexpected noise or interference. From this we could determine whether we will need to switch a particular instrument off when we are making a very sensitive measurement with another one."

In addition, system validation tests were performed from the mission control facility at ESOC in Germany, to validate the planned flight operations procedures.

Thanks to the continuing hard work and 24 hour, seven days per week shifts of the ground teams, Rosetta (with its payload and lander in flight configuration) was finally ready to be shipped to the launch site in Kourou, French Guiana.

Meanwhile, the expansion of the ground segment was completed with the construction and testing of the 35 meter dish antenna at New Norcia in Australia. Simulations were carried out to train the operators at ESOC, and by the end of the year all elements of the ground network were ready for the launch.

4.9   FAREWELL EUROPE, HELLO SOUTH AMERICA

Rosetta was lifted into a huge, protective metal container on 2 September 2002, ready for its journey to the launch site on the far side of the world. As a final precaution, engineers tested the spacecraft's fuel tanks to ensure they were completely leak-proof, then Rosetta was sealed inside its nitrogen-purged box.

"This room-sized container held the complete spacecraft, including the lander and the large dish of the high-gain antenna," said Walter Pinter-Krainer, the mission's principal assembly, integration and verification engineer. "Only the folded solar arrays, which were shipped to the Kourou spaceport a few weeks earlier, were not attached to the orbiter."

On 10 September, the container was transferred to a trailer for a slow overnight drive to Schiphol airport near Amsterdam. Illuminated by the flashing lights of the police escort, the precious cargo passed along deserted country roads and onto the motorway, on its way to the airport.

Later that day, the container and ground support equipment – 62 tonnes in all – were gingerly loaded aboard a Russian Antonov 124 transport plane, ready for transportation to the South American launch center. After a 16½-hour journey, which included a stop-over on the Cape Verde Islands, the air freighter touched down at Cayenne airport, near the Kourou spaceport, in the early evening.

The Antonov was immediately unloaded and the spacecraft was taken to the S1 building, one of the specialist payload facilities at Kourou, where engineers from ESA, Astrium Germany and Alenia had installed the electrical links, internet communications and miscellaneous other equipment that had been shipped to the launch center on an Arianespace supply vessel in late August.

"With the move from Europe to Kourou, we have now entered the most exciting phase of the Rosetta program so far – the launch campaign," noted Walter Pinter-Krainer. "Everything is looking good and we are on schedule for the launch."

4.10   LAUNCH PREPARATIONS

Over the next three months Rosetta was subjected to a rigorous, step-by-step program of flight preparations which would conclude by mating the spacecraft with the upper stage of the Ariane 5 launch vehicle.

The first step was to reassemble the 3-tonne Rosetta orbiter inside the spacecraft preparation facility at Kourou. Once Rosetta had been set up in its launch configuration (without the solar arrays and high-gain antenna) and passed a leak test of its propulsion system, the ground team installed the final versions of its flight software and validated the electrical systems.

This was followed by the Final Acceptance Test, which involved engineers in the payload preparation facility working around the clock for eight days on checks to verify the spacecraft was responding as it should.

Fig. 4.9: The Rosetta spacecraft and the lander at the Center Spatial Guyanais (CSG) facilities in Kourou, French Guiana, in September 2002. The high-gain antenna is visible on the upper face, and instruments can be seen at the left. The solar arrays have yet to be attached. (ESA-Service Optique CSG)

"This was the last full functional test of the spacecraft, including its payload, before lift-off," said Claude Berner, Rosetta payload and assembly, integration and verification manager. "It was a critical moment, but the test was successfully completed on schedule. We will not carry out another full functional test until after the launch."

A complex system validation test was also conducted with colleagues at ESOC in Germany. During a transatlantic link that lasted almost non-stop for 96 hours, specialists verified their ability to communicate with Rosetta in real time.

"The test was extremely successful," reported Paolo Ferri, operations manager at ESOC for Rosetta. "We tried to operate the spacecraft in a realistic manner. Members of the experiment teams were present in ESOC to study the results as we commanded the experiments – both in sequence, one after the other, and then several at a time."

One of the key aspects of the long-distance trial was a simulation of the operations that would take place promptly after launch.

"Immediately after they manually triggered the separation switches (simulating the spacecraft separation from the Ariane launcher) we took control and ran through the same sequence that we will follow on 13 January," said Ferri. "It was a very real simulation of the first day and a half of Rosetta's mission."

Finally, the 2.2 meter diameter high-gain dish antenna was integrated with the orbiter and the solar arrays were stacked against the orbiter's side.

On 13 November, after carefully reviewing all aspects of the mission, the Flight Readiness Review Board gave the 'green light' for the final stages of the launch campaign. The team of experts declared itself fully satisfied with the state of the spacecraft, the payload, the lander, and the ground segment.

The Board – which comprised about 40 representatives of ESA and participating countries – was co-chaired by Professor David Southwood, ESA's Science Director, and Rene Bonnefoy, the ESA Inspector General. The scientific principal investigators also attended the review to report on the ground calibration of their instruments.

"I am delighted to say that the Board identified no 'show stoppers' that will cause us to delay the launch," declared John Ellwood, the project manager. "There are still some areas relating to preparation of the launch vehicle that are a little behind schedule, but we are confident that they will be completed in the next few weeks."

In mid-December, the orbiter was loaded with approximately 1,060 kg of oxidizer (nitrogen tetroxide) and 660 kg of fuel (monomethylhydrazine). This was more than half of the spacecraft's lift-off mass. Every precaution had to be taken to ensure that there was no spillage or exposure to the toxic fuel.

Wearing special pressurized SCAPE protective suits, a team of six engineers from Astrium UK worked in shifts over three days to fuel the orbiter – two shifts for each propellant tank, with two men on each shift and two more in the control room. Each shift was restricted to a maximum of 4.5 hours because that was the lifetime of the team's radio batteries.[3]

"We breathe normal air, but the temperature of the suit is controlled by an air conditioner," explained Chris Smith, the team leader. "The suit weighs about 40 pounds (18 kg) and the helmet is very heavy, so it presses on your shoulders and gives you a backache. While we were working, everyone was cleared from the site for a distance of one-third of a mile (0.5 km) and blast walls were in place in the building. 'Sniffers' monitored the air in the room to detect any release of vapors. It all went very smoothly. It is always difficult connecting the fuel lines to the spacecraft while wearing a SCAPE, but there was only one leaky valve and that was spotted immediately."

The success of this fueling operation completed the hectic three-month prelaunch campaign at Kourou for the engineers from ESA and industry. Rosetta now rested inside building S5B, awaiting the go-ahead for mating with its launch vehicle.

On 19 December, the Ariane 5 for Flight 158 was transferred from the spaceport's launcher integration building to the final assembly facility. During the transfer, the giant rocket was moved on its launch table along the dual rail track that connected the various sites within the ELA-3 launch complex. The rocket's central core stage had previously been mated to its twin solid rocket boosters. The vehicle equipment bay housing the guidance, telemetry and control systems was added atop the core stage, and then the EPS upper stage.

The scheduled date of lift-off for the Rosetta mission was a few weeks away, but it was not to be. On the night of 11-12 December, an updated version of the Ariane 5 failed on its maiden flight. This more powerful Ariane 5 ECA variant was to have put two satellite payloads into geosynchronous transfer orbit.

A Board of Inquiry named by Arianespace, ESA and the French space agency CNES was set up to investigate the cause of the "anomaly" and its potential impact on preparations for the upcoming Rosetta mission.

"This was a very critical period for ESA, and I remember that this was a very intense time," recalled Gerhard Schwehm later. "Most of the project team had taken leave to get some rest before the final preparation for launch."

On 30 December, as the Board was deliberating, irreversible operations linked to Rosetta's launch were suspended. The launch would have to be postponed at least several days beyond the targeted date of Sunday, 12 January. A new launch date was expected to be announced no earlier than 11 January 2003.

---

[3] SCAPE stands for Self-Contained Atmosphere Protective Equipment.

REFERENCES

ROSETTA Comet Rendezvous Mission, ESA SCI(93)7, September 1993

Rosetta – ESA's Planetary Cornerstone Mission, G. Schwehm and M. Hechler, ESA Bulletin No. 77, p. 7-18, February 1994

ESA's Report to the 30th COSPAR Meeting, ESA SP-1169, May 1994

Beyond This World, Nigel Calder, ESA BR-112, May 1995

ESA's Report to the 32nd COSPAR Meeting, ESA SP-1219, July 1998

ESA's Report to the 33rd COSPAR Meeting, ESA SP-1241, July 2000

Rosetta Rendezvous Mission with Comet 67P/Churyumov-Gerasimenko: https://directory.eoportal.org/web/eoportal/satellite-missions/content/-/article/rosetta

The International Rosetta Mission, ESA Bulletin 93, p. 38-50, February 1998, http://www.esa.int/esapub/bulletin/bullet93/VER.pdf

Rosetta Hardware Design Review, Daimler Chrysler Aerospace, 1999

Industrial *Involvement in the Rosetta Mission*, http://sci.esa.int/rosetta/54180-industrial-involvement-in-the-rosetta-mission/

Rosetta: ESA's Comet Chaser, ESA Bulletin 112, p. 10-17, November 2002a

Exploring A Cosmic Iceberg, ESA Bulletin 112, p. 18-23, November 2002

A Race Against Time, ESA Bulletin 112, p. 24-29, November 2002

Rosetta arrives in South America, 13 September 2002b: https://sci.esa.int/web/rosetta/-/30527-rosetta-arrives-in-south-america

Rosetta – Starting from Scratch, Paolo Ferri, Room, Issue #1(7) 2016 https://room.eu.com/article/rosetta-starting-from-scratch

5

Anatomy of a Mission

5.1   THE ROSETTA ORBITER

ESA had little experience of sending spacecraft far beyond Earth orbit, so it was not surprising that, in some ways, the design of the Rosetta orbiter was fairly traditional. It was based upon the Eurostar platform used by many European communications satellites and consisted of a box-shaped central frame and aluminum honeycomb platform whose dimensions were 2.8 × 2.1 × 2.0 meters.

"One reason that we based it on the Eurostar platform had to do with the lifetime of the mission," recalled Rosetta project scientist Gerhard Schwehm. "We wanted to rely on designs for telecom satellites, which usually have a design lifetime of 12-plus years. However, it was a big struggle in the beginning for some guys in the project team to get used to a three-axis stabilized spacecraft design."

Table 5.1: Spacecraft Vital Statistics

Size: main structure	2.8 × 2.1 × 2.0 meters
span of solar arrays	32 meters
Launch mass	
- total	3,000 kg (approx.)
- propellant	1,670 kg (approx.)
- science payload	165 kg
- lander	100 kg
Solar array output	850 W at 3.4 AU, 395 W at 5.25 AU
Propulsion subsystem	24 × 10 Newton bi-propellant thrusters
Communications	S-band and X-band. One 2.2 meter diameter high-gain antenna, one medium-gain antenna, two low-gain antennas
Operational mission	12 years

© Springer Nature Switzerland AG 2020
P. Bond, *Rosetta: The Remarkable Story of Europe's Comet Explorer*,
Springer Praxis Books, https://doi.org/10.1007/978-3-030-60720-3_5

The spacecraft was built around a vertical thrust tube having a diameter of 1.194 meters. This matched the interface with the Ariane 5 launcher. The primary structure consisted of a central cylinder of aluminum that housed the propellant tanks. Shear panels connected the side plates on which the spacecraft and payload equipment were mounted.

Including the 100 kg lander and 165 kg of science instruments the spacecraft's overall launch mass was approximately 3,000 kg, some 56% of which was propellant. The spacecraft's upper mass limit was determined by the capability of the Ariane 5 launcher for the intended mission, which allowed for a maximum 'wet' (spacecraft and fuel) mass of about 3 tonnes.

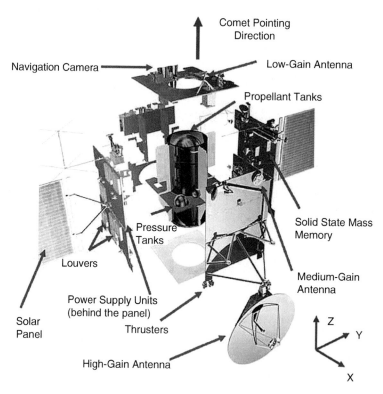

Fig. 5.1: An exploded view of the Rosetta orbiter. (ESA/AOES Medialab)

The orbiter comprised two distinct sections: the Payload Support Module which carried the scientific instruments, and the Bus Support Module with the major support subsystems. The scientific instruments that would face the comet during the operational phase of the mission were mounted on top of the bus.

On one side of the orbiter was a 2.2 meter diameter communications dish – the steerable high-gain antenna. The lander was attached to the opposite face.

The requirement for continuous comet pointing meant that the orbiter needed rotatable solar arrays and a two-axis steerable high-gain antenna. In the vicinity of the comet, the instruments would almost always be facing the nucleus, while the solar arrays and antennas were pointing toward the Sun and Earth respectively.[1]

In contrast, the orbiter's side and back panels would be in shade for most of the mission. As these panels usually received little sunlight, they were the ideal location for the radiators and louvers employed for thermal control. They would also face away from the comet, so damage from dust ejected by the nucleus and in the coma would be minimized.

Propulsion

Rosetta's main propulsion system consisted of 24 bi-propellant thrusters (arranged in a dozen redundant thruster blocks) for trajectory and attitude control. Four pairs of thrusters were for changes in velocity ('delta-V' burns). This reaction control system was also used to adjust the spacecraft's attitude.

Each of these thrusters propelled the spacecraft with a force of 10 Newtons, or about the same as experienced by someone holding a large bag of apples. The thrusters, made to a design that was widely used in spacecraft, were provided by DASA (later Astrium) in Germany.[2]

Mounted around the vertical thrust tube were two large, equally sized, propellant tanks for the fuel and oxidizer. The top tank was filled with MMH (monomethylhydrazine), a volatile form of hydrazine with the chemical formula $CH_3(NH)NH_2$ which was commonly used in bi-propellant engines because it is hypergolic with a range of oxidizers.[3] The lower tank held the N_2O_4 (nitrogen tetroxide) which would chemically interact with the MMH in the combustion chamber. The propellants were pressurized by helium in a pair of 68 liter, high pressure tanks.

The spacecraft carried 1,719.1 kg of propellant at launch: 659.6 kg of fuel and 1,059.5 kg of oxidizer, contained in two 1,108 liter, grade 5 titanium alloy tanks. The thrusters were able to provide delta-V of at least 2,300 m/s over the course of the mission.

During Rosetta's long trek to Comet 67P, a leak in the reaction control system that appeared in 2006 meant the tanks could not be re-pressurized, so the flight operations team had to learn to use the thrusters at a slightly reduced efficiency and lower pressure than planned.

---

[1] Of course, at large distances from the Sun, the Earth and Sun lie in roughly the same direction.

[2] Astrium was subsequently renamed Airbus Defence & Space.

[3] A hypergolic fuel ignites spontaneously when mixed with the oxidizer. This simplifies the propulsion system by eliminating the need for mechanical ignition.

Power System

Rosetta's appearance was dominated by the huge solar arrays, one on each side, that were the primary source of electricity. They were made by Dutch Space, a Netherlands-based company that specialized in making solar arrays for satellites. Officine Galileo Milan was responsible for the design and layout of the solar array photovoltaic assemblies for both the lander and the orbiter.

Fig. 5.2: Extension of one of Rosetta's 14 meter long solar arrays during a ground test in June 2002. Each wing consisted of five 2.25 × 2.736 meter panels. The spacecraft's overall span was 32 meters. (ESA)

The wings were fitted to the spacecraft body with a yoke and drive mechanism that permitted ±180 degrees of rotation in order to catch the maximum amount of sunlight. Both sides of the solar arrays were electrically conductive to avoid the build up of electrostatic charges.

Rosetta was the first deep space mission to rely exclusively on solar arrays for its electrical power generation. Previous missions that ventured to the orbit of Jupiter and beyond had been equipped with nuclear radioisotope thermal generators (RTGs) that produced power from the heat released in the radioactive decay of plutonium. But Europe did not have the capability to design and manufacture such power sources.

To generate sufficient power at distances of more than 5 AU (800 million km) from the Sun, where the amount of sunlight was only 4% of its intensity at Earth, 64 square meters of solar cell surface were necessary. Each wing comprised five 2.25 × 2.736 meter folding panels and spanned 14 meters.

The arrays were covered with more than 22,000 non-reflective solar cells that were specially developed by German company ASE. The individual solar cells were 200 microns thick and each measured 61.95 × 37.75 mm.

The very high efficiency, low intensity, low temperature (LILT) silicon cells were made in Germany by DASA, under contract to ESA. Their top surfaces employed pyramids to maximize the capture of sunlight. At a temperature of −130°C and an intensity of 40 Watts per square meter, they could provide a sunlight conversion efficiency of 25%.

Spacecraft power was controlled by a redundant Terma power module that had previously been incorporated into ESA's Mars Express spacecraft.[4]

While the spacecraft was in the vicinity of Earth, the solar arrays could provide up to 8,700 W of power. When close to Jupiter's orbit they would generate 440 W, which was just sufficient to keep the orbiter alive. At 3.4 AU (510 million km from the Sun), the panels would generate the 850 W required to initiate scientific observations of the comet.

In order to survive 31 months in hibernation, Rosetta's solar panels would be rotated toward the Sun, so that they could receive as much sunlight as possible. Meanwhile, almost all of the electrical systems would be switched off, with the exception of the radio receivers, command decoders and power supply.

Back-up power for the orbiter, required during periods of eclipse or darkness, was originally to have been provided by rechargeable nickel-cadmium batteries, but these were replaced by lithium-ion batteries developed by AEA Technology, which were smaller, lighter, more reliable and more flexible.

Temperature Control

Temperature control was particularly challenging for Rosetta, owing to a 25-fold variation in incoming solar radiation during the course of the mission. At perihelion, the spacecraft would be less than 150 million km from the Sun and actually inside Earth's orbit, but at aphelion it would be some 800 million km from the Sun.

Designers had to cater for two opposing scenarios: a cold environment (−130°C), where the priorities were preserving heat and keeping the spacecraft systems warm, and a much warmer environment in which heat rejection was necessary in order to maintain equipment within the allowable temperature range.

To minimize the power consumption during its treks through the asteroid belt and beyond, it was important to limit leaks of heat to space and minimize the

---

[4] Mars Express was originally meant to fly after Rosetta, so a lot of design features and subsystems developed for Rosetta were used in Mars Express and its sister craft, Venus Express.

radiator surface areas. At such times, the power consumption of Rosetta's equipment was estimated to be 93 W, so, in order to balance the anticipated heat loss to space of 222 W, it was necessary to provide 132 W of power to heaters at strategic points.

Unlike most spacecraft that had previously ventured into deep space, Rosetta avoided the use of radioisotope heater units to offset the power budget, but this strategy was only possible by imposing tight control of heat leakage to space.

The most significant sources of heat loss at aphelion were instrument apertures and external spacecraft equipment (26%), the radiators (22%), the side panels (26%) and appendages. Also important were leaks through the thermal blankets (16%) and the launch vehicle interface ring (10%).

Fig. 5.3: One of the adjustable louvers that were fitted over the radiators to enable different amounts of heat to escape when the spacecraft was exposed to warm and cold environments. (ESA/A. Le Floc'h)

The orbiter was fitted with 14 polished, metallic louver panels – rather like hi-tech Venetian blinds – which covered an area of 2.25 square meters. They were installed over the radiators across the side and back of the spacecraft. The opening and closing of the louvers was fully passive, requiring no power. Instead, they worked on the bi-metallic thermostat principle. The 16 blades on each louver were moved by coiled springs made up of three different metals that expanded and contracted at differing rates. Designed by Spain's SENER company, they were designed to rotate precisely when required.

The primary method of reducing heat loss was to wrap the spacecraft in blankets of high performance multi-layer insulation (MLI), each of which comprised two distinct sections.

The Dunmore Corporation of Pennsylvania in the USA supplied the Kapton polyimide-based films. These were layered together with protective separation films to create blankets ranging from 10 to 30 layers thick, depending upon the type and level of protection required for each location on the spacecraft. The films were stacked, cut into complex shapes in order to cover each component of the spacecraft, then sewn together and secured with polyimide tapes. The resulting MLI not only absorbed and reflected radiant energy but also gave protection against impact from micrometeoroids.

The external foil was 1 mm carbon-filled black Kapton with VDA on the interior side. This material had stable properties over a long lifetime, was electrically conductive (to prevent the build-up of static electricity), and could be handled without risk of degradation.

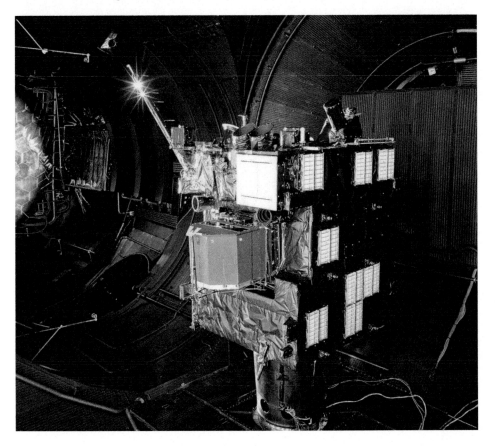

Fig. 5.4: The Rosetta spacecraft without thermal blankets, prior to vibration testing on the shaker at ESTEC in the Netherlands. (ESA/Anneke Le Floc'h)

Fig. 5.5: Technicians installing multi-layer insulation on Rosetta in November 2003, during the second launch campaign at Kourou spaceport. (ESA)

Care was taken in the design and attachment of the MLI to ensure that there were interleaved overlaps, no through-stitching, no local compression, and no stand-offs terminating outside.

Battery-powered heaters were fitted at strategic locations, such as propellant tanks, pipework, thrusters and some equipment in order to guard against temperatures sinking too low. During the prolonged hibernation periods, primary heaters would always be turned on and redundant lines thermostatically controlled.

These measures were confirmed by rigorous checks on the Structural Thermal Model of the spacecraft tested in a thermal-vacuum chamber in the Large Space Simulator at ESTEC in the Netherlands.

Attitude and Control

Rosetta was three-axis stabilized for normal operations, but a spin stabilized mode was used when most of the systems were shut down during the prolonged hibernation in deep space.

Data fed into the attitude and control system (AOCS) came from a pair of navigation cameras (NAVCAMs), a pair of star trackers, four Sun acquisition sensors, and three laser gyro inertial measurement packages.

The 5 degree field of view and 12 bit, 1,024 × 1,024 pixel resolution of the NAVCAMs was sufficient to enable ground controllers to determine the spacecraft's position and orientation during the approaches to asteroids and the comet. Supporting data came from monitoring the radio signals sent back by the spacecraft.

Rosetta's attitude was controlled by the reaction control system of bi-propellant thrusters and four reaction wheel assemblies (RWA). Over time, the RWAs were vulnerable to friction, so they incorporated a method for lubrication. They were also fitted with heaters for protection against the low temperatures encountered in deep space.

In the qualification program, an existing life test wheel was frozen at −35°C for 30 days to demonstrate that a mechanical wheel could survive the experience and then be restored to an operational state for the main observational period of the mission.

When two of the four reaction wheels started to show signs of friction and vibrations at their normal operating speed in flight, the ground controllers learned to operate the wheels at lower speeds than originally intended, and, as a precaution in case the two faulty wheels were to fail completely, new software was uploaded to enable Rosetta to operate using only two wheels.

Communications

Rosetta's communications suite included a 2.2 meter diameter, steerable high-gain parabolic dish antenna, a 0.8 meter diameter medium-gain antenna installed in a fixed position, and two omnidirectional low-gain antennas. It utilized an

S-band telecommand uplink, and S-band and X-band for downlinking telemetry and science data.

For nominal communications, the high-gain antenna with its 0.5 degree half-cone angle beam width was the primary communications link for uplinks and downlinks in S-band and X-band. Additional performance capability and communication robustness was added by the medium-gain antenna, which served as back-up for the primary dish. The low-gain antennas with their hemispherical coverage supported emergency operations in S-band.

During the mission, the rate at which data could be sent from Rosetta varied from 8 bit/s to 64 kbit/s, depending on the distance from Earth and the radio frequency used. However, real time communications were not always possible.

Owing to Earth's rotation, the spacecraft was only visible from ESA's New Norcia antenna in Australia for an average of 12 hours per day.[5]

In addition, there were several periods of communication blackout when the spacecraft was passing behind the Sun. To overcome these breaks in communication, the science data could be stored temporarily in the onboard 25 Gbit solid-state memory for downlink at a convenient time.

To deal effectively with the limited downlink bandwidth, the data were compressed. Coding and modulation schemes were adopted that were optimum for a power limited system; e.g. a new signal coding concept known as Turbo Code that gave an improvement in the link budget of 2.7 dB. This coding scheme software could be updated in flight as appropriate.

The communication equipment included a 28 Watt X-band Traveling Wave Tube Amplifier (TWTA) and a dual 5 Watt S/X-band transponder. Onboard heaters kept the equipment from freezing when the spacecraft was far from the Sun.

Autonomy

One of the great challenges of the Rosetta mission was ensuring that the spacecraft survived the hazards of traveling through deep space for more than a decade.

It would spend much of its outward journey in a state of hibernation to limit consumption of power and fuel, minimize operating costs, and allow the mission teams to have a less intense work program. There would also be communication blackouts whenever the Sun was close to the line of sight (solar conjunctions would up add to 8 months in total). The prolonged periods of inactivity would be interrupted by relatively short spells of intense activity, in particular for the encounters with Mars, Earth, and two asteroids on the way to the target.

Furthermore, the large distances involved meant that signal travel times to and from Rosetta would last for up to 45 minutes each way, so the mission could not

---

[5] Two additional deep space antennas would be inaugurated by ESA before Rosetta encountered Comet 67P.

be operated by real time command and monitoring from the ground. And, as noted, daily coverage by ground stations would be intermittent, with the duration of a given link lasting no more than 12 hours.

As a result, it was essential to ensure that Rosetta could function autonomously and safely in contingency situations, taking corrective actions and, if possible, continuing its programmed scientific observations without ground intervention.

Autonomy and redundancy was ensured by installing four identical computers based on the MA-31750 processor, any two of which were necessary for operating the data management system and the attitude and orbit control system. In the event of an anomaly, Rosetta would enter the relevant 'safe mode' and, if necessary, switch over to a redundant system.

The exception to this was the deep space hibernation mode, scheduled to last about 2.5 years, when the seven electrical systems were essentially switched off. For that time the spacecraft would be spinning at 1 rpm with its solar arrays facing the Sun and only the radio receivers, command decoders and power supply remained active. All of this equipment was redundant, so in the hibernation phase there was no need to have an active autonomous fault management system.

5.2 MISSION CONTROL

The Rosetta ground segment had to meet both the scientific objectives and the challenges of a deep space mission. These challenges included long turnaround times for signals (up to 100 minutes), low bit rates (8 bps), low signal power, spacecraft hibernation for 2.5 years, and gravity assist maneuvers at Mars and Earth.

In addition, personnel had to deal with a mission duration exceeding 12 years, and the related issues of maintaining staff expertise and experience while minimizing the overall cost.

The mission was directed from the Mission Control Center at the European Space Operations Center (ESOC) in Darmstadt, Germany, which was responsible for all mission operations, in particular:

- Mission planning, monitoring and control of the spacecraft and its payload.
- Determination and control of the spacecraft trajectory.
- Distribution of the scientific data from the spacecraft to the Rosetta scientific community and the principal investigators.

The payload data were received, pre-processed, archived, and issued to the mission scientific community via a data disposition system with remote access.

The Flight Dynamics System provided attitude and orbit determination and prediction, and it prepared slew and orbit maneuvers and evaluated spacecraft dynamics and navigation.

Fig. 5.6: The Rosetta control room at ESOC in Darmstadt. (DLR)

Operational capabilities for performing telemetry, telecommand and tracking operations at S- and X-band frequencies were supported by a specially built ground station at New Norcia in Australia, which was equipped with a 35 meter dish antenna and cryogenically cooled, low noise amplifiers to receive the spacecraft's weak radio signals. The Kourou ground station in French Guiana provided additional support during the launch, parking orbit, and near-Earth phases of the mission.

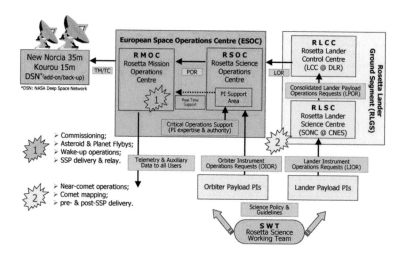

Fig. 5.7: The Rosetta space operations network. (ESA)

The Science Operations Center (SOC) was initially planned to be co-located with ESOC, but it was moved in 2006 to the European Space Astronomy Center (ESAC), in Villanueva de la Cañada, near Madrid, Spain. ESAC supported scientific mission planning and preparation of experiment command requests, prior to their submittal to the operations facility. One key role was to ensure that the science instruments operated as required to satisfy the science goals of the mission. The SOC was also responsible for pre-processing the incoming science data and making the archived data available to the scientific community.

The operations of the lander were coordinated through the German Aerospace Center (DLR) control facility in Cologne, with support from the CNES Lander Science Center in Toulouse, France. The Rosetta Lander Ground Segment supported operations of the lander, in particular before and after completion of the landing and relay phase.

5.3   DOWN TO EARTH

During Rosetta's prolonged interplanetary expedition, reliable two-way communication links between the spacecraft and the ground were essential. All of the scientific data collected by the instruments on the spacecraft were sent to Earth using a radio link. The operations facility, in turn, remotely controlled the spacecraft and its instruments via the same radio link.

Signals were transmitted and received in two radio frequency bands: S-band (2 GHz) and X-band (8 GHz). At the speed of light, the signals could take up to 50 minutes to travel between the spacecraft and Earth or vice versa.

Radio communications between Rosetta and Earth relied upon ESA's three newly developed deep space ground stations at New Norcia, near Perth in Western Australia, Cebreros, near Madrid (Spain), and Malargüe in Argentina. In addition, support was provided by NASA's Deep Space Network (DSN), including the 34 meter and 70 meter antennas at Goldstone in California, Canberra in Australia, and Madrid in Spain.

Construction of ESA's first deep space tracking station at New Norcia started in 2002, not long before the original launch date for Rosetta. It entered service in March 2003, in time to support the comet chaser's delayed launch. It was the primary ground station for the Rosetta mission, providing extended support throughout the cruise phase, planetary fly-bys, asteroid encounters and arrival at, landing on, and orbiting of Comet 67P.

ESA's second ground station at Cebreros in Spain was inaugurated in September 2005. The third station at Malargüe in Argentina was added in 2013, shortly prior to Rosetta's arrival at 67P. Together, they provided 360 degree coverage for ESA's growing fleet of missions exploring the Solar System.

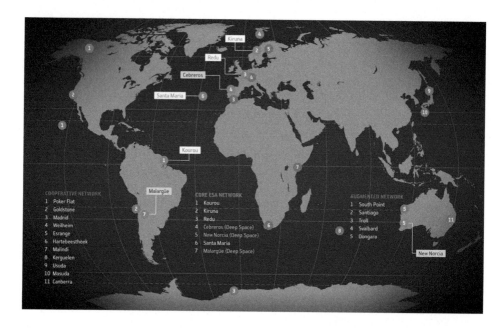

Fig. 5.8: A map of the ESA tracking (ESTRACK) ground station network. Blue shows core ESA-owned stations operated by the Network Operations Center (NOC) at ESA's European Space Operations Center (ESOC) in Darmstadt, Germany. Orange indicates the Augmented ESTRACK stations, procured commercially and operated by commercial entities on ESA's behalf. Green indicates Cooperative ESTRACK stations that were owned and operated by external agencies, notably NASA, and regularly provided services to ESA on an exchange basis. (ESA)

Fig. 5.9: The 35 meter diameter antenna at New Norcia, near Perth, in Western Australia. It was specially built for ESA to ensure reliable communications for deep space missions such as Rosetta. (ESA)

All three antennas were remote controlled for routine operations from ESOC in Germany. They incorporated state-of-the-art technology, including a control system that provided the highest possible pointing accuracy under the prevailing environmental, wind, and temperature conditions.

The Malargüe station featured low noise amplifiers that were cooled to −258°C in order to receive the ultra-weak signals from deep space, while atomic clocks were used for the precise radiometric measurements that were required to accurately locate and guide vehicles in deep space.

All three ground stations provided routine tracking support, including the transmission of telecommands to Rosetta and its lander, monitoring their status and health, and receiving the precious scientific data.

Other ground stations in Hawaii, Kourou and Santa Maria Island in the Atlantic took part in the post-launch phase of the mission.

NASA's DSN stations provided essential support throughout, including serving as prime or back-up links during critical mission phases, when commands had to be sent without delay or interruption, or when a prime station was unavailable. DSN support was especially important in 2014 when Rosetta awoke from its 31 months in hibernation, during its arrival at 67P, and during the lander's descent to the nucleus. The DSN stations also played a crucial role when the orbiter concluded its mission by making an unprecedented descent and touchdown on the nucleus.

Table 5.2:   Ground Stations Used During the Rosetta Mission

Location	Latitude	Longitude	Antenna Diameter
Kourou, French Guiana (ESA)	5.25° N	52.80° W	15 m
New Norcia, Australia (ESA)	31.03° S	116.11° E	35 m
DSN Goldstone, USA (NASA)	36.69° N	116.87° W	34 m / 70 m
DSN Madrid, Spain (NASA)	41.15° N	4.25° W	34 m / 70 m
DSN Canberra, Australia (NASA)	35.40° S	148.96° E	34 m / 70 m
Cebreros, Spain (ESA)	40.45° N	4.37° W	35 m
Malargüe (Argentina)	35.76° S	69.38° W	35 m

5.4   STAFF TURNOVER

The remarkably long duration of the Rosetta mission was a major headache for ESA's project team and the engineers and scientists involved in the development and operation of the entire program. The timeline covered some three decades, with the first design proposals in the late 1980s, official mission approval in 1993, launch in 2004, arrival at 67P in 2014, and the final touchdown in 2016.

There was considerable turnover of the ESA project managers and project scientists during that time (see Appendix 2). Several months were set aside for each succeeding team leader to learn the ropes and familiarize themselves with the mission's status and requirements, as well as the current members of the team.

It was also anticipated that, by the time Rosetta and its lander were carrying out critical near-comet operations, most of the engineers and scientists who designed, developed and tested the spacecraft and its scientific instruments 15 years earlier would no longer be available to offer support – many would be committed to other projects and a fair number would have reached retirement age.

Special attention had to be paid to preserving the design knowledge of the spacecraft, along with the skills and expertise, throughout the lengthy cruise phases. This required proficiency training for the operations personnel, refreshment of skills and motivation, and cross-training to guarantee back-up support as appropriate.

A high fidelity spacecraft simulator and the spacecraft electrical model, both located at ESOC during the mission, were vital for training, for tests and simulations, for reproducing in-flight anomalies, and for devising solutions to problems.

5.5 PLANETARY PROTECTION

Planetary protection – the avoidance of biological contamination of a pristine Solar System environment – was not a consideration for Rosetta. The spacecraft was built in a clean room in accordance with accepted international rules, but sterilization was not considered necessary because although comets may well possess prebiotic molecules, it appeared unlikely that they would host living microorganisms.

On the other hand, the spacecraft had to be extremely clean in order to reduce the possibility of instruments detecting pollutants carried from Earth by instruments such as the ROSINA gas mass spectrometer. Although the spacecraft surpassed the ROSINA requirements, during calibrations in flight the instrument detected outgassing from surfaces which were normally in shadow but temporarily exposed to sunlight.

5.6 ROSETTA'S SCIENTIFIC INSTRUMENTS

Rosetta's scientific payloads were supplied by consortia of institutes in Europe, Canada and the United States. In addition to eleven experiments on the orbiter, there was a lander carrying its own experiments.[6]

In addition, four interdisciplinary scientists provided wider expertise in cometary science:

- Marco Fulle, Osservatorio Astronomico di Trieste, Italy.
- Marcello Fulchignoni, Laboratoire d'Etudes Spatiales et d'Instrumentation en Astrophysique, Observatoire de Paris, France.
- Eberhard Grün, Max-Planck-Institut für Kernphysik, Heidelberg, Germany.
- Paul Weissman, Planetary Science Institute, Pasadena, California, USA.

---

[6]The UK initially intended to provide two gas analyzers – one called Ptolemy for the lander, and Berenice for the orbiter. The development program ran into technical difficulties and in summer 2000 it became clear that the UK team had neither the time nor the resources to make both instruments, so it was decided to develop only Ptolemy for the lander.

Apart from the lander, all of the orbiter's experiments were located on its 'top' because this would face the comet throughout the main scientific phase of the mission. Until its release to descend to the nucleus, the Philae lander was carried on the side of the orbiter opposite to the high-gain dish antenna.

Table 5.3:   Rosetta Orbiter Scientific Experiments

Experiment Name	Purpose	Original Principal Investigator
ALICE	UV imaging spectrometer	Dr. Alan Stern Southwest Research Institute, Boulder, Colorado
CONSERT	Radio sounding of nucleus interior	Prof. Wlodek Kofman Ecole Nationale Superieure d'Ingenieurs, CNRS, Grenoble, France
COSIMA	Dust mass spectrometer	Dr. Jochen Kissel Max-Planck-Institut für Extraterrestrische Physik, Garching, Germany
GIADA	Dust velocity and impact measurement	Prof. Luigi Colangeli Osservatorio Astronomico di Capodimonte, Naples, Italy
MIDAS	Micro-imaging dust analysis	Prof. Willi Riedler Space Research Institute, Graz, Austria
MIRO	Microwave spectrometer	Dr. Samuel Gulkis Jet Propulsion Laboratory Pasadena, California
OSIRIS	Optical, spectroscopic and infrared imaging	Dr. Horst Uwe Keller Max-Planck-Institut für Aeronomie, Katlenburg-Lindau, Germany
ROSINA	Neutral gas and ion mass spectrometer	Prof. Hans Balsiger University of Bern, Switzerland
RPC (Rosetta Plasma Consortium)	Plasma measurements to monitor comet activity and its interaction with solar wind	Dr. Rolf Boström Swedish Inst. of Space Physics, Uppsala, Sweden. Dr. James Burch Southwest Research Institute, San Antonio, Texas. Prof. Karl-Heinz Glassmeier Technische Universität, Braunschweig, Germany. Prof. Rickard Lundin Swedish Institute of Space Physics, Kiruna, Sweden Dr. Jean-Gabriel Trotignon LPCE/CNRS, Orleans, France
RSI (Radio Science Investigation)	Use Rosetta radio signals to study characteristics of nucleus and inner coma	Dr. Martin Pätzold Universität Köln, Germany
VIRTIS	Visible and infrared thermal imaging spectrometer	Dr. Angioletta Coradini Istituto di Astrofisica Spaziale, CNR, Rome, Italy

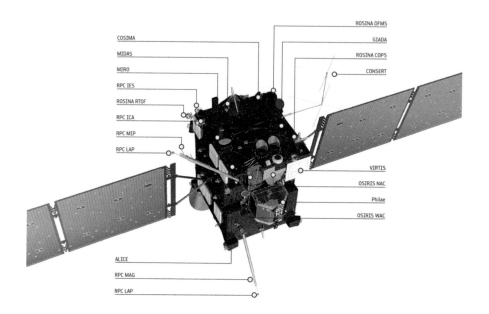

Fig. 5.10: An artist's impression of the Rosetta orbiter with scientific instruments labeled. (ESA)

The orbiter experiments can be grouped according to the types of measurements they were to carry out.

Remote Sensing

OSIRIS (Optical, Spectroscopic, and Infrared Remote Imaging System)

The primary imaging system, OSIRIS comprised a Wide-Angle Camera (WAC) and a Narrow-Angle Camera (NAC).

The WAC was fitted with 14 filters at wavelengths optimized for mapping the gas and dust in the vicinity of the comet.

The NAC had 12 filters for the wavelength range 250-1,000 nanometers and was intended for high-resolution mapping of Mars, the asteroids visited by Rosetta, then the comet's nucleus. During operations at the comet, the NAC would be able to capture photos with resolutions as sharp as two centimeters per pixel. In order to capture images over a wide variety of ranges, it had focusing plates that enabled it to perform both far-focus (2 km to infinity) and near-focus (1 to 2 km) imaging. Its imagery would be used to determine the volume, shape, bulk density, and surface properties of the asteroids and the comet's nucleus.

With a total mass of 35 kg the OSIRIS instrument was developed by six European countries, under the leadership of principal investigator Horst Uwe Keller at the Max-Planck-Institut for Aeronomy, Katlenburg-Lindau, Germany.

ALICE (Ultraviolet Imaging Spectrometer)

ALICE was designed to characterize the composition of the comet's nucleus and to analyze gases in its coma and tail. It had a resolution of 0.6 degrees, and analyzed ultraviolet light in the 70-205 nanometer wavelength range.

At 67P, it measured the abundances of noble gases (helium, neon, argon, krypton and xenon) in the coma to investigate the thermal history of the comet. It also determined the production rates, variability, and structure of water, carbon monoxide, and carbon dioxide gas (molecules that are plentiful in comets and vital for life) and the abundance of carbon, hydrogen, oxygen, nitrogen and sulfur atoms within the comet's coma. It also provided data during the fly-bys of Mars and the asteroids Steins and Lutetia.

The instrument mass was 3 kg, and its principal investigator was Alan Stern of the Southwest Research Institute, Boulder, Colorado.

VIRTIS (Visible and Infrared Thermal Imaging Spectrometer)

VIRTIS was an imaging spectrometer that studied the comet through the use of spectroscopy across visible and infrared wavelengths. Its resolution was 0.1×0.5 degrees. It had a mass of 2.7 kg, and combined three data channels into one instrument. Two of the data channels were in the Mapper (-M) optical subsystem that performed spectral mapping at 0.25-5 micrometers with medium spectral resolution. The High resolution (-H) optical subsystem's single channel was solely for high resolution spectroscopy across the 2-5 micrometer range. The optics were passively cooled to –143°C and the detector was actively cooled to –203°C.

The primary scientific objectives were to study the comet's nucleus and its environment. This involved mapping and studying the nature of the solids and the temperature on the surface of the nucleus. It also identified comet gases and characterized the physical conditions within the coma. Secondary objectives included assisting with selecting the best landing sites and giving support to other instruments.

It was developed by DLR with partners Kayser Threde/Munich, Astrium GmbH/Munich, IB Ulmer of Frankfurt on Oder and the Institute of Planetary Research, Laben, Italy. The original principal investigator was Angioletta Coradini of the Istituto di Astrofisica Spaziale, CNR, Rome, Italy. After Coradini died in 2011, she was succeeded by Fabrizio Capaccioni (INAF).

MIRO (Microwave Instrument for the Rosetta Orbiter)

MIRO was the first microwave instrument to be sent into space to study a Solar System body. Resembling a miniaturized, ground-based radio telescope, it consisted of two heterodyne radiometers, one operating at millimeter wavelengths (190 GHz, ~1.6 mm) and the other at submillimeter wavelengths (562 GHz, ~0.5 mm). Both were configured with a broadband continuum detector to measure the brightness temperature of the comet nucleus and the fly-by asteroids. The submillimeter receiver also served as a very high resolution spectrometer. The resolution at a range of 2 km from the nucleus was 15 meters at millimeter wavelengths and 5 meters at submillimeter wavelengths.

As a combined spectrometer and radiometer that could both sense temperature and identify chemicals, it was used to study the composition, velocity and temperature of gases on, or near the surface of 67P, and to measure the temperature of the nucleus down to a depth of several centimeters. Its data established how the comet changed in temperature as it approached the Sun. MIRO was designed to study four of the ten most abundant molecules usually found on a cometary nucleus: water, carbon monoxide, methanol and ammonia. By detecting the ratios of isotopes, it enabled scientists to determine the conditions in which the comet was formed.

In addition to studying Mars during the Rosetta fly-by, MIRO measured the subsurface temperatures of the asteroids encountered and searched for gases around them.

The MIRO principal investigator was Samuel Gulkis of NASA's Jet Propulsion Laboratory in California. The science and engineering teams were from the U.S., France and Germany. The instrument mass was 19.9 kg.

Composition Analysis

ROSINA (Rosetta Orbiter Spectrometer for Ion and Neutral Analysis)

ROSINA consisted of two mass spectrometers and a pressure sensor. The mass spectrometers determined the composition of 67P's atmosphere and ionosphere, measured the temperature and bulk velocity of the gas and electrified gas particles (ions), and the reactions in which they participated. The pressure sensor could measure both total and ram pressure, and was used to determine the gas density and rate of gas flow around the comet. It also investigated possible asteroid outgassing.

The original principal investigator was Hans Balsiger of the University of Bern, Switzerland. The hardware for the mass spectrometers was provided by a consortium from four European countries and the USA. The instrument mass was 36 kg.

COSIMA (Cometary Secondary Ion Mass Analyzer)

By performing microanalysis of individual dust particles emitted by the comet and collected by the dust collector, COSIMA was Rosetta's primary in-situ dust analyzing instrument.

It consisted of a secondary ion mass spectrometer equipped with dust collectors, an ion gun, and an optical microscope for grain localization. Dust from the near-comet environment was collected on a target. The target was then moved under a microscope, where the positions of any dust particles were determined. The cometary dust particles were then bombarded with pulses of indium ions from the primary ion gun, and the resulting secondary ions were sent to a time-of-flight mass spectrometer which provided the data to determine their elemental and isotopic composition.

The main objective of COSIMA was to study the physical and chemical composition of the comet's dust environment, as well as processes that occurred in and around the nucleus. In particular, it was able to identify the composition of dust particles that were released from different regions of the nucleus during various levels of activity, and to compare the mineralogy, chemistry and composition of dust grains.

The original COSIMA principal investigator was Jochen Kissel of the Max-Planck-Institute for Extraterrestrial Physics, Garching, Germany. It was built by a consortium that included laboratories and academic centers in France, Finland, Germany and Austria. The instrument's mass was 19.1 kg.

MIDAS (Micro-Imaging Dust Analysis System)

Being capable of imaging the smallest of cometary dust particles in-situ, MIDAS was the first instrument of its type to be launched into space. It studied the dust environment around the comet and the asteroids that were visited by Rosetta. In particular, it performed microanalysis of cometary dust particles and determined their 3D structure, size and texture.

In order to collect particles, MIDAS used a funnel that protruded beyond the spacecraft's exterior and which usually pointed towards the comet. Behind this was a wheel with sticky targets mounted around its circumference. After dust had been collected, the wheel would be rotated to present the sample to an atomic force microscope for analysis. A tiny needle was scraped across the surface of a sample, with the deflection of the needle being used to build an image of the particle's size. By scanning over the dust grain, line by line, MIDAS would build a 3D picture. This technique was expected to enable features as small as 4 nanometers (2,500 times smaller than the average width of a human hair) to be studied.

The original principal investigator was Willi Riedler from the Space Research Institute (IWF) Graz, Austria, and hardware was contributed by IWF Graz, Austria, the ESA-ESTEC Science Payload and Advanced Concepts Office, Universität Kassel of Germany and Technische Univeristät Vienna, Austria. The instrument's mass was 8 kg.

Large-scale Nucleus Structure

CONSERT (Comet Nucleus Sounding Experiment by Radiowave Transmission)

CONSERT was designed to transmit radio signals from the Rosetta orbiter to the lander after the latter was on the comet's nucleus. The experiment – on both the orbiter and the lander – consisted of a transmit/receive antenna and a transmitter and receiver housed in one box. The mass of the instrument was 3.1 kg on the orbiter and 1.9 kg on the lander.

When the geometry was right, the signals from the orbiter would pass through the nucleus to the lander, which would extract data and send a signal back to the orbiter, where the principal experimental data collection would occur. Each transmission of the 90 MHz radio signal lasted about 25 microseconds. The measurement cycle was about 1 second. Rosetta did not travel far during the measurement cycle, so the signal returned to the orbiter along essentially the same path as it was sent. As the radio waves traveled through different parts of the nucleus, being slowed, reflected and scattered, variations in their propagation time and amplitude could be used to determine various properties of the internal material and thereby undertake a form of internal mapping.

When the launch was delayed, and another comet was selected as the target, there was some concern about whether the radio transmissions would be able to penetrate the larger nucleus. However, after the lander suffered an unexpected bounce across the surface, the instrument assisted in identifying the location of the lander. By making measurements of the distance between Rosetta and Philae at times of direct visibility between the orbiter and lander, as well as measurements through the core, the experiment team was successfully able to narrow down the search.

The original principal investigator was Wlodek Kofman from the Ecole Nationale Superieure d'Ingenieurs, CNRS, Grenoble, France. The experiment was designed and built in France by the Laboratoire de Planétologie of Grenoble and by the Service d'Aéronomie in Paris, with the Max-Planck-Institut for Aeronomy in Lindau-Katlenburg, Germany.

GIADA (Grain Impact Analyzer and Dust Accumulator)

GIADA measured the number, mass, momentum and velocity distribution of dust grains in the near-comet environment. These included grains that traveled directly from the nucleus to the spacecraft, and those that arrived from other directions after their ejection momentum had been altered by solar radiation pressure.

It combined three different detection subsystems. The Grain Detection System (GDS) could detect individual dust particles by using a laser curtain, without affecting particle dynamics. The Impact Sensor (IS) consisted of five piezoelectric (PZT) sensors that were connected to an aluminum plate in order to measure the momentum released by an impacting particle. The Micro-Balances System (MBS) was composed of five Quartz Crystal Microbalances (QCM), which measured the cumulative mass of particles less than 5 micrometers in size.

The original principal investigator was Luigi Colangeli from the Astronomical Observatory of Capodimonte in Naples, Italy. The instrument was developed by Selex ES, a subsidiary of Finmeccanica. Its mass was 6.35 kg.

Comet Plasma Environment and Solar Wind Interaction

RPC (Rosetta Plasma Consortium)

The RPC comprised five sensors: an Ion Composition Analyzer (ICA), an Ion and Electron Sensor (IES), a dual Langmuir Probe (LAP), a Fluxgate Magnetometer (MAG) and a Mutual Impedance Probe (MIP). Its overall mass was 8 kg.

These sensors measured the physical properties of the comet nucleus, examined the structure of its inner coma, and monitored the activity of the comet and how it interacted with the solar wind.

The ICA measured the 3D velocity distribution and mass distribution of positive ions with a resolution that could differentiate between the major particle species such as protons, helium, oxygen, molecular ions, and heavy ion clusters (dusty plasma).

The IES consisted of two electrostatic analyzers – one for electrons and one for ions – which shared a common entrance aperture, and it simultaneously measured the flux of electrons and ions in the plasma surrounding the comet over an energy range from around one electron volt up to 22,000 electron volts.

The LAP had two spherical sensors mounted at the tip of deployable booms to measure the density, temperature and flow velocity of the cometary plasma. The sensors could measure the current-voltage characteristics of the intervening plasma, to determine the electron number density and temperature.

The MAG measured the magnetic field in the region where solar wind plasma interacted with the comet. It had a pair of three-axis fluxgate magnetometer sensors mounted on a deployable boom 1.5 meters long which pointed away from the comet's nucleus. One sensor was near the far end of the boom and the other was part way along it. In cooperation with the ROMAP magnetometer experiment on the Rosetta lander, the instrument also looked for any magnetic field associated with the nucleus.

The MIP derived the electron gas density, temperature, and drift velocity in the comet's inner coma by measuring the frequency response of the coupling impedance between two dipoles. It also investigated the spectral distribution of natural waves in the 7 kHz to 3.5 MHz frequency range and monitored the dust and gas activity of the nucleus.

The Swedish Institute of Space Physics (IRF) provided two of the sensors – the ICA from Kiruna and the dual LAP from Uppsala. The original investigators for these sensors were Rolf Boström (LAP) and Rickard Lundin (ICA). The other RPC investigators were James Burch of the Southwest Research Institute in San Antonio, Texas (LAP), Karl-Heinz Glassmeier of the Technical University of Braunschweig, Germany (MAG), and Jean-Gabriel Trotignon of the Laboratory of Physics and Chemistry of the Environment and Space in Orleans, France.

RSI (Radio Science Investigation)

The RSI made use of the communication system that linked Rosetta with ground stations on Earth. One-way or two-way radio links could be used for these investigations. In the one-way case, the signal produced by an ultra-stable oscillator on the spacecraft was received on Earth for analysis. In the two-way case, the signal transmitted by a ground station was received by the spacecraft and transmitted back to Earth. In either case, the downlink could be performed at X-band frequencies, or in both X-band and S-band frequencies.

RSI investigated frequency (Doppler) shifts and dispersive frequency shifts (induced by the ionized propagation medium), signal power, and the polarization of the radio carrier waves. Variations in these parameters provided information about the motion of the spacecraft, the perturbing forces acting upon it, and the medium through which the signal propagated.

Shifts in the spacecraft's radio signals were used to measure the mass, density and gravity of the nucleus, to define the comet's orbit, and to study its inner coma. The RSI was also used to measure the mass and density of asteroids visited by Rosetta, and to study the solar corona at times when the spacecraft was passing behind the Sun as viewed from Earth.

The principal investigator was Martin Pätzold from the Rhenish Institute for Environmental Research at the University of Cologne, Germany.

5.7   THE ROSETTA LANDER (PHILAE)

Although the initial concept of Rosetta as a Comet Nucleus Sample Return mission had fallen by the wayside (see Chapter 3), the mission kept its capability to make in-situ measurements of the comet's nucleus with the addition of a small lander. The lander would provide 'ground truth' data that would verify the orbiter's measurements of the comet's composition, surface and subsurface conditions. It was to be the first spacecraft ever to make a soft landing on the surface of a comet.

The lander, named Philae, was supplied by a European consortium under the leadership of the German Aerospace Center (DLR). Other members of the consortium were ESA, CNES, and institutes from Austria, Finland, France, Hungary, Ireland, Italy and the UK.

The Max Planck Society was the most important DLR partner, responsible for building the landing module. Among other things, the Institute for Solar System Research developed the landing gear, the mechanism that would eject the lander from the spacecraft, and the evolved gas analyzer instrument (COSAC).

The Max-Planck-Institute for Extraterrestrial Physics in Garching built the harpoons to anchor the lander to the comet, while the Max-Planck-Institute for Chemistry in Mainz contributed an instrument (APXS). Additionally, the universities of Münster, Mainz and Braunschweig were involved alongside international partners.

The DLR lander project was managed by Stephan Ulamec. The original principal scientific investigator was Berndt Feuerbacher, the Director of the Institute für Raumsimulation, but he was soon succeeded by Helmut Rosenbauer from the Max-Planck-Institut for Aeronomy.

Although the European Space Operations Center (ESOC) in Darmstadt, Germany, had overall responsibility for operating the Rosetta orbiter, the lander was monitored by the Microgravity User Support Center (MUSC) in Cologne.

Table 5.4: The Lander

The Philae platform comprised 8 subsystems:
- The carbon fiber structure
- Thermal control system
- Power management system
- Flight electronics (main memory, main clock, Telemetry/Telecommand management, etc.)
- S-band telecommunication system (two patch antennas on the upper face)
- Inertial wheel for stabilization and attitude control during descent
- Cold gas thruster propulsion system for counteracting touchdown recoil
- Deployable tripod landing gear (three legs, three feet with rigging screws, telescopic tube to cushion impact)

Shape: Hexagonal
Dimensions: $1 \times 1 \times 0.8$ meter
Mass: 98 kg, including 21 kg of instruments
Energy/power from primary battery (1,300 Watt-hour) and secondary rechargeable batteries (140 Watt-hour) fed by solar panels
Onboard memory: 2×16 Mb
Communication rate with the orbiter: 16 kb/s
Anchoring system based on two harpoons fired into nucleus after touchdown

The hexagonal Rosetta lander measured about 1 meter across and 80 cm in height, and had a mass of approximately 100 kg. The lander was carried in piggyback fashion on the side of the orbiter until its release following arrival at the comet. The main structure was made of carbon fiber.

The lander's power system comprised surface-mounted solar panels and two sets of batteries. The low intensity, low temperature solar cells covered 2.2 square meters and were designed to deliver up to 32 W at a distance of 3 AU (450 million km) from the Sun.

A non-rechargeable primary battery comprised four strings, each of which had eight lithium thionyl chloride (Li/SOCl$_2$) cells.[7] The 1,300 Wh battery was intended to provide power for the first 60 hours on the comet's surface, then the secondary 140 Wh Li-ion batteries would become the prime energy source for the

---

[7]This type of battery has the highest energy density of any lithium battery. Although not rechargeable, they have an extremely long shelf-life that makes them ideal for low power consumption devices which run for years.

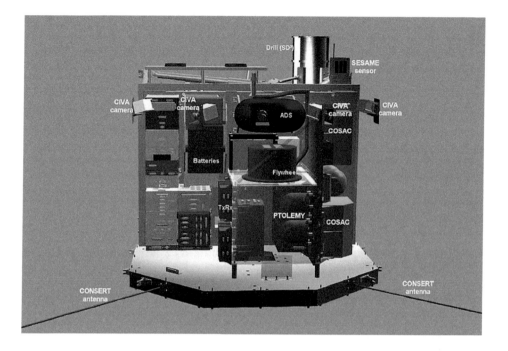

Fig. 5.11: A cutaway side view of the Philae lander compartment showing the location of COSAC and Ptolemy systems, the CONSERT antennas, the SESAME dust sensor and the ÇIVA cameras. (Philae Team)

long term operations on the comet. These consisted of two blocks, each of 14 Li-ion cells. They were rechargeable via the solar panels or, during the cruise, using current lines from the Rosetta orbiter.

Once a suitable landing site was identified from mapping of the nucleus, the orbiter would be aligned appropriately, ground control would command the lander to self-eject, then unfold its three legs for a gentle touchdown at the end of its ballistic descent.

The legs were designed to damp out most of the kinetic energy from the impact to minimize the likelihood of bouncing. They could also rotate, lift or tilt to return the lander to an upright position if it were to settle on its side or topple over.

Immediately after each foot touched the surface, its momentum would drive an ice screw into the ground to prevent lateral motion during the touchdown and help to fix the lander in place.

The touchdown signal from the landing gear would trigger an anchoring process – essential to prevent the lander from bouncing back into space. The process was to start with the firing of a cold gas thruster positioned on the top of the lander to push it against the surface.

After several seconds, two harpoons were to be ignited, one after the other, and fired into the ground at 70 m/s (250 km/h). After they had come to a halt, attached ropes were to be reeled in to lock the lander in position.

The orbiter would function as a relay for the lander. With a clear line of sight, communication between the two in S-band was possible at data rates of up to 16 kbit/s across ranges of a few hundred meters to 150 km. The high performance, low cost transceiver had an overall power consumption of 6.5 W (1.5 W for the receiver and 5 W for the transmitter) and a total mass of 1 kg.

The Lander project managers were:

- Dr. Stephan Ulamec, DLR, Köln Porz-Wahn, Germany.
- Prof. Denis Moura, CNES, Toulouse, France.
- Dr. R. Mugnuolo, Italian Space Agency, Matera, Italy.

Lander Scientific Experiments

The Rosetta lander carried nine experiments with a total mass of about 21 kg. It also carried a drilling system to obtain samples of subsurface material. The instruments were designed to study in great detail the composition and structure of the nucleus material.

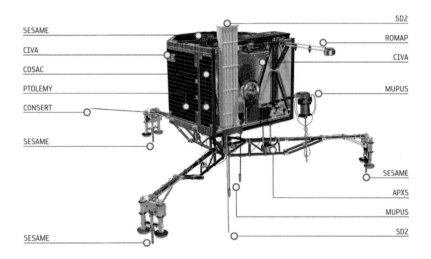

Fig. 5.12: The scientific instruments on the Philae lander. (DLR)

The minimum mission target for science operations was 60 hours, which was the expected life of the primary battery, but there was a slight possibility they could continue for many months.

German institutes were heavily involved in providing the scientific payload. In particular, DLR contributed three of the 10 experiments on the lander: the ROLIS camera used during the landing phase, the SESAME experiment for seismic analysis of the comet core, and the MUPUS system to measure the thermal and material properties of the comet.

Table 5.5: Rosetta Lander Scientific Experiments

Instrument Name	Purpose	Principal Investigator
APXS	Alpha proton X-ray spectrometer	Dr. Rudolf Rieder Max-Planck-Institute for Chemistry, Mainz, Germany
ÇIVA	Panoramic and microscopic imaging system	Dr. Jean-Pierre Bibring Institut dAstrophysique Spatiale, Université Paris Sud, Orsay, France
CONSERT	Radio sounding, nucleus tomography – see also the orbiter instruments	Prof. Wlodek Kofman Laboratoire de Planetologie, Grenoble, France
COSAC	Evolved gas analyzer – elemental and molecular composition	Dr. Helmut Rosenbauer Max-Planck-Institute for Aeronomy, Lindau, Germany
MODULUS Ptolemy	Evolved gas analyzer – isotopic composition	Prof. Colin Pillinger / Dr. Ian Wright Open University, Milton Keynes, UK
MUPUS	Measurements of surface and subsurface properties	Dr. Tilman Spohn University of Münster, Germany
ROLIS	Imaging system	Dr. Stefano Mottola DLR, Berlin, Germany
ROMAP	Magnetometer and plasma monitor	Dr. Hans-Ulrich Auster Technical University of Braunschweig, Germany.
Sampling, Drilling and Distribution Subsystem (SD2)	Surface drilling and sample retrieval	Dr. A. Ercoli-Finzi Polytechnic of Milan, Italy
SESAME	Surface electrical, acoustic and dust impact monitoring	Dr. Dietrich Möhlmann DLR, Cologne, Germany Dr. Harri Laakso Finnish Meteorological Institute, Helsinki, Finland Dr. Istvan Apathy KFKI, Budapest, Hungary

Some experiments required more time than others, e.g. a long period of time was necessary to analyze the chemical and physical characteristics of the comet's surface. A significant amount of time was also required to study the characteristics of the temperature, thermal conductivity, and electrical conductivity of the nucleus. Medium duration measurements included acoustic and seismic investigation of the surface.

There were also experiments that were less time-dependent, such as the Sampling Drilling and Distribution (SD2) device that was to collect samples from different depths and present them for chemical analysis.

Other instruments would investigate the elemental, molecular, and mineralogical composition and the morphology of the nucleus's material.

Scientists were particularly interested in the results from Philae's investigations because the material had been preserved in the nucleus since its formation 4.5 billion years ago, the same era that saw the creation of the planets.

ROLIS (Rosetta Lander Imaging System)

Consisting of a miniaturized CCD camera, ROLIS was one of two imaging systems carried by Rosetta's lander. Its primary objective was to obtain images during the descent to the nucleus, providing increasingly high resolution images of the landing site. After touchdown, it was to focus at an object distance of 30 cm and document the surface beneath the lander, as well as the areas sampled by other instruments.

Multispectral imaging was achieved through an illumination device consisting of four arrays of monochromatic light-emitting diodes in bands at 470, 530, 640 and 870 nanometers. The drill sites and the targets for in-situ analyzers were to be imaged to provide context to assist in interpreting the results.

After the drilling operation, the borehole was to be inspected for morphology and any signs of stratification. The lander's ability to adjust its position would enable stereoscopic imaging to assist with analyzing the immediate surroundings and identifying features.

The principal investigator was Stefano Mottola of DLR, Berlin. The instrument was designed and built in the DLR Institute of Planetary Research.

COSAC (Cometary Sampling and Composition experiment)

As one of two evolved gas analyzers on the lander, COSAC detected and identified complex organic molecules from their elemental and molecular composition. Material from the surface of the comet would be fed into it by the SD2 instrument, combusted, and the resultant gas sent to the analysis section consisting of a gas chromatograph and a mass spectrometer.

COSAC was the instrument chosen to receive the single drill sample that would be available while running on the primary battery. It was hoped this data would help to determine whether some of the organic material on Earth was delivered by comets.

COSAC was developed and built under the leadership of the Max-Planck-Institute for Solar System Research. The original principal investigator was Helmut Rosenbauer. The partners were the Laboratoire Inter-universitaire des Systèmes Atmosphériques (Paris), the Laboratoire Atmosphères, Milieux, Observations Spatiales (Paris), and the University of Giessen.

MODULUS Ptolemy

Ptolemy was an evolved gas analyzer that was designed to obtain accurate measurements of isotopic ratios of light elements. The first example of a new concept in space instrumentation, it was to make in-situ isotopic measurements of the comet's nucleus. The instrument concept was known as MODULUS (Methods Of Determining and Understanding Light elements from Unequivocal Stable isotope compositions) and the name honored Thomas Young, the initial translator of the Rosetta stone.

The scientific goal of MODULUS was to understand the geochemistry of light elements such as hydrogen, carbon, nitrogen and oxygen by determining their nature, distribution and stable isotopic compositions. The size of a small shoe box and with a mass of under 5 kg, Ptolemy used gas chromatography/mass spectrometry methods to investigate the comet's surface and subsurface.

Material from the SD2 sampling and drilling system was to be delivered to one of four ovens dedicated to Ptolemy, mounted on a circular, rotatable carousel. (The carousel had a total of 32 ovens, with the remainder allocated for use by COSAC and ÇIVA.) Of the four Ptolemy ovens, three were for solid samples collected and delivered by SD2 and the fourth contained a gas-trapping substrate that would be used to collect volatiles from the near-surface cometary atmosphere. Samples would be heated and the resultant gas purified, quantified and sent to the mass spectrometer.

However, as a result of the non-optimal landing conditions, Ptolemy was only operated in its 'sniffing' mode, analyzing material released from the comet during the lander's approach and rebounds.

Ian Wright from the Open University in Milton Keynes in the UK was the lead scientist for the Ptolemy instrument, after it was originally proposed for the mission by Colin Pillinger. It was designed and developed by the Open University, in conjunction with CCLRC Rutherford Appleton Laboratory in the UK.

MUPUS (Multi-Purpose Sensors for Surface and Subsurface Science)

MUPUS used sensors on the lander's anchor, probe, and exterior to measure the thermal and mechanical properties of the surface.

The instrumentation included 16 temperature sensors attached to the interior of a 35 cm long penetrator that was to be deployed at a distance of about 1.5 meters from the lander's body. The hollow rod of the penetrator was to be inserted into the ground using an electromagnetic hammering mechanism. As it was driven into the ground, the progress per hammering stroke and the temperature of the subsurface were to be measured for an indication of how resistant the surface was to penetration and how the temperature changed with depth.

The lander team attempted to deploy MUPUS after the off-nominal landing, but it is possible the penetrator broke during the analysis.

In addition to the sensors on the penetrator, MUPUS also had two heat sensors inside the anchoring harpoons that did not fire. Finally, an infrared sensor known as the thermal mapper, mounted on one of the lander's struts, could measure the heat emitted from the surface of the comet across a small area.

The experiment was designed and built by an international consortium led by the principal investigator, Tilman Spohn, of the Institute for Planetology (IfP) at the University of Münster, with major contributions from IfP, SRC Warsaw in Poland, DLR Berlin and IWF Graz.

ROMAP (Rosetta Lander Magnetometer and Plasma Monitor)

ROMAP comprised a magnetometer and a plasma monitor. It was to determine the interaction of the comet with the solar wind and measure any residual magnetic field on the comet.

The sensors were positioned on a 60 cm long retractable rod. The fluxgate magnetometer was designed to measure the magnetic field while the plasma monitor measured the abundance of the ions and electrons in the plasma environment of the comet. This involved the development of a digital magnetometer with low mass and power requirements, and was the first time that a magnetic sensor would operate within a plasma sensor.

ROMAP was built by a consortium headed by the principal investigator, Hans-Ulrich Auster of the Technical University of Braunschweig (Germany) and István Apáthy from AEKI in Budapest (Hungary). It also included the Max-Planck-Institute for Solar System Research (Göttingen, Germany) and the Austrian Academy of Sciences (Graz, Austria).

SESAME (Surface Electrical, Seismic and Acoustic Monitoring Experiments)

SESAME comprised three instruments designed to measure various properties of the comet's outer layers.

The Cometary Acoustic Sounding Surface Experiment (CASSE) was to study the surface and subsurface using sensors incorporated into the lander's feet. They were intended to listen for noise produced within the nucleus by sources such as expansion and contraction from heating and cooling, impacts, and seismic events. In this regard, it was similar to a seismometer on Earth. It could also generate sound and use the echoes for information about the material that the signal passed through to gain information on any layering within the comet, as well as the presence of voids and other irregularities. In this active mode it was similar to a sonar.

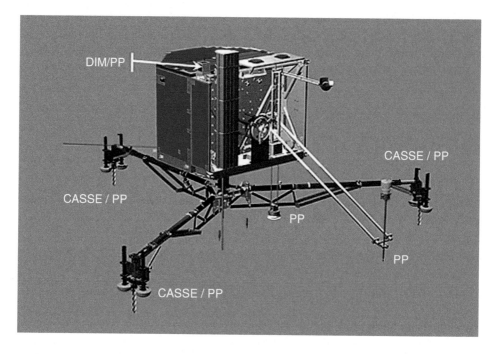

Fig. 5.13: The locations of the different SESAME instruments: CASSE, PP and DIM. (DLR)

The Permittivity Probe (PP) was to study the electrical characteristics of the nucleus. It used five electrodes incorporated into parts of the lander: three transmitters on one of the feet of the lander, the APXS instrument, and the MUPUS penetrator, and two receivers in the remaining two feet of the lander.

The transmitter electrodes were to send an electric signal into the surface to determine the electrical conductivity of the surface down to a depth of about 2 meters in order to detect the presence and abundance of water in the surface.

The Dust Impact Monitor (DIM) was to measure the dust environment close to the surface. In particular, it was to measure the impact of particles that had been released from the comet at insufficient velocity to escape its gravity. Mounted on one of the upper surfaces of the lander, it would measure the impact of these particles from three directions to determine the energies with which they were ejected from the surface.

APXS (Alpha Proton X-ray Spectrometer)

Lowered to within 4 cm of the ground, APXS was to provide information on the elemental composition of the comet's surface, notably all elements from carbon to nickel. This involved a source of alpha particles irradiating surface material to enable its elemental composition to be determined from backscattered alpha particles and alpha-induced X-ray radiation. The data would be compared with known meteorite compositions and contextualized by data collected from other instruments on both the orbiter and lander.

The principal investigator was Rudolf Rieder from the Max-Planck-Institute for Chemistry, Mainz, Germany. Contributing institutions included the Institute for Inorganic and Analytical Chemistry of Johannes Gutenberg-University, Mainz, Germany; the Center d'Etude Spatiale des Rayonnements, Toulouse, France; and the Department of Physics at the University of Guelph, Ontario, Canada.

CONSERT (Comet Nucleus Sounding Experiment by Radiowave Transmission)

CONSERT probed the internal structure of the nucleus by transmitting radio waves from the orbiter right through the nucleus to the lander, where a transponder returned the signal to the orbiter. (See Rosetta Scientific Instruments above)

ÇIVA (Comet Infrared and Visible Analyzer)

This comprised two separate experiments.

ÇIVA-P consisted of five identical 'mono' cameras plus two cameras which were aligned for stereoscopic imaging. They were to produce a panoramic image of where the lander had come to rest, in order to characterize the site and map the topography and albedo (reflectivity) of the surface.

The second experiment, ÇIVA-M, combined a visible light microscope and a coupled infrared spectrometer. They were mounted on the base plate of the lander and would analyze samples delivered by SD2 to determine the composition, texture and albedo of samples collected from the surface.

The principal investigator was Jean-Pierre Bibring from the Institut d'Astrophysique Spatiale, Université Paris Sud, Orsay, France. The cameras were developed by the Swiss Center for Electronics and Microthechnology in Neuchâtel, Switzerland.

SD2 (Sampling, Drilling and Distribution Device)

The principal purpose of SD2 was to obtain subsurface comet material for analysis by ÇIVA, COSAC and Ptolemy. It would interface with COSAC and Ptolemy to study the geochemistry of the comet (including complex organic molecules) and with ÇIVA to image and investigate the composition, texture and albedo of samples.

It comprised a carbon fiber toolbox and a drill that contained a retractable sampling tube. It was capable of boring to a depth of 230 mm and collecting samples, and included a rotatable carousel with 26 ovens. The entire system weighed about 5 kg and the average power needed for drilling was about 10 W.

After the drill reached the desired depth, the sampling tube would extend from the drill bit to collect the sample. The drill would then move back to its home position. Once the drill was in position, the carousel would rotate to locate the assigned oven under the drill. The drill would position the sampling tube on the oven's opening and push the sample into the oven, then the carousel would rotate to deliver the sample to the appropriate instrument.

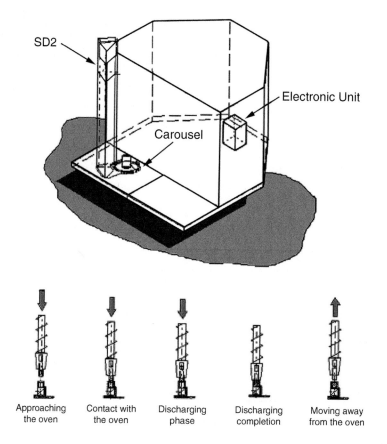

Fig. 5.14: Top: A simplified cutaway view showing SD2 and the rotatable carousel. Bottom: The process of discharging a drill sample into one of the ovens on the carousel. (Pierluigi Di Lizia/SD2 instrument team)

There were 10 medium temperature ovens (giving a maximum temperature of 180°C) and 16 high temperature ovens (800°C). By progressively heating a sample, different gases could be released to identify the composition of the material. Alternatively, the samples could be used to measure characteristics such as density, texture, strength and thermal properties.

It was anticipated that more than four samples would be collected, each measuring some 20-30 cubic millimeters. In the event, the system was able to deliver a sample only to COSAC in the time before the lander exhausted its primary battery.

The principal investigator was Amalia Ercoli-Finzi from the Polytechnico of Milan, Italy, and the hardware was developed by Tecnospazio SpA of Italy, under the leadership of the Italian Space Agency (ASI).

5.8   ROSETTA'S DISK

What have a comet-chasing spacecraft, a 2,200-year-old piece of basalt and a global language archive got in common? The answer is that not only are they all named 'Rosetta', but all three provide an enduring link spanning millennia.

In homage to the role played by the Rosetta Stone in deciphering the mysteries of Egyptian hieroglyphics, ESA agreed to carry the Rosetta Disk aboard its pioneering comet explorer as a modern equivalent of the original Rosetta Stone.

At a pre-launch press event in Kourou, French Guiana, on 18 November 2002, this unique disk was attached to the orbiter where it would be shielded by the thermal blankets.

On the 7.5 cm nickel disk was a cultural archive provided by the San Francisco-based Long Now Foundation. It featured pages of text with 1,000 different languages, each miniaturized and etched as an image. All that would be required to read the pages would be a microscope. This simple technique guarded against the threat of changing technologies that could make a digital disk unreadable by computers in the future.

"The Long Now Foundation is trying to preserve the world's languages for future generations and we are happy to carry the disk on the Rosetta spacecraft in order to ensure that the archive survives for posterity," said John Ellwood, ESA's Rosetta project manager.

At the time, it was anticipated that Rosetta would continue orbiting the Sun for thousands of years, preserving Earth's linguistic heritage for possible retrieval by future generations, rather like the Rosetta Stone. In practice, however, the orbiter's mission ended with a landing on the comet and the Rosetta Disk is now circling the Sun on the surface of an object which formed more than 4 billion years ago.

Will an expedition from Earth ever land on Comet 67P to retieve the disk? Only time will tell. Quite possibly this record of our current linguistic diversity will remain long after many of the languages have been forgotten.

5.9   THE ARIANE 5 LAUNCHER

Rosetta was designed to be delivered to orbit by Europe's Ariane 5 launcher, one of the few rockets in the world with the capability to propel a 3 tonne spacecraft towards a distant comet.

In order to operate Ariane 5, ESA built a new launch site (ELA-3) at Europe's spaceport near the town of Kourou in French Guiana, as well as facilities to integrate the solid boosters and to manufacture propellant. The rocket hardware was shipped in by sea, but the spacecraft was delivered by air in a special container.

Fig. 5.15: The fully assembled Ariane 5G+ on the launch pad at Kourou on 1 March 2004,   showing   its   core   stage   and   twin   solid-fueled   boosters. (ESA-CGS-CNES-Arianespace)

The Ariane 5 was developed by prime contractor EADS Launch Vehicles – a merger of the French company Aerospatiale with DASA of Germany.[8] The central core stage is flanked by a pair of solid-fueled rocket boosters.

The core stage contained hydrogen and oxygen as cryogenic liquids for its single Vulcain engine. This motor would be ignited first, followed 7 seconds later by the two boosters, which would deliver most of the thrust needed to get the massive rocket off the ground. After two minutes 19 seconds the boosters would burn out and be jettisoned, and some 50 seconds later the protective fairing that cocooned the spacecraft would be discarded. By the time the core stage engine shut down, 9 minutes 50 seconds into flight, the rocket would have achieved an altitude of almost 170 km and a velocity of almost 8,100 m/s.

The basic version of the Ariane 5 (the Ariane 5G with the 'G' standing for Generic) was the original choice for Rosetta, but this was grounded after a launch failure of a newly introduced version (Ariane 5 ECA) on 11 December 2002 (see Chapter 6). This resulted in an intensive investigation, as well as a re-evaluation of the Rosetta mission. In addition to the prospect of switching to a Russian Proton rocket, it assessed several different comets as potential targets for a new launch date.

When Comet 67P was eventually chosen as Rosetta's new destination, it was decided to use a slightly modified variant known as the Ariane 5 Generic-Plus (Ariane 5G+).[9]

For this key mission, the complete upper stage – including all systems such as engine, tanks and feeder systems – underwent specific testing and re-qualification.

On a mission designed to put satellites into geostationary transfer orbit the upper stage would ignite very soon after separation from the core stage, but this time (and for the first time) the new upper stage coasted unpowered in an approximately 4,000 × 200 km orbit for 1 hour 46 minutes.[10] On nearing perigee (the closest point to Earth) it was fired up for 16 minutes to attain the 37,476 km/h needed to escape Earth's gravity and enter circumsolar orbit. This was the greatest velocity ever achieved by an Ariane 5.

Rosetta was released by the spent EPS some 2 hours 13 minutes after lift-off, and started its decade-long trek toward Comet 67P. Despite the powerful initial boost given by the Ariane, the spacecraft required gravitational assistance from fly-bys of Mars (2007) and Earth (2005, 2007, 2009) to enable it to rendezvous with the comet passing through the outer Solar System in 2014.

---

[8] EADS was the European Aeronautic Defense and Space Company.

[9] The modifications included lengthening the tanks in the upper stage to accommodate additional propellants.

[10] To keep temperatures onboard the upper stage under control during its 105 minutes of coasting, it was rotated around its longitudinal axis in 'barbeque mode'. In addition, the storable propellants were pre-heated to deliver proper ignition.

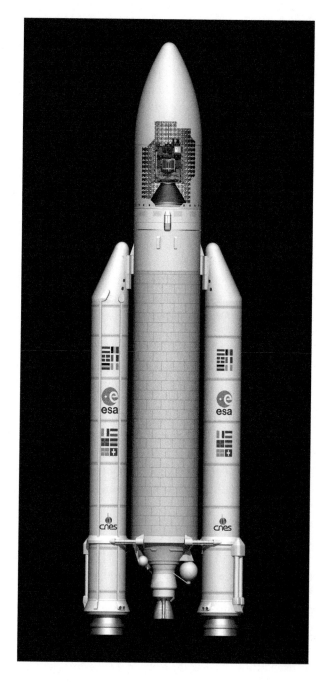

Fig. 5.16:   An artist's impression of Rosetta on top of its Ariane 5G launcher. (ESA)

Table 5.6:   Ariane 5G+ Vital Statistics

Overall length: 46.1 m
Core stage length: 30.53 m
Core stage diameter: 5.46 m
Core stage engine: Vulcain, using 158.1 tonnes of cryogenic propellants (liquid oxygen and liquid hydrogen)
EPS restartable upper stage: height 3.36 m
EPS engine: Aestus using 9.7 tonnes of storable propellant – monomethylhydrazine (MMH) and nitrogen peroxide (N_2O_4)
Two solid propellant rocket boosters: height 27.1 meter, diameter 3.05 meters. 238 tonnes of powdered propellant (each)
Payload fairing: 12.73 meters high; internal volume 115 cubic meters
Launch mass: 749 metric tons

REFERENCES

Rosetta Rises to the Challenge, ESA Bulletin 112, p. 30-37, November 2002a

Inorbit Maintenance Of The Rosetta Reaction Wheels (RWAs), Paul McMahon et al: http://esmats.eu/esmatspapers/pastpapers/pdfs/2017/mcmahon.pdf

Philae (spacecraft): https://en.wikipedia.org/wiki/Philae_(spacecraft)

Philae lander fact sheets: https://www.dlr.de/rd/Portaldata/28/Resources/dokumente/rx/Philae_Lander_FactSheets.pdf

The Evolution of Philae, 20 January 2014: http://blogs.esa.int/rosetta/2014/01/20/the-evolution-of-philae/

Industrial *Involvement In The Philae Lander*: http://sci.esa.int/rosetta/54181-industrial-involvement-in-the-philae-lander/

Rosetta Lander (Philae) Investigations, J. P. Bibring et al, January 2007a: https://www.researchgate.net/publication/41625911_Rosetta_Lander_Philae_Investigations

OSIRIS – The Scientific Camera System Onboard Rosetta, H. U. Keller et al, 2004: https://pdssbn.astro.umd.edu/holdings/ro-a-osiwac-3-ast1-steinsflyby-v1.4/document/osiris_ssr/osiris_ssr.pdf

The Rosetta *ALICE UV Spectrometer:* http://alice.boulder.swri.edu/pages/introduction

VIRTIS: http://sci.esa.int/rosetta/35061-instruments/?fbodylongid=1646

MIRO: http://sci.esa.int/rosetta/35061-instruments/?fbodylongid=1641

ROSINA: http://sci.esa.int/rosetta/35061-instruments/?fbodylongid=1650

The MIDAS Experiment For The Rosetta Mission, H. Arends et al, 2001: http://esmats.eu/esmatspapers/pastpapers/pdfs/2001/arends.pdf

GIADA performance during Rosetta mission scientific operations at comet 67P, Advances in Space Research, Vol. 62, Issue 8, 15 October 2018, pp 1987-1997: https://www.sciencedirect.com/science/article/pii/S0273117717305410

CONSERT: https://www2.mps.mpg.de/en/projekte/rosetta/consert/technical_description.html

RPC: The Rosetta Plasma Consortium, C. Carr et al, Space Science Reviews, February 2007, Volume 128, Issue 1-4, pp 629-647: https://link.springer.com/article/10.1007/s11214-006-9136-4

Rosetta Radio Science Investigations (RSI), Martin Pätzold et al, Space Science Reviews, February 2007b, Volume 128, Issue 1-4, pp 599-627: https://link.springer.com/article/10.1007%2Fs11214-006-9117-7

Industrial Involvement *In The Philae Lander*, http://sci.esa.int/rosetta/54181-industrial-involvement-in-the-philae-lander/

Robot Technology for the Cometary Landing Mission Rosetta, R. Mugnuolo et al: http://www.esa.int/esapub/pff/pffv7n2/pozzv7n2.htm

Lander instruments: http://sci.esa.int/rosetta/31445-instruments/

Ion Trap Mass Spectrometry On A Comet Nucleus: The Ptolemy Instrument And The Rosetta Space Mission, John F. J. Todd et al, Journal of Mass Spectrometry, 12 December 2006, https://doi.org/10.1002/jms.1147 https://onlinelibrary.wiley.com/doi/full/10.1002/jms.1147

CASSE – Cometary Acoustic Surface Sounding Experiment: https://www.dlr.de/pf/en/desktopdefault.aspx/tabid-5326/8936_read-17316/

The Rosetta Alpha Particle X-Ray Spectrometer (APXS), G. Klingelhöfer et al, Space Science Reviews, February 2007, Vol. 128, Issue 1-4, pp 383-396: https://link.springer.com/article/10.1007%2Fs11214-006-9137-3

The Rolis Experiment on the Rosetta Lander, Stefano Mottola et al, Space Science Reviews, February 2007, Volume 128, Issue 1-4, pp 241-255: https://link.springer.com/article/10.1007/s11214-006-9004-2

Philae Blog: https://www.mps.mpg.de/3086295/Philae-Blog

Introducing SD2: Philae's Sampling, Drilling and Distribution instrument, 9 April 2014: http://blogs.esa.int/rosetta/2014/04/09/introducing-sd2-philaes-sampling-drilling-and-distribution-instrument/

Rosetta – Starting from Scratch, Paolo Ferri, Room, Issue #1(7) 2016: https://room.eu.com/article/rosetta-starting-from-scratch

ESTRACK, ESA's Deep Space Tracking Network: https://download.esa.int/esoc/estrack/esa_estrack_brochure_2015_EN.pdf

New Norcia ground station: http://www.esa.int/nno

Ground Stations' Last Week With Rosetta, 28 September 2016: https://blogs.esa.int/rosetta/2016/09/28/ground-stations-last-week-with-rosetta/

ESA Reaches Out Into Deep Space From Spain, ESA Bulletin 118, pp 16-20, May 2004: http://www.esa.int/esapub/bulletin/bulletin118/chapter2_bul118.pdf

Ariane Flight VA158 Press Kit: http://www.arianespace.com/wp-content/uploads/2017/06/04_feb_26-en.pdf

Rosetta: The Great Escape: https://www.ariane.group/en/commercial-launch-services/ariane-5/rosetta/

Rosetta Disk Goes Back to the Future, 4 December 2002: http://sci.esa.int/rosetta/31242-rosetta-disk-goes-back-to-the-future/

6

Switching Comets

During the mid-1990s, when the details of the Rosetta rendezvous mission were still being refined, mission planners were considering several possible comets as targets. The favored candidate was periodic comet 46P/Wirtanen, but potential back-ups included periodic comets 73P/Schwassmann-Wachmann 3 and 15P/Finlay.

As time went by, Wirtanen was confirmed as the ideal objective for Rosetta and the mission was planned with it in mind. The baseline plan targeted launching on a European Ariane 5 in January 2003. In order to rendezvous with the comet in November 2011, the spacecraft would require a gravity assist from Mars in August 2005 and two assists from Earth in November 2005 and November 2007. Wirtanen was selected because of its fairly active nature, modest size, and orbital path, which meant that it would be in the right place at the right time for a rendezvous with Rosetta. Upon arrival, the spacecraft would fly alongside the nucleus. Full payload operations would start at a heliocentric (solar) distance of 3.25 AU (488 million km) in August 2012 and continue to July 2013, when the comet was at perihelion (closest point to the Sun).[1]

6.1   COMET 46P/WIRTANEN

Comet Wirtanen was discovered by chance on 15 January 1948 by Carl Wirtanen while he was examining photographic plates at the Lick Observatory in California.

---

[1] One astronomical unit (AU) represents the average radius of Earth's orbit around the Sun; some 150 million km or 93 million miles.

P. Bond, *Rosetta: The Remarkable Story of Europe's Comet Explorer*, Springer Praxis Books, https://doi.org/10.1007/978-3-030-60720-3_6

Like many periodic comets that have been captured or influenced by the power-ful gravity of the largest planet in the Solar System, collectively known as the 'Jupiter family', Wirtanen commutes between the orbits of Jupiter and Earth.

Wirtanen's elliptical orbit is susceptible to change by gravitational interactions with the planets. In particular, close approaches to Jupiter in 1972 (at a distance of 0.28 AU or 41.9 million km) and 1984 (0.46 AU or 68.8 million km) shortened its orbital period from 6.71 years to 5.46 years.

By the close of the 20th century, Wirtanen's perihelion was just outside Earth's orbit, so the amount of heating during its inward passage was quite modest. At perihelion its heliocentric distance was 159 million km (1.06 AU) and when far-thest from the Sun (aphelion) it was 768 million km (5.13 AU, near the orbit of Jupiter). The inclination between the orbit of the comet and that of Earth was moderate, at less than 12 degrees.

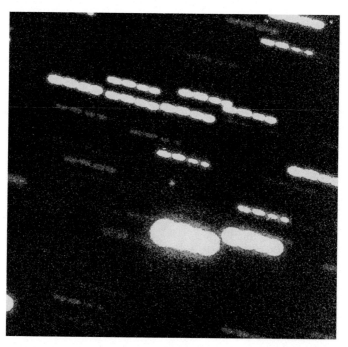

Fig. 6.1: A false-color composite image of Comet 46P/Wirtanen, based on four expo-sures recorded on 9 December 2001 by the 8.2 meter VLT YEPUN telescope. The tele-scope was tracking the motion of the comet, so stars are seen as four consecutive trails. The star-like image of the comet's nucleus shows no surrounding gas or dust. The brightness indicates a diameter of roughly 1 km. The comet's distance from Earth at that time was approximately 534 million km. (ESO)

With the exception of 1980, Wirtanen had been observed during every approach to the Sun since its discovery. It was particularly closely monitored during a coor-dinated observational campaign in 1996-1997, and again following its selection as the primary target for Rosetta.

Despite such intensive observations, little was known about the comet's size, shape, mass or rotation period. Usually, its faint image was drowned in a sea of stars, making ground-based studies extremely difficult. Although it released little dust or gas near aphelion, it was too far away to study in detail. During its brief ventures into the inner Solar System, the warmth of the Sun prompted ices on its surface to sublimate and jets of gas to blast dust grains into the surrounding space – characteristics that led scientists to favor it as the target for the Rosetta mission. Unfortunately, although this enveloping coma increased its brightness, it also hid the nucleus from view.

Assuming Wirtanen's nucleus to be very dark, reflecting 3% of incoming sunlight, as was the case for most other comets, its brightness implied a diameter of approximately 1.1 km. If the reflectivity were higher, then of course the nucleus would be smaller. Ground-based studies identified water, oxygen, carbon dioxide, and various compounds of nitrogen, hydrogen and carbon.

6.2   ASTEROID FLY-BY OPPORTUNITIES

Since opportunities to investigate asteroids at close quarters were few and far between, ESA planners wanted Rosetta to visit two rocky objects on the way to Comet Wirtanen.

In 1995, ESA announced that Rosetta's baseline mission would have opportunities to fly past main belt asteroids (3840) Mimistrobell and (2530) Shipka, after the spacecraft had made its first and second fly-bys of Earth, respectively.

Although Wirtanen remained as Rosetta's prime target, further studies of possible candidates resulted in changes to the asteroid fly-bys. By 1997, the planners were considering whether to visit a different S-class asteroid, (2703) Rodari, instead of Shipka (see Table 6.1).

Table 6.1: Summary of Planned Major Events for Rosetta's Mission to Comet Wirtanen with Fly-bys of Asteroids Mimistrobell and Rodari

Event	Date	Days	Object Distance (km)	Earth Distance (km or AU)
Launch from Earth	21 Jan 2003	0	0	0 km
Mars gravity assist	26 Aug 2005	948	200	0.69 AU
First Earth gravity assist	26 Nov 2005	1,040	3,332	3,332 km
Mimistrobell fly-by	15 Sep 2005	1,333	600	2.34 AU
Second Earth gravity assist	26 Nov 2007	1,770	2,315	2,315 km
Rodari fly-by	4 May 2008	1,930	1,580	1.46 AU
Orbiting Wirtanen	24 Aug 2011	3,136	4-18	4.81 AU
Delivery of RoLand	22 Aug 2012	3,443	1	2.60 AU
Shutdown of systems	10 July 2013	3,768	0	1.06 AU

As the launch came closer, the fly-by targets were changed to another pair of contrasting objects: (140) Siwa, which would be the largest asteroid yet encountered by a spacecraft, and (4979) Otawara, the smallest apart from Dactyl, the tiny satellite discovered by the Galileo spacecraft during a fly-by of (243) Ida.

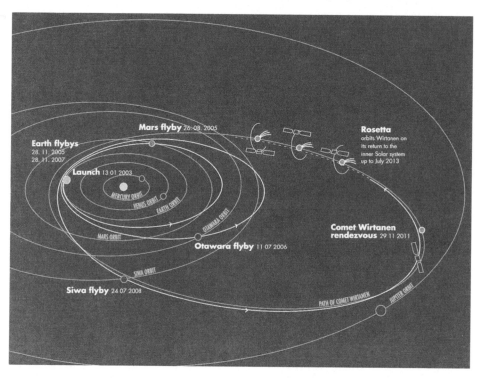

Fig. 6.2: The mission plan for the Rosetta mission with one Mars fly-by, two Earth fly-bys, and encounters with main belt asteroids Otawara and Siwa on the way to Comet Wirtanen. (ESA)

The Otawara fly-by (see Fig. 6.2) was to occur 1.89 AU from the Sun, on 11 July 2006. The spacecraft would pass its sunlit side at a range of about 1,595 km and a relative velocity of 10.63 km/s.

Apart from its orbit, little was known about Otawara until it became the subject of a ground-based program of studies undertaken by telescopes in France, Chile and the USA.

Otawara was believed to be a stony object rich in the minerals pyroxene and/or olivine, but it was also possible it belonged to an asteroid family named after its largest member, (4) Vesta. Presuming Otawara to be dark, its diameter was likely 2.6-4 km. Its density was estimated at 2-2.5 times greater than water, suggesting a substantial rocky component. A study of changes in its reflected light (its light curve) indicated that it rotated once every 2.7 hours, which was faster than any asteroid visited by spacecraft to that time. This would be an advantage during a fly-by, as it would enable the spacecraft's instruments to image the surface and measure its characteristics at high resolution during one complete rotation.

In contrast, with a diameter of 110 km, Siwa was much larger than any asteroid previously examined by spacecraft. Spectral studies indicated that it was a primitive, very black, carbon-rich object. Estimates for its rotation period varied between 18.5 hours and 22 hours. Rosetta was to obtain images and high-resolution data as it flew within 3,000 km of Siwa on 24 July 2008. It would approach the sunlit side at 17.04 km/s and see a crescent phase as it withdrew. Siwa would be 2.75 AU from the Sun and 3.11 AU from Earth, so signals from the spacecraft would take 26 minutes to reach ground stations.

Table 6.2:   Vital Statistics of Rosetta's Asteroid Targets (pre-2003)

	Otawara	Siwa
Average distance from Sun (million km)	324	409
Orbital period (years)	3.19	4.51
Estimated size (km)	2.6-4	110
Estimated rotation period (hours)	2.7	18.5
Orbital inclination (degrees)	0.91	3.19
Orbital eccentricity	0.144	0.215
Asteroid type	V or SV	C
Date of discovery	2 August 1949	13 October 1874
Name of discoverer	K. Reinmuth	J. Palisa

6.3   A DRASTIC CHANGE OF PLAN

After years of planning for a launch to Comet Wirtanen in 2003, a major spanner was thrown in the schedule on 11 December 2002, when the Ariane 5 ECA rocket exploded during its maiden flight, with the loss of its payload of two communication satellites.

An Inquiry Board appointed by Arianespace, ESA and CNES (the French Space Agency) was established to investigate the cause of the "anomaly". Meanwhile, Arianespace, the operator of the Ariane 5, decided to create a Review Board to offer advice regarding the launch date of the next payload on the manifest: the Rosetta mission.

On 5 January 2003, the key participants announced that all irreversible operations involved in the Rosetta launch must be suspended. This would result in a launch postponement of at least several days beyond the targeted date of Sunday, 12 January (Kourou time).

On 7 January, the Board announced that the cause of the explosion early in the flight was a fault in the main rocket motor. The investigation blamed a leak in the cooling system of the nozzle of the Vulcain 2 engine. The overheating and deterioration of the nozzle produced an imbalance in the thrust of the engine which resulted in loss of control over the trajectory.

Although Rosetta was to be launched on a basic Ariane 5G – which differed from the ECA version by using tried-and-tested Vulcain 1 engines – the Review Board decided to play safe and recommended postponing the ground-breaking comet mission. Arianespace, ESA, and all of the other interested parties accepted this recommendation, and began a long consultation to decide arrangements for the earliest possible launch of Rosetta, and how it might differ from the planned mission.

Meanwhile, there was a thorough re-examination of the system qualification procedures for the Ariane 5 program to prevent the recurrence of such a mishap.

As for the spacecraft, it had to be moved and stored in flight-ready condition in a clean room at Kourou while its next launch campaign was decided. Long term storage involved removing its batteries, removing the harpoons from the lander, and draining the propellant tanks.

"The same care that went into building the spacecraft will now be applied to storing it and making sure it will be in perfect shape for us to launch it when the date comes," said John Ellwood, the project manager.

After the initial shock and disappointment of the last minute postponement, the Rosetta team set about redefining the entire mission profile. The overall sentiment was one of defiance and a determination to succeed, despite the significant setback.

As Rosetta's project scientist, Gerhard Schwehm, put it, "During the decade it has taken us to develop and build Rosetta, we have faced many challenges and overcome them all. This new challenge will be met with the same energy, enthusiasm and, ultimately, success."

6.4   THE PROJECT MANAGER'S VIEW

In November 2019, John Ellwood, Rosetta project manager at the time of the Ariane 5 ECA disaster, commented in an email about its impact on the comet mission:

> I was actually in Paris in mid-December when we heard of the problems with the Ariane launch. The spacecraft was in Kourou being filled with fuel and oxidizer, which we planned to do before the Christmas break to be ready for launch in January 2003.
>
> We took no immediate action until Arianespace could assess the problem. I went out to Kourou just after Christmas and had many discussions with Jean

Jacques Dordain, who was in Europe during this time. He was Director of Launchers, just about to become DG (ESA Director General). I remember the actual call – when we decided to postpone the mission – was when I had taken the day off with my family to go to visit Devil's Island. I was sitting on the side deck of a large catamaran in the hot sun, difficult to hear my mobile due to the gentle breeze, taking part in this decision-making process!

There was no real choice in this decision – Arianespace/ESA/Europe could not risk another failure and there was not really enough time to demonstrate what had gone wrong and how to make the Rosetta launch safe.

At first the team were pretty devastated – we had had a pretty long and hard launch campaign and were almost at the climax. We also did not know what were the back-up scenarios. The scientists had continuously told us that this was the unique opportunity.

The immediate decisions were what to do with the spacecraft and what were our future options. We managed to defuel the spacecraft but could not take the oxidizer out for fear of potential technical problems – this had never been done before. We therefore decided to leave the spacecraft in Kourou and the immediate tasks were to organize the logistics of this.

We then started to have discussions with the scientists and ESOC about what other possible target comet opportunities there were. There were none with comets around the same size and with the same journey time with a launch in the near future.

Someone proposed that there could be a possibility to launch to Wirtanen later in the year, using the slightly more powerful Proton launcher. I then embarked on a crash action with the Russians to address this possibility. There were all sorts of political and financial implications with it, but we started with looking at the technical possibilities. Although it was technically possible from an interface and orbit viewpoint, the main problem was moving the spacecraft from Kourou to Baikonur – we could not this do with oxidizer on board, and eventually we judged it too risky to the reaction control system to try and take it out.

We were back to looking at other opportunities and then the scientists hit on 67P/C-G, which was larger than Wirtanen and would take, I think, another two years' journey time.

The mission team had quite a frustrating 2003, but there were many things to do. After the business with the Russians, we had to prepare the new scenario. It also gave us a bit of time to sort out other issues. (See Chapter 4.)

6.5   A NEW DESTINATION

One obvious issue was that Rosetta could no longer reach its original target, Comet Wirtanen, in the planned time frame. The mission team was tasked to identify any suitable comets that it could reach if launched within the next two-and-a-half years.

There were three overriding criteria: the potential for maximum scientific return, minimizing the technical risks to the spacecraft, and minimizing the additional expenses, estimated at that time as likely to be in the range €50-100 million.

Fortunately, the Rosetta team was able to start with the list of potential targets for the Comet Nucleus Sample Return mission developed by the ESOC mission analysis team, headed by Martin Hechler (see Chapter 3). During the early study phase for CNSR they had developed extended lists of launch date and comet target combinations, along with calculations of the spacecraft mass that could be delivered to each target by an Ariane launcher.

Nevertheless, the search for Rosetta's new destination proved problematic, and the shortlist presented to ESA's Science Program Committee (SPC) on 25-26 February 2003 provided no easy answers.

The options included:

- Keeping Comet Wirtanen as the target. The spacecraft, and especially its lander, were designed to explore this comet with its small nucleus. However, waiting for the next easy launch window to Wirtanen was undesirable because that would require keeping Rosetta in storage until the comet returned to the inner Solar System after completing another 5.5 year orbit.
- A fly-by of Venus that could sling the spacecraft to a 2012 rendezvous with Wirtanen after launches in October 2003 or April 2004. However, sending Rosetta to Wirtanen by utilizing a Venus gravity assist was impossible because Rosetta's design was only qualified to go within 0.9 AU of the Sun. The greater intensity of solar radiation near Venus would potentially damage many spacecraft systems – unless a major redesign was undertaken, which was undesirable.
- Comets Tempel 2 or Howell could be reached without a fly-by of Venus. However, Rosetta would still require an approach within 0.8 AU of the Sun. Furthermore, the nucleus of Tempel 2 was far too large at 16×8 km. In the stronger gravity field the lander would crash onto the surface.
- A launch to Wirtanen in January 2004, using a more powerful rocket than the Ariane 5G. This might be done using an Ariane 5 ECA, but there was a doubt over whether this version would be ready in time. The only qualified rocket that was suitable was Russia's Proton DM, but Rosetta was 40 cm too

big for the Proton's payload fairing, which would have to be modified and qualified within the next 10 months.

- A launch to another familiar periodic comet – 67P/Churyumov-Gerasimenko. This seemed to be the easiest solution. Using an Ariane 5G+ rocket to launch in February 2004 and taking advantage of fly-bys of Earth and Mars, the mission could reach the comet in 2014. On the down side, the nucleus was thought to measure about 5 km in diameter, which was somewhat larger and more massive than the lander's designers had envisaged.

Of the nine mission scenarios studied by the Rosetta Science Working Team, three survived and were presented to the delegations of the ESA Member States during the Science Program Committee meeting on 25-26 February 2003. Two of the scenarios would see Rosetta launch to 67P/Churyumov-Gerasimenko in February 2004 or 2005 using either an Ariane 5 hybrid or a Proton. The alternative was to use a Proton to launch it to Comet Wirtanen in January 2004.

To better inform the comet selection process, intensive efforts were made to learn as much as possible about the potential targets using facilities that included the Hubble Space Telescope and the Very Large Telescope of the European Southern Observatory in Chile.

Having discussed the suitability and viability of the options, the SPC announced its decision on Rosetta's new baseline mission at its meeting on 13-14 May 2003.

The revamped mission was to be launched in February-March 2004 by an Ariane 5G+ for a rendezvous with Comet 67P/Churyumov-Gerasimenko in November 2014. Mission planners were to study a launch to the same target one year later as a back-up.

Even then, the revamped mission did not immediately receive the all-clear, owing to financial constraints. The cost of the proposed postponement was estimated at €80 million.

In January 2003, ESA's Director of Science, Professor David Southwood, had been confident that the additional commitment could be absorbed by the existing science budget. However, since then a number of other unexpected financial challenges had arisen – notably the need to inject €70 million into the development of instruments for two other high profile astronomy missions: Herschel and Planck.

"I am not asking for more money overall, but for help in cash flow," explained Southwood. "We in ESA are sure that we will find the necessary sensitivity, understanding and, ultimately, solidarity from the [ESA] Council. Europe paved the way to comet science with Giotto and it is a matter of great pride that the ultimate comet explorer will be European."

He gained a sympathetic hearing at the June meeting of the ESA Council, which decided the money to save Rosetta would be found through some immediate "financial flexibility" at Agency level.

6.6 COMET 67P/CHURYUMOV-GERASIMENKO

Like Comet Wirtanen, Rosetta's new target was a regular visitor to the inner Solar System. 67P/Churyumov-Gerasimenko – hereafter abbreviated to 67P – was a member of the Jupiter family of comets whose orbits were modified by close approaches to the giant planet. Both of these comets were thought to have originated in the Kuiper Belt (see Chapter 1) and been deflected into the inner Solar System.

Comet 67P was discovered in 1969, when several astronomers from Kiev were visiting the Alma-Ata Astrophysical Institute to undertake a survey of comets. On 20 September, while studying photographs of 32P/Comas Solá taken by Svetlana Gerasimenko, Klim Churyumov found a comet-like object near the edge of one plate. He assumed that the faint object was the expected comet, but further analysis established it to be a new one.

Table 6.3: Comet 67P/Churyumov-Gerasimenko (2003 data)

Diameter of nucleus (km)	5 × 3
Orbital period (years)	6.57
Perihelion distance from Sun	194 million km (1.29 AU)
Aphelion distance from Sun	858 million km (5.74 AU)
Orbital eccentricity	0.632
Orbital inclination (degrees)	7.12
Year of discovery	1969
Discoverers	Klim Churyumov, Svetlana Gerasimenko

The comet's orbital history is particularly interesting. Until 1840, it never approached the Sun closer than 4 AU and was thus completely unobservable from Earth.

That year, a fairly close encounter with Jupiter caused the orbit to move inward, producing a perihelion of 3 AU. Over the next century, this was gradually decreased to 2.77 AU. Then, in 1959, another Jupiter encounter reduced it to a mere 1.28 AU. The orbit continued to evolve and, after another perturbation from Jupiter in 2007, the perihelion at the time of the Rosetta encounter in 2014 was expected to be 1.24 AU.

At the time that 67P was selected as Rosetta's target, the comet was making one circuit of the Sun every 6.57 years. It had been observed from Earth on six apparitions – 1969 (discovery), 1976, 1982, 1989, 1996 and 2002 – and was unusually active for a short period object, with a diffuse coma of dust and gas surrounding the solid nucleus and often producing a tail when at perihelion. At the 2002-2003 apparition, the tail was up to 10 arcminutes long, with a bright central condensation in a faint extended coma. Even seven months after perihelion its tail was very well developed, although it then rapidly faded.

Fig. 6.3: A composite of 15 images of the nucleus of Comet 67P (the central point of light), taken on 26 February 2004 using the 3.5 meter New Technology Telescope of the European Southern Observatory. The telescope was tracking the comet, so the stars appear as streaks. The comet's nucleus appears almost star-like, indicating it to be surrounded by a very small amount of gas or dust. The comet was approximately 600 million km from Earth. (ESO)

The comet typically reached a magnitude of 12, with outbursts around perihelion on its 1982-1983, 1996-1997 and 2002-2003 apparitions. Despite being a relatively active object, even at the peak of outburst the rate of dust production was estimated to be some 40 times lower than for Halley's Comet. Nevertheless, 67P was classed as a dusty comet. In 2002-2003, the peak dust production rate was approximately 60 kg/s, and values as high as 220 kg/s were reported in 1982-1983. The gas to dust emission ratio was approximately two.

The Wide Field Planetary Camera 2 on the Hubble Space Telescope (HST) took 61 images of Comet 67P on 11-12 March 2003. The HST's sharp vision enabled astronomers to isolate the comet's nucleus from the surrounding coma. The images showed that the nucleus measured 5 × 3 km and had an ellipsoidal (rugby ball) shape. It rotated once in approximately 12 hours.

"Although 67P is roughly three times larger than the original Rosetta target, its elongated shape should make landing on its nucleus feasible – now that measures are in place to adapt the lander package to the new configuration before next year's launch," said Philippe Lamy of the Laboratoire d'Astronomie Spatiale in France.

6.7   LANDING ON A LARGER COMET

In May 2003, engineers were presented with a new challenge when ESA's SPC announced that 67P would replace Wirtanen as Rosetta's objective.

The most obvious challenges were the different orbits and dates of arrival in the inner Solar System. However, the team from ESA, industry and academia would also have to prepare the Rosetta lander for a hazardous descent onto a much larger cosmic iceberg than was initially envisaged.

With time of the essence, the team began to study the implications of exploring 67P and the modifications that the fragile lander might require. After months of studies and simulations, engineers were confident that everything possible had been done to ensure the success of the first soft touchdown on such a pristine surface.

As Philippe Kletzkine, ESA's manager for the Rosetta lander, explained:

Churyumov-Gerasimenko is a much bigger comet than Wirtanen. It is about four times the diameter and its gravity could be at least 30 times greater. This means that the landing speed will increase from 0.2-0.5 meters per second to 0.7-1.5 meters per second.

In the case of Wirtanen, our biggest problem was avoiding a rebound – the spacecraft only had to bounce slightly and the momentum would overcome the weak gravitational hold of the comet.

Now, we also have to worry about absorbing the shock from a faster landing and the stability of the lander upon touchdown. In the worst case scenario of a 'hard' comet surface, rough terrain and relatively high gravity, it was possible that the lander could topple over. In order to prevent this, we decided to modify the landing gear.

Reluctant to remove the landing gear or eliminate the entire lander from the Rosetta orbiter, the team considered options for something that would be small, light, and easy to fit. Their solution was a bracket, called a tilt limiter, that could be attached to the bottom of the lander.

Jean-Christophe Salvignol, the Rosetta lander mechanical engineer, explained the issue:

By restricting the angle at which the landing gear can flex on touchdown to only 3-5 degrees, we improve the damping effect on touchdown and reduce the possibility of a rebound.

The limiter was designed by Astrium GmbH in collaboration with ourselves and the Max-Planck-Institute in Lindau. During pendulum tests with a model of the landing gear, we simulated landing on a wall at different angles of approach, and verified that the spacecraft could successfully touch down at speeds of up to 1.5 meters per second on a 10 degree slope, or up to 1.2 meters per second on a 30 degree slope.

In parallel, computerized simulations of landings were run by the Max-Planck-Institute to better determine the landing performances for various surface characteristics, impact velocities and lander attitudes.

On 30 September 2003 the tilt limiter was delivered to Kourou and installed on the Rosetta lander.

Fig. 6.4: The location of the tilt limiter on the Rosetta lander. (Max-Planck-Institute/ ESA)

"This excellent collaboration between ESA, industry and MPAe has enabled us to adapt to the new mission very quickly and efficiently," said Salvignol.

No major changes were envisaged for the lander's descent profile. However, under the new mission scenario, there would be more time available for the orbiter's instruments to map the nucleus in detail, in order to facilitate the selection of a safe touchdown location for the 100 kg lander.

The historic touchdown on the icy nucleus of Comet 67P was expected to occur sometime in November 2014.

6.8   THE NEW FLIGHT PLAN

During the decade-long journey to reach 67P – two years longer than the planned mission to Comet Wirtanen – Rosetta would have to travel as far from the Sun as Jupiter's orbit. Since no launch vehicle was capable of sending it there directly, the plan required the spacecraft to gain energy from gravitational assists during one fly-by of Mars in 2007 and three fly-bys of Earth in 2005, 2007 and 2009.

It was anticipated that the amount of science that could be conducted during this deep space cruising would be similar to that expected in the original flight plan. The Wirtanen flight plan had included observations of Mars and two very unusual asteroids. However, with the revised mission scenario, Rosetta would experience an eclipse during the Mars fly-by, and this would restrict the science activities that could be performed there.

As before, Rosetta would pass twice through the asteroid belt – where it was hoped to make close-up observations of at least one of these primitive objects. A number of candidate targets were identified, but the final selection would not be made until after launch, once the amount of fuel that was available had been verified by mission engineers.

Since Rosetta would be launching to 67P with the same amount of oxidizer and fuel that was available for the Wirtanen target, the mission team had to examine the spacecraft's propellant margins very carefully. Of particular concern was the extended thruster firing that would be required to rendezvous with the comet.

"We do not have too much fuel to spare," explained John Ellwood, Rosetta project manager. "Our capability to target one or more asteroids will depend on the efficiency of the launch and how much fuel we will need to conduct orbital maneuvers and course corrections, so no decision will be made until after lift-off."

Despite the modifications and unknowns mentioned above, many aspects of the expedition to explore some of the most primitive objects in the Solar System remained very similar to those originally planned.

The Ariane 5 launch from Kourou in February-March 2004 would put the upper stage into a 4,000 × 200 km orbit of Earth. About two hours later, the rocket's upper stage would ignite once more to send the comet chaser on its way.

The first gravitational 'slingshot' was to take place in March 2005, when Rosetta returned to the vicinity of Earth. Three years into the mission, it would pass Mars. The second encounter with Earth would occur in November 2007. With its orbit now more elliptical, Rosetta would penetrate the asteroid belt for the first time, prior to its third and final visit home in November 2009. Only then would it have sufficient velocity to set course for the comet. After its second passage through the asteroid belt it was to be placed in a state of hibernation.

Finally, after more than seven years of interplanetary travel, Rosetta was to cross the orbit of Jupiter, some 800 million km from the Sun, and fire its thrusters to alter course and intercept 67P. Handicapped by the reduced sunlight (25 times

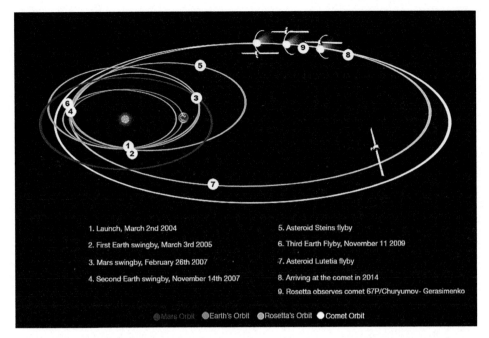

Fig. 6.5: Launching in March 2004, Rosetta would take 10 years to rendezvous with Comet 67P using gravitational assists from three Earth fly-bys and a Mars fly-by. Along the way, it would twice cross the main asteroid belt, where there would be opportunities to conduct two asteroid fly-bys. (NASA-JPL)

less intense than on Earth) the spacecraft would be running on minimal electrical power and relying heavily on its huge solar panels to capture every photon. But the power levels would gradually rise as it started to head sunward and close in on 67P. By the second rendezvous maneuver in May 2014, the electricity supply would be adequate to enable operation of the suite of 10 scientific instruments.

When the target's position was pin-pointed, Rosetta would edge towards the speeding comet and, in August 2014, maneuver into orbit around it. Once the nucleus had been surveyed, and a safe landing site selected, the spacecraft would release its lander to slowly fall toward the black, pristine surface.

"We may separate at a lower altitude, since this means less acceleration," explained Philippe Kletzkine. "We anticipate a maximum separation speed of just half a meter per second, so the overall descent time is likely to be between 30 minutes and one hour. We anticipate a landing on the 'summer' side of the nucleus, where there is maximum illumination."

Over a period of several weeks, a treasure trove of data from the lander's nine instruments would be sent back to Earth via the orbiter.

Meanwhile, the orbiter would continue to watch the dramatic changes in the nucleus during its headlong plunge toward the inner Solar System.

Despite its generally more active nature, scientists reckoned that the dust environment close to the nucleus of 67P would be only marginally more hazardous to the spacecraft than would have been so for Wirtanen. This was because 67P's larger perihelion distance meant that its nucleus was heated less strongly by the Sun, potentially limiting the output of dust that could threaten the orbiter.

Rosetta's unique odyssey of exploration was expected to end in December 2015, six months after the comet passed perihelion and was retreating to the frigid regions of deep space. After a saga lasting almost 12 years, the curtain would finally fall on the most ambitious European scientific mission ever launched.

Table 6.4: The Revised Mission Plan – The Voyage to Comet 67P

Launch from Kourou	2 March 2004 (UT)
1st Earth gravity assist	4 March 2005
Mars gravity assist	25 February 2007
2nd Earth gravity assist	13 November 2007
Asteroid Steins fly-by	5 September 2008
3rd Earth gravity assist	13 November 2009
Asteroid Lutetia fly-by	10 July 2010
Enter deep space hibernation	8 June 2011
Exit deep space hibernation	20 January 2014
Major comet rendezvous maneuver	May 2014
Arrive at comet	August 2014
Philae lander delivery	11 November 2014
Perihelion passage	13 August 2015
Mission end	31 December 2015

6.9   PREPARING FOR LAUNCH, ROUND 2

After deciding to ground Rosetta in January 2003, weeks before its launch campaign was due to complete, the priority was to ensure that the orbiter and its attached lander could be stored in a completely safe, clean environment until a new launch date could be agreed.

Once the spacecraft was carefully moved to the empty S3B clean room at Kourou, a number of safety precautions were undertaken, including the removal of the needle-sharp explosive harpoons, the high-gain antenna, and the huge solar arrays. The mission team also decided to exploit the delay by removing and refurbishing five of the orbiter's instruments.

One of the main questions was how to deal with the fully fueled spacecraft. Eventually, it was decided to offload the 660 kg of toxic, corrosive, monomethyl-hydrazine (MMH) fuel. This dangerous and time-consuming procedure was eventually completed on 7 May. However, it was decided to leave the nitrogen tetroxide oxidizer on board, with the system pressurized. Experience with other spacecraft had indicated that, after offloading this oxidizer, the residual nitric oxide acid had the potential to corrode the titanium tank.

After the mission was retargeted to explore 67P, the Rosetta ground team was able to begin preparations. One of their first tasks was to update the software to satisfy the requirements of the revised mission.

As Jan van Casteren, Rosetta's systems engineering manager, explained, "We had already prepared some software for uplink to Rosetta in May, four months after its planned launch, so we decided to take advantage of the delay to include additional functionality and put the new software on board the spacecraft while it is on the ground."

Other modifications were made to allow for the fact that Rosetta would at various times move closer to the Sun, or farther away from it, than previously planned during its prolonged trek to its target.

"We put reflective surfaces on the exterior of some thermal blankets to prevent overheating," explained van Casteren. "We also had to analyze the potential impact of spending longer in space during a period of maximum solar activity. By accumulating a larger overall dose of radiation, there was a likelihood that the solar arrays would be degraded more quickly, so we carefully investigated the power situation to ensure that we would have a sufficient margin throughout the mission. This gave us confidence that Rosetta will have enough power at all stages of its mission, even when it is beyond the orbit of Jupiter."

John Ellwood, Rosetta's project manager, said, "Although we were all disappointed by the delay, we've been able to take advantage of the additional time on the ground to ensure that Rosetta is in perfect health for its exciting new mission."

The Second Launch Campaign

After verifying system functionality in August-September, the Launch Preparation Readiness Review Board gave the go-ahead to initiate Rosetta's second launch campaign. Once the new flight profile and fly-by targets had been identified, the way was clear for the Center Spatiale Guyanais (CSG) in Kourou to formally start the new campaign on 24 October 2003.

After the postponement of the original launch, some pieces of hardware were removed from the Rosetta orbiter, prior to its storage so the first steps on the road to mission recovery were to reinstall these appendages.

By 3 November, the Alenia assembly, integration and verification (AIV) team had successfully reattached the high-gain antenna to the spacecraft. On 28 November, it passed its deployment test and was then returned to the stowed position required for launch.

In parallel, experts from Dutch Space were carefully inspecting the solar arrays whilst they were still dismounted and suspended beneath the solar array rig. Once this was finished, the electrical and mechanical connections for the arrays were undertaken by a joint Dutch Space and Alenia team.

When the solar arrays were reinstalled, a final deployment test was carried out on each wing. This involved six sequential firings of the thermal knives on each wing, enabling each panel to open, supported by the deployment rig. Then the wings were restowed in readiness for the launch.

Another task included finalization of the multi-layer insulation. In November, personnel from Austrian Aerospace, assisted by staff from Alenia, Astrium and ESA, carefully sowed these blankets back into position.

By late November, the two PROM (Programmable Read-Only Memory) cassettes had been successfully integrated and verified, including tests that were made remotely from ESOC in Germany. The final activities for the GIADA instrument were also undertaken, including the cleaning of internal optical surfaces and laser system health checks.

Rosetta activities in Kourou were closed down for the rest of the year on 3 December, after which the spacecraft was "baby-sat" by a small team of engineers.

The next major pre-launch milestone in Kourou took place on 27-28 January 2004, when the orbiter was refueled with MMH propellant and then pressurized by a team from Astrium Ltd.

Fig. 6.6: MMH propellant being loaded into the Rosetta orbiter by a team from Astrium Ltd., on 27-28 January 2004. (ESA-CNES-Arianespace)

January also saw the completion of the ground checkout activities of the Rosetta lander. The flawless Cruise Abbreviated Functional Test demonstrated that all subsystems and payloads were fully operative. To round off these tests, the electrical configuration was finalized for launch, the primary battery was checked, and the secondary battery was charged. Finally, the harpoons to anchor the lander to the surface of the comet were reinstalled, still fitted with tip protectors.

Fig. 6.7: The Rosetta orbiter and its lander (center) in the clean room at Kourou in January 2004. Note the folded solar arrays on either side of the orbiter. (ESA)

On 5 February, the DLR-led team and ESA announced that the pioneering comet lander had been named 'Philae'.

Philae is an island in the river Nile where an obelisk was found with a bilingual inscription which included the names of Cleopatra and Ptolemy in Egyptian hieroglyphs. This gave the French researcher Jean-François Champollion the final clues that he required to decipher the hieroglyphs of the Rosetta Stone, and thereby unlock the secrets of the civilization of ancient Egypt (see Chapter 3).

On 10 February, the Ariane 5G+ (V158) launcher, minus its payload, was transferred on its mobile table from the Launcher Integration Building (Batiment d'Integration des Lanceurs, BIL) to the Final Assembly Building (Batiment d'Assemblage Final, BAF) with a temporary dome in place to protect the EPS upper stage and vehicle equipment bay.

That same day, the finely choreographed campaign continued with the transfer of the Rosetta spacecraft from the fueling hall, followed by its integration with the mechanical and electrical launcher interfaces on the cone-shaped launch adapter (Adapteur de Charge Utile, ACU) that would attach it to the top of the launcher.

Whilst the spacecraft underwent a day of electrical health tests, the power supply rack for the spacecraft was installed in the bottom of the rocket's launch table. This would supply power to Rosetta until several minutes before launch.

After its transfer from the S3B building to the Final Integration Building, Rosetta was placed on top of its launcher on 16 February. This maneuver required it to be lifted about 40 meters to the top of the BAF, then moved sideways, lowered onto the Ariane 5, and secured in place by nearly 200 bolts.

Fig. 6.8: On 16 February 2014, Rosetta and its payload adapter were lifted about 40 meters in the Final Assembly Building, ready for mating with its Ariane 5 launch vehicle. (ESA)

Fig. 6.9: Rosetta was secured on top of the Ariane 5 launch vehicle by almost 200 bolts. The upper stage is visible, as is one of its side boosters on the right, beneath the platform. (ESA)

The next day, the orbiter's batteries were connected and charged to full capacity. At the same time, its protective covers were removed. This cleared the way for the aerodynamic fairing to be installed that would protect the spacecraft on its final days on the launch pad and the early part of the ascent.

Fig. 6.10: Rosetta with all of its protective covers removed, shortly before the aerodynamic fairing was attached to the launcher on 17 February. The lander is visible in the foreground. (ESA)

A large hose connected to the fairing provided a continuous airflow of 3,400 cubic meters per hour in order to keep the satellite in a clean, temperature controlled environment until launch.

The final activity involving the spacecraft took place on the evening of 23 February, when the lander's harpoons were armed and their protective covers removed – a delicate operation that involved a team member entering the fairing with what was referred to as a "diving board".

Fig. 6.11: A technician working inside the payload fairing to arm the lander's harpoons and remove their protective covers. (ESA)

At 15:30 local time on 24 February the Ariane 5G+ rocket moved along the 2.8 km rail line from the Final Assembly Building to the ELA-3 launch zone.

Although the available launch window lasted from 26 February until 17 March, the launch time was unusually precise due to Rosetta's unique mission profile, with the lift-off of Ariane Flight 158 scheduled for 07:36:49 UT on 26 February.[2]

All seemed to be going as planned, but with only 20 minutes to go the launch was postponed due to strong winds at high altitude above the launch site. Both the Ariane launch vehicle and its payload were put in a safe mode. Arianespace Chief Executive Officer Jean-Yves Le Gall announced that the next opportunity would occur at the same time on the following day.

However, unexpected problems arose once again on 27 February, causing the countdown to be stopped for a second time. On this occasion, the launch vehicle was the cause. Prior to the start of the cryogenic stage's fueling, a visual inspection of the core stage's exterior revealed that a 10×15 cm piece of insulation was missing. The thermal protection was necessary to insulate the cryogenic oxygen/hydrogen in the core stage from the environment at the tropical launch site.

Arianespace and ESA announced that the lift-off would be delayed by several days while the insulation was repaired. To the frustration of all concerned, the rocket had to be moved back to the Final Assembly Building, where a new block of thermal protection would be installed. The adhesive would require some 36 hours to dry and cure.

Resumption of the countdown was postponed until the beginning of the following week. In the meantime, the 'safed' spacecraft remained inside the fairing.

The third attempt to launch Rosetta was scheduled for 07:16 UT on 2 March, at the start of that day's window, with an opportunity at 07:36 UT if the weather intervened again.

Lift-off!

After the postponement of the two previous launch attempts, it proved to be third time lucky. The final countdown resumed at 19:47 UT on 1 March. At 23:47 UT the ground team carried out a check of all electrical systems. Early the next morning, the filling of the main cryogenic stage with liquid oxygen and hydrogen took place, followed by the chill-down of the Vulcain engine.

At 06:07 UT on 2 March, checks of the communication links between the launcher and the telemetry, tracking, and command systems were successfully completed, and 63 minutes later mission control announced "all systems go" and initiated the synchronized launch sequence.

---

[2] Flight 158 had two precise launch slots on 26 February: one at 7:16:49 UT and the other at 7:36:49 UT. These times would change a little if the actual launch date were to be changed. The actual launch time on 2 March was 07:17:44 UT. The overall launch window began on 26 February and lasted for 21 days. If this window had been missed, the mission would have been postponed.

Fig. 6.12: The Ariane 5G+ rocket lifts off from the ELA-3 launch site at 07:17:44 UT on 2 March 2004 carrying the 3 tonne Rosetta spacecraft. (Arianespace)

Fig. 6.13: The spent boosters jettisoned, the core stage of the Ariane 5 discards the fairing on its way to Earth orbit. (ESA/J. Huart)

At 07:17:44 UT, Ariane 5 Flight 158 roared off the ELA-3 launch pad at Kourou, its core stage and solid rocket boosters leaving a trail of fire as it rose into the black sky and headed east across the Atlantic Ocean. Three minutes later, the side booster rockets were successfully jettisoned, followed 50 seconds later by the fairing. At 07:29 UT the core stage exhausted its fuel and was discarded. The upper stage and its Rosetta payload were placed in an elliptical parking orbit that ranged between 250 km and 4,000 km.

Regular checks on the spacecraft's status were provided by ground tracking stations in Natal, Dongara (Australia) and Hawaii.

As this was the first time an Ariane 5 was being used to put a spacecraft on an Earth escape trajectory, involving a long delay prior to the final ignition of the upper stage, the European Space Operations Center (ESOC) in Darmstadt, Germany, monitored progress with mounting tension.

At 09:16 UT, the rocket's upper stage was ignited at an altitude of around 550 km. About 17 minutes later, the upper stage motor shut down on schedule, at an altitude of around 1,200 km. By the end of the burn, Rosetta's velocity had been boosted from 7,500 m/s to around 10,250 m/s. The tracking station at Kourou

Table 6.5: The Planned Sequence of Key Launch Activities

07:17:44 UT – Lift-off.

07:20 UT – The two solid rocket booster stages are successfully jettisoned, followed by the fairing.

07:25 UT – Acquisition by tracking station at Natal, Brazil.

07:29 UT – Ariane 5 has moved into a ballistic phase. Having burnt all its fuel, the main cryogenic stage was jettisoned. Rosetta and the Ariane 5 upper stage entered an elliptical parking orbit around Earth ranging between 250 km and 4,000 km.

08:05 UT – Acquisition of signal at the Dongara tracking station in Australia.

09:04 UT – Acquisition of signal at the South Point Tracking Station in Hawaii.

09:16 UT – Ignition of the EPS upper stage at an altitude of around 550 km and a velocity of 7,500 m/s. By the end of the 17 minute engine burn Rosetta is traveling at over 10,000 m/s (36,000 km/h).

09:28 UT – Signal acquisition at Kourou tracking station confirms an altitude of 750 km and a velocity of 10,180 m/s.

09:32 UT – Shut down of the EPS stage on schedule at an altitude of around 1,200 km and a velocity of around 10,250 m/s.

09:33 UT – Successful separation of the spacecraft. Rosetta is now on an escape trajectory and traveling out into the Solar System at a speed of about 3.4 km/s.

09:37 UT – ESOC in Germany takes over control of the mission and begins communicating with the spacecraft.

13:30 UT – The solar panels have been deployed successfully and the spacecraft is receiving power through them.

confirmed that Rosetta was looking good for separation and soon afterward the spacecraft set off at the start of its independent voyage into deep space.

After many years of trials and tribulations, Europe's comet chaser was finally escaping the grip of Earth's gravity and starting its 10-year trek to Comet 67P. The adventure had begun.

REFERENCES

Rosetta Status Report Archive: https://sci.esa.int/web/rosetta/status-report-archive

Rosetta: Europe's Comet Chaser, Peter Bond, ESA BR-179, September 2001, and ESA BR-215, December 2003.

Rosetta Rises to the Challenge, ESA Bulletin 112, p. 30-37, November 2002

Arianespace Flight 157 – Inquiry Board submits findings, 8 January 2003: http://www.esa.int/Our_Activities/Space_Transportation/Arianespace_Flight_157_-_Inquiry_Board_submits_findings2/

Rosetta Launch Postponed, 14 January 2003: http://www.esa.int/Our_Activities/Space_Science/Rosetta/Rosetta_launch_postponed

ESA'S New Challenge With Rosetta, 21 January 2003: http://www.esa.int/our_activities/space_science/rosetta/esa_s_new_challenge_with_rosetta

The Cosmic Mirror, Daniel Fischer, 16 February 2003: http://www.astro.uni-bonn.de/~dfischer/mirror/249.html

Rosetta overcomes major setback, Paolo Ferri, Room, Issue #2(8) 2016: https://room.eu.com/article/rosetta-overcomes-major-setback

Hubble assists Rosetta comet mission, 5 September 2003: http://sci.esa.int/hubble/33778-hubble-assists-rosetta-comet-mission-heic0310/

Rosetta's *Target, Comet 67P/Churyumov-Gerasigmenko*: http://sci.esa.int/rosetta/14615-comet-67p/

New Destination For Rosetta, Europe's Comet Chaser, 29 May 2003: http://www.esa.int/our_activities/space_science/rosetta/new_destination_for_rosetta_europe_s_comet_chaser

ESA confirms Rosetta's new target but identifies financial challenge for the mission, 2 June 2003: http://sci.esa.int/rosetta/32331-esa-confirms-rosetta-s-new-target-but-identifies-financial-challenge-for-the-mission/

Replanning Europe's mission of the decade: the backstory, 15 April 2014: http://blogs.esa.int/rosetta/2014/04/15/replanning-europes-mission-of-the-decade-the-backstory/

Second Rosetta Launch Campaign, Journal Archive: http://sci.esa.int/rosetta/31522-journal-archive/?farchive_objecttypeid=31&farchive_objectid=30928

Rosetta Ready to Explore A Comet's Realm, 12 January 2004a: http://sci.esa.int/rosetta/34479-rosetta-ready-to-explore-a-comet-s-realm/

Rosetta Begins its 10-year Journey to the Origins of the Solar System, 2 March 2004b: http://www.esa.int/Our_Activities/Space_Science/Rosetta/Rosetta_begins_its_10-year_journey_to_the_origins_of_the_Solar_System

Rosetta On Its Way, 2 March 2004c: http://sci.esa.int/rosetta/34784-02-march-2004-launch/

Ariane 5 Sends Rosetta on its way to Meet a Comet, Arianespace, 2 March 2019: http://www.arianespace.com/mission-update/ariane-5-sends-rosetta-on-its-way-to-meet-a-comet/

Ariane 5 V158 press kit: http://www.arianespace.com/wp-content/uploads/2017/06/04_feb_26-en.pdf

Svetlana Gerasimenko, co-discoverer of Comet 67P: https://sci.esa.int/web/rosetta/-/54597-svetlana-gerasimenko

Klim Churyumov, co-discoverer of Comet 67P: https://sci.esa.int/web/rosetta/-/54598-klim-churyumov

7

The Long Trek

7.1 EARLY COMMISSIONING

Although Rosetta was rapidly leaving Earth behind after launch at 07:17:44 UT on 2 March 2004, there were many checks to be carried out before the mission team could celebrate this achievement.

First, the venting and priming of the propulsion system was completed at 09:44 UT. The reduction of the spacecraft's initial spin rate and the Sun acquisition phase also proceeded very smoothly. Once the deployment of the huge solar arrays was completed at 10:11 UT, power started to flow. Next was switching on and testing the two star trackers required for stellar navigation and attitude control. The launch locks of the Philae lander were released successfully at the end of the first ground station pass. It would remain firmly attached to the spacecraft by the cruise latches until its release at the comet.

Another major milestone was passed at 00:34 UT on 3 March when the high-gain antenna was deployed, starting with firing the pyros of the launch locks. This was followed by three rotations, first in elevation, then in azimuth, and finally an azimuth-elevation slew that ended with the 2.2 meter dish aimed at Earth.

Later that morning, Rosetta completed its first trajectory correction maneuver – a test that changed its velocity by 1 m/s. After the spacecraft had adopted the requisite orientation, the 7 minute burn was initiated at 11:49 UT. The thrusters and the attitude and orbit control system (AOCS) performed flawlessly.

The mission operations specialists at ESOC completed their calculations and confirmed that the launch vehicle had put Rosetta on a highly accurate trajectory. Owing to the spacecraft's excellent performance and the progress of planned activities, it was decided to bring forward commissioning activities for the platform and payload.

© Springer Nature Switzerland AG 2020
P. Bond, *Rosetta: The Remarkable Story of Europe's Comet Explorer*,
Springer Praxis Books, https://doi.org/10.1007/978-3-030-60720-3_7

Activation of S-band transmission using the high-gain antenna got underway at 23:16 UT on 3 March. Successful commissioning of the S-band uplinks and downlinks using the low-gain and high-gain antennas proceeded throughout the night. Configuration of the high throughput X-band link using the high-gain antenna was accomplished by a downlink signal received by both the ground stations at Kourou and Madrid at 13:07 UT on 4 March. After the S-band uplink was terminated at 13:20 UT, the X-band uplink was established at 13:35 UT. When this was terminated at 14:30 UT, uplink communication was re-established using S-band and the high-gain antenna.

These tests successfully demonstrated the nominal performance of the major communication systems, which would be critical for the mission's success. However, the results had shown that a maximum data rate of 22 kbits per second was sustainable using the high-gain antenna.

Meanwhile, owing to the rapidly increasing distance from Earth, the possible telemetry data rate using the low-gain antenna had already decreased to 7.8 bits per second and the antenna would soon become redundant until the next Earth fly-by.

During this time the attitude control system was subjected to various tests, such as gyroscope calibrations and determination of the friction in the reaction wheel system. This included, for the first time, switching on all four reaction wheels simultaneously. Substantial disturbance of the spacecraft's orientation caused by outgassing occurred during the first few days, but these torques gradually decreased to nominal levels.

Full configuration of the 25 Gbit solid-state mass memory took place on 4 March in order to support routine operations, creating data stores for all instruments and storing redundant files of application software. Activation of all memory modules for the mission was successfully completed.

Commissioning of the power subsystem took place at the end of the Madrid pass on 4 March. All checks were successful and the power subsystem behaved as expected.

The drive mechanisms of the solar arrays were employed during the early days of the flight in order to keep the solar cells perpendicular to the Sun as the spacecraft rotated – thus providing maximum power.

The Launch and Early Orbit Phase (LEOP) was completed successfully on 5 March, clearing the way for the Mission Control Team at ESOC to move from the Main Control Room to the Rosetta Dedicated Control Room, in readiness for commencement of the initial phase of the commissioning program. From now on, primary communications links were provided solely by the ESA ground station at New Norcia, Australia, which could provide coverage for 10-12 hours per day.

7.2   INITIAL PAYLOAD TESTING

As Rosetta headed into deep space and closer to the Sun, it maintained the 'tilted' orientation achieved during LEOP and was slewed to +X Sun-pointing only when this was required for payload operations. This kept the high-gain antenna mechanism cooler, after an unexpectedly high temperature was measured in LEOP. This had no immediate impact on operations, apart from imposing additional attitude slews, which in turn meant more reaction wheel off-loading cycles.

Attention then turned to the commissioning of Rosetta's scientific payload, scheduled to take place throughout March, April and May. The instruments were activated one at a time, with experts from the instrument teams present at ESOC during the daily ground contacts with the spacecraft.

The first three instruments to be activated were COSIMA, CONSERT and OSIRIS, and their first commissioning activities were successful. On 10 March, the CONSERT antenna was the first payload appendage to be deployed.

For five days in mid-March, mission controllers concentrated on commissioning the Philae lander.

Rosetta instrument commissioning resumed with the RPC on 18-19 March. On the second day, spacecraft booms carrying the RPC MIP (Mutual Impedance Probe) and LAP (Langmuir Probe) instruments were successfully deployed using the primary systems. As the instruments were active for the deployment, the science telemetry confirmed the success of the operation. However, in order to analyze a failure that was identified in the redundant power supply, the remaining RPC activities were postponed to a later date.

The following week saw ROSINA, ALICE and VIRTIS activated for the first time, with successful completion of their initial commissioning activities.

At the end of March, the RSI was commissioned during four ground passes, followed by the initial activation and commissioning of the MIRO instrument. This coincided with several software maintenance activities on the spacecraft. In early April, MIRO commissioning was followed by the commissioning of GIADA and MIDAS. With all of the scientific instruments having been successfully activated at least once since the start of the mission, this marked an important milestone.

Meanwhile, a new attitude pointing mode (known as GSEP Earth-pointing) was initiated and maintained for most of the year. Periodic re-pointings to prevent the Sun coming too close to the +X spacecraft axis were planned, typically once per month.

When priority use of the New Norcia ground station was temporarily assigned to support the main science phase of ESA's Mars Express mission, Rosetta's daily sessions were reduced to about 7 hours.

7.3   ASTEROID SELECTION

Rosetta's scientific goals always included the possibility of studying one or more asteroids at close range, but only after Rosetta's accurate insertion into an inter-planetary orbit could the ESA mission managers assess how much fuel was actu-ally available for fly-bys. Information from ESOC enabled the Rosetta Science Working Team (SWT) to select a pair of asteroids of high scientific interest in orbits between those of Mars and Jupiter that were reachable within the fuel budget.

On 11 March, the SWT made its final selection of (2867) Steins and (21) Lutetia. These were contrasting objects, both in terms of size and physical properties.

Steins was known to be relatively small, with a diameter of a few kilometers. The fly-by was set for 5 September 2008, during Rosetta's first excursion into the asteroid belt, at a range of just over 1,700 km. The encounter would be at the rela-tively low speed of about 9 km/s.

Ground-based studies had shown Lutetia to be a much bigger object with a diameter of about 100 km, but little was known of its composition. Rosetta would pass within 3,000 km of it at a speed of 15 km/s on 10 July 2010, during its second venture into the asteroid belt.

In addition to close-up images of these primordial objects, Rosetta would sup-ply data on their mass and density as clues to their composition, measure their temperature, and search for any nearby gas and dust.

7.4   COMMISSIONING, PHASE 2

The second phase of lander and instrument commissioning was carried out during April. One minor glitch was the failure of the prime explosive pyro to fire on the first attempt to open the door on the ALICE detector. After an investigation, the redundant pyro was fired successfully and the door safely opened.

On 23-25 April, the spacecraft was rotated to point its payload toward Earth for the first time. This gave MIRO and VIRTIS the opportunity to take calibration measurements, with Earth as their target.

There was some concern about the effects of increasing temperature as Rosetta continued to cruise nearer to the Sun. An unplanned slew was executed to change the spacecraft's attitude when the –Z thrusters reached the highest permitted tem-perature. Monitoring of the thermal environment resulted in a further restriction on the range of allowed spacecraft attitudes. The Sun would be restricted to –4 and +50 degrees from the +X axis within the X/Z plane, at least until after perihelion on 25 May.

The second slot of commissioning OSIRIS was achieved on 25-30 April, with the exception of the planned Earth-Moon imaging, which was postponed as a result of the limits imposed in response to the high thruster temperatures that occurred during the Earth-pointing period for the MIRO instrument.

Rosetta's first scientific observations took place on 30 April, when MIRO, ALICE, VIRTIS and OSIRIS took spectra and images of Comet C/2002 T7 (LINEAR), which was making its first (and only) visit to the inner Solar System. The coma and tail were studied at wavelengths ranging from the ultraviolet to microwave portions of the spectrum, from a distance of about 95 million km. The presence of water molecules around the comet was identified. Follow-up observations by ALICE and MIRO were made on 17 May.

Fig. 7.1: An image of Comet C/2002 T7 (LINEAR) taken in blue light by the OSIRIS camera on Rosetta. It shows the comet's nucleus from a distance of about 95 million km, with the tail extending over 2 million km. (ESA/MPGH/H. Uwe Keller)

Even though the calibrations of the instruments remained to be completed, the data from the remote sensing observations confirmed their excellent performance.

As the insolation increased, the thruster modules on the +X side of the spacecraft continued to warm up, and by 7 May the hottest module, number 7, had reached 65°C.

Preparations for the first deep space maneuver – a long burn of the thrusters – commenced on 6-7 May by firing a dozen pyro valves to enable pressure regulators to link the high-pressure helium tanks to the propellant tanks.

Rosetta's first major propulsion burn was completed successfully on 10 May. This was the most critical spacecraft activity executed since LEOP, the maneuver produced a change in velocity (delta-V) of 152.8 m/s. This was achieved through a continuous burn of the four axial thrusters over a period of about 3.5 hours. The performance of the spacecraft was excellent, within about 0.05% of nominal.

A 'touch-up' maneuver was made on 16 May, as part of the trajectory targeting activities around perihelion. This burn of less than 17 minutes gave a delta-V of 4.989 m/s. Having achieved the desired trajectory, the reaction control subsystem was isolated on 24 May.

Rosetta reached perihelion at 17:00 UT on 24 May. Its minimum distance from the Sun was 132.6 million km (0.886 AU). At the end of the last New Norcia pass on 28 May, Rosetta was 40.3 million km from Earth and the one-way signal travel time was 2 minutes 14 seconds.

With Rosetta now beginning to move away from the Sun and follow a circuitous path back to the vicinity of Earth, the main commissioning activities got under way.

The Cruise 1 Phase formally began on 7 June 2004, and was expected to last until the start of the second (and final) commissioning slot in September-October 2004. It required the parallel activation of all onboard instruments, and special pointing activities designed to calibrate the remote sensing payload. At the same time, the period of non-contact with the spacecraft was gradually increased, culminating in weekly passes by mid-July.

During June, the attitude and orbit control system was checked and updated, and the medium-gain antenna was brought online as a precaution, in case the loss of the high-gain link should prompt the spacecraft to adopt 'survival mode'.

Starting in the fourth week of the Cruise 1 Phase, Rosetta was configured for 'quiet cruising' for the first time and the payload remained off, with no test, maintenance or characterization activities taking place.

The uplink and activation of new onboard avionics software (version 7) was undertaken over 16-23 July. The mission team judged this "probably the most critical operation since launch". This successfully concluded four months of intense system testing at ESOC.

The activation of the new software involved commanding a reboot of the Data Management and Attitude and Orbit Control Systems on 22 July. This automatically put the spacecraft into safe mode. It was the first time that safe mode had been entered in flight, but the automatic reconfiguration operations and six hours of manual recovery controlled from ground went as planned. Once verification of the new software was completed on 2 August, it took over full control of the spacecraft.

Following verification of the low-gain and high-gain antennas, the second part of the main commissioning phase began on 6 September. This incorporated payload software updates and full functional testing of each instrument. The commissioning was completed in mid-October, after which the baseline attitude of the spacecraft was changed to +X Earth pointing until it was time to prepare for the first Earth flyby.

Rosetta entered cruise mode again on 17 October. Apart from routine flight operations and monitoring tasks, the mission team used this time to perform some troubleshooting activities. These included a health check of the star trackers and activation of the OSIRIS instrument on 20 January 2005. OSIRIS received a software upload designed to solve an issue with its door mechanism that was discovered during the commissioning phase, but follow-up tests showed the update had not solved the problem. Nevertheless, the instrument was able to take pictures of Comet Macholtz as part of the verification of its operations.

7.5 EARTH FLY-BY #1

Fly-bys make use of the gravitational attraction of planets to modify a spacecraft's trajectory and achieve the orbital energy needed to reach the final target.

Rosetta's first Earth swing-by would be an essential gravity boost in its 10-year, 7.1 billion km trek to Comet 67P/Churyumov-Gerasimenko. Several small tweaks to its trajectory were made in preparation for the Earth fly-by. These culminated in TCM-6, on 17 February 2005, which was so precise that a back-up maneuver planned for 24 February was cancelled.

The spacecraft was gradually configured for the gravity assist, including activating its fourth reaction wheel on 25 February, switching its radio communications from X-band to S-band on 27 February, and shifting from the high-gain antenna to the low-gain antenna on 2 March.

Rosetta was given ground coverage priority from 27 February to 10 March, with daily New Norcia passes that supported the various maneuvers and provided the tracking data needed to accurately calculate the fly-by geometry.

On 1 March, the first payload instruments were activated: RPC and ROMAP on the lander. These were followed by MIRO and VIRTIS on 4 March. Meanwhile, SREM, the radiation environment monitor, was active in the background. Another lander instrument, ÇIVA, was operated for three hours around closest approach. The OSIRIS imager was not able to operate due to unresolved problems with its instrument cover.

Fig. 7.2: Earth viewed by Rosetta's navigation camera from a distance of about 250,000 km after the fly-by on 4 March 2005, featuring South America and Antarctica. (ESA)

The first Earth fly-by took place at 22:09 UT on 4 March 2005. The spacecraft approached from the night side, and the closest point of approach was above the daylight hemisphere. During the passage, the spacecraft seemed to fly to the west and disappeared from view on the day side. ESA's closest-ever Earth fly-by up to that time saw the spacecraft pass above the Pacific Ocean, just west of Mexico, at an altitude of 1,954.74 km and a relative velocity of 38,000 km/h.

Fig. 7.3: An image of the Moon rising over the Pacific Ocean at 22:06 UT on 4 March 2005, taken by a navigation camera just three minutes prior to the point of closest approach. (ESA)

The Earth swing-by included various tracking tests with the navigation cameras, using the Moon as a target, as Rosetta approached on 4 March. All operations were successfully completed – with the exception of a problem in configuring the link between Camera B and the spacecraft's mass memory that prevented the pictures from this camera being stored.

The spacecraft was commanded at 01:00 UT on 5 March into Asteroid Fly-by Mode, using the navigation camera aimed at the Moon to control its attitude. In this special mode, Rosetta made use of the NAVCAM to steer in such a way that the payload platform was kept pointed towards the target. This was the first and only in-flight test opportunity for this mode, which was solely intended to be used operationally during the actual fly-bys of the asteroids Steins in 2008 and Lutetia in 2010. The test lasted 9 hours and was fully successful. After the end of the test, the spacecraft was pointed back Earthward to enable the payload instruments and the navigation cameras to take black-and-white Earth pictures and other measurements from a rapidly growing distance of about 250,000 km.

Fig. 7.4:   Rosetta's VIRTIS instrument took this composite infrared view of the Moon on 4 March 2005, prior to closest approach to Earth, from a distance of 400,000 km. The spatial resolution is 100 km per pixel. It combines images at three different infrared wavelengths (2,225, 3,000 and 3,650 nanometers). Red represents the basaltic plains known as 'maria', while blue represents the ancient highlands. (ESA)

On 4-5 March, prior to closest approach to Earth, VIRTIS took a series of images in visible and infrared light at a distance of 400,000 km from the Moon. From its perspective, the Moon was a crescent with only a small portion of the surface illuminated (ranging between 19% and 32%). The infrared images detected both thermal (heat) radiation from the lunar surface and reflected solar radiation. The spatial resolution was low but it was still possible to distinguish Oceanus Procellarum, Kepler crater and Mare Humorum (see Fig. 7.4 and Fig. 7.5). Spectral analysis distinguished the mineralogical differences between the flat 'seas' (in Latin 'maria') and the highlands. For instance, it was possible to see marked differences in the abundance of two kinds of rocks known as pyroxene and olivine.

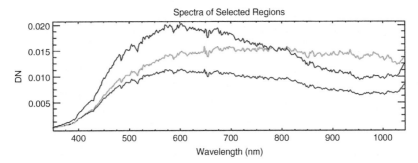

Fig. 7.5: VIRTIS infrared spectra of Kepler crater (blue), Mare Humorum (green), and Oceanus Procellarum (red) on the Moon. The spectra are uncorrected for the phase angle. Despite the low spatial resolution of this instrument's images, the mineralogical differences between the highlands and the basaltic plains ('seas' or 'maria') were evident. (ESA)

On 5 March, after the closest approach to Earth, VIRTIS took a series of images of our planet in visible and infrared light from a distance of 250,000 km. Only 49% of it was illuminated, but the instrument was able to detect infrared radiation emitted from both the day and night hemispheres.

Fig. 7.6: A composite red-green-blue image (left) and stretched false-color image (right) of Earth in visible light from the VIRTIS instrument. It is possible to distinguish Argentina (a) and the Andes mountain chain (b). (ESA)

Fig. 7.7: An infrared image of Earth from the VIRTIS instrument taken from a distance of 250,000 km on 5 March 2005, after Rosetta's closest approach. It shows the distribution of carbon dioxide in the atmosphere, with higher concentrations in the green areas. The spatial resolution is 62 km per pixel. (ESA)

VIRTIS was switched off on 5 March, followed by MIRO. Next was the lander's ROMAP and RPC on 7 March. ALICE was activated for various slots on 8-9 March. The radio link was configured back to S-band using the high-gain antenna on 6 March, and a new Earth-pointing attitude was initiated on 8 March. Downlinking the science data from the fly-by was completed on 10 March.

At the Science Working Team's meeting at ESOC on 8-9 March, the principal investigators presented the preliminary results from their instruments during the fly-by, and expressed their satisfaction.

The final observation during the fly-by phase was on 26 March, when ALICE made another observation of the Moon.

7.6   ACTIVE CRUISE AND DEEP IMPACT

A passive checkout of the payload in late March found an anomalously high electrical current in MIDAS, so the instrument was shut down for several weeks until functional tests could be arranged.

During much of April and the first half of May, the mission team carried out the first in-flight test of the Near Sun Hibernation Mode (NSHM), with the spacecraft spending many days in a special low activity mode, during which the gyroscopes and reaction wheels were inactive and the attitude was controlled by the star tracker and thrusters only. This hibernation capability was required in order to enable the spacecraft to operate with minimal hardware for periods of up to six months during the long 'quiet cruise' phases, with ground contact typically taking place just once per month.

The next major event, which had been added to the original mission timeline, was the monitoring of the after-effects of a collision between a projectile fired from NASA's Deep Impact spacecraft and the nucleus of Comet 9P/Tempel 1.

At NASA's request, Rosetta's initiation of quiet cruising was postponed by several months to enable it to observe the collision from a viewpoint about 80 million km from the comet, with an angle of 90 degrees between the Sun and the comet. On 28 June, the spacecraft slewed to point its four remote sensing instruments at the comet. At 05:52 UT on 4 July, the 362 kg copper projectile smashed into the nucleus at high speed.

ALICE, MIRO and OSIRIS operated continuously from 29 June to 14 July, using a complex pointing profile designed to ensure that the best scientific data were returned from each of the instruments. VIRTIS was operated only for a few hours around the predicted time of impact. Rosetta downlinked an average of 60 Mbytes of scientific data every day to the New Norcia ground station.

OSIRIS observed the nucleus and coma of the comet both prior to and after the impact, using different filters to study the extensive dust cloud that resulted. By measuring the water vapor content and the cross-section of the dust cloud, scientists calculated that some 4,500 tonnes of water were blasted into space by the impact, and, rather surprisingly, that even more dust was ejected (see Fig. 7.8).

The results indicated that comets are composed more of dust held together by ice, rather than being ice contaminated with dust. They are more like 'icy dirtballs' than the 'dirty snowballs' previously believed (see Chapter 1). The scientists did not find evidence of enhanced activity from Comet 9P/Tempel 1 in the ensuing days, suggesting that, in general, impacts with small bodies are not the cause of prolonged cometary outbursts.

The observation campaign was very successful, with all four instruments operating very well. Some problems occurred with the timing of commands for OSIRIS, and with MIRO, but both instruments were recovered within 24 hours, with minor impact on the overall operations and data return.

This exercise was the first scientific planning and operations scenario on a large scale and an extended period of time for the Rosetta mission. It provided an important learning experience that would assist in designing future operations.

With the conclusion of the Deep Impact scientific operations, Rosetta slewed to the Near Sun Hibernation Mode entry attitude on 25 July, and the next day the Passive Cruise Mode was entered, including transition to NSHM and configuration of all subsystems for the new mode. This was the first operational use of this mode, and the intention was to employ it for most of the rest of the journey to Mars.

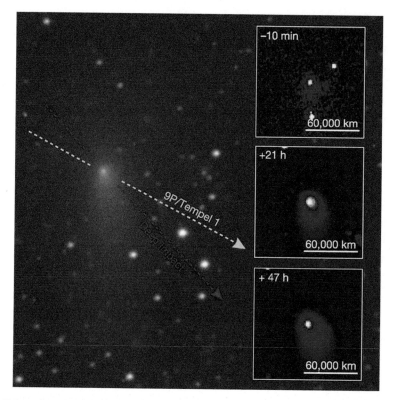

Fig. 7.8: Images taken by the OSIRIS-NAC on Rosetta show the expanding dust cloud that resulted from the collision between Deep Impact's copper projectile and Comet Tempel 1 on 4 July 2005. The dotted lines indicate the projected orbits of the comet (blue) and the Deep Impact spacecraft (red) from the vantage point of Rosetta. The insets show the dust coma and the expanding debris cloud around the time of impact, +21 hours, and +47 hours. The Sun is to the top. A scale line is also included. The outward flow of the impact debris was estimated from the expansion rate of the cloud to be up to 300 m/s. (Kuppers et al, Nature, 13 October 2005)

The spacecraft was to remain in this low activity, low bit rate mode for the next two months, but this changed when an unexpected fuel over-consumption of 20 grams and an unexpected delta-V of 2.5 mm per second were spotted off-line in data that was obtained on 4 August.

An unplanned tracking pass on 8 August provided spacecraft telemetry that indicated that the fuel consumption had returned to normal and the acceleration had disappeared. Nevertheless, analysis of earlier telemetry confirmed that anomalous spacecraft behavior had occurred at the start of August, so it was decided to return to Active Cruise Mode as soon as possible. This transition was completed on 19 August, when the spacecraft was reconfigured to employ the high-gain antenna in X-band for maximum telemetry bit rates. As a safety measure, mission controllers once again began monitoring the spacecraft on a weekly basis.

Another major unexpected event was a solar flare on 8-9 September that hit Rosetta at the beginning of the weekly non-coverage period. When the signal was acquired for the weekly contact on 15 September, mission control found that the spacecraft's active star tracker had crashed, and the second star tracker (not used for attitude control) was in standby mode.

For that time, Rosetta had been relying only on the attitude and orbit control system (AOCS) to determine its attitude using just its gyroscopes, which resulted in a drift of about 0.7 degrees and a small offset in the pointing of the high-gain antenna. Nevertheless, it remained possible to maintain a radio link with Earth.

Recovery activities were implemented immediately and, by the end of that day's ground station pass, both star trackers were in tracking mode and the attitude was restored. Later checks verified that no long-term damage had been caused by the solar flare. It was decided to leave the radiation environment monitor (SREM) on for the long journey out to Mars when Rosetta resumed Passive Cruise Mode.

Rosetta was then used to support a tracking campaign for the validation of the new ESA deep space ground station in Cebreros, Spain. Three passes were conducted successfully on 26, 27 and 28 September.

Apart from regular monitoring and minor maintenance activities, there were few interruptions to this 'quiet' routine. Attitude guidance was changed to +X Earth-pointing on 14 December 2005 in order to minimize the disturbance torques experienced by the spacecraft and the fuel which was then consumed in offloading the reaction wheels. Contact with the spacecraft was via weekly New Norcia passes.

On 10-12 March 2006, Rosetta turned to aim its OSIRIS imaging instrument towards asteroid Steins, its final opportunity to observe the asteroid prior to the fly-by in September 2008. The images collected over a period of about 24 hours were downlinked to Earth on 15-16 March. Although the brightness of the asteroid was comparable to that of a candle from a distance of about 2,000 km, OSIRIS was able to measure variations with an accuracy of better than 2%.

The 'light curve' for Steins implied a spin period of slightly over six hours, in agreement with previous Earth-based studies. The asymmetry of the curve suggested an irregular shape, but OSIRIS found no evidence for either a 'tumbling' motion of the asteroid or the presence of a satellite. By combining the OSIRIS observations with ground-based data, scientists attempted to determine the orientation of the spin axis.

On 15 March, the spacecraft was configured for the first solar conjunction of the mission, in preparation for when it would pass very close to the Sun in the sky, as viewed from Earth. In practice, this meant that its angular distance from the Sun would be less than 5 degrees from mid-March to mid-May 2006, with the minimum of about one third of a degree on 13 April. Such conjunctions cause major communication problems between a spacecraft and its ground stations. To accommodate the significant degradation of the uplink and downlink, the level of activity on the spacecraft was minimized and the S-band transmitter was used in parallel with the nominal X-band link.

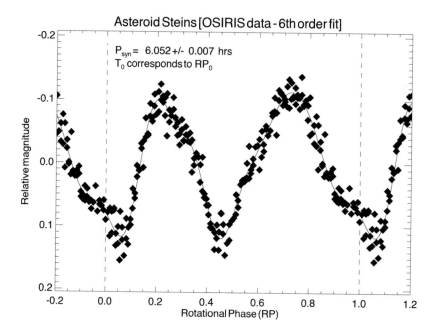

Fig. 7.9: The OSIRIS instrument measured the 'light curve' of asteroid Steins, showing its variation in brightness over one day. The light curve is ~23% brighter at maximum than at minimum. The asymmetry of the curve suggested an irregular shape. One rotation lasted slightly over six hours. (Stefano Mottola/DLR, OSIRIS team)

On 30 March, the telemetry bit-rate was reduced, as planned, to 3.5 kbps in order to help deal with the worsening signal disturbance from the Sun. On 6 April, a telecommand link test was successfully performed and all commands successfully decoded on board, although the uplink signal received by the spacecraft was significantly affected by the Sun.

To measure the performance of the communications link during the solar conjunction (and to support studies of the solar corona) daily ground station passes were initiated, although most of the New Norcia passes were brief (4 hours) and only for tracking, without establishing any telemetry or telecommand connections. The stored telemetry was downlinked when Rosetta emerged from solar conjunction.

On 24 May, Rosetta resumed the Near Sun Hibernation Mode attitude that it was to maintain until the Mars fly-by.

The RPC suite of instruments was activated on 4 July for five days, as Rosetta crossed the tail of Comet Honda. The science observations were stored until they could be downloaded at the next opportunity in August. Preliminary analysis of the data indicated that the magnetometer had been able to detect the tail far downstream from the nucleus. As a result, such campaigns were requested for further potential crossings of comet tails.

7.7   MARS FLY-BY

On 26 July 2006, Rosetta was reconfigured to Active Near Sun Cruise Mode. This meant the spacecraft resumed two ground station passes per week for telemetry recovery and spacecraft maintenance operations.

The Mars Swing-by Phase formally started on 28 July. The fourth passive payload checkout took place in the second half of August, when all instruments except GIADA and ROSINA were activated and checked out in sequence.

Rosetta's second Deep Space Maneuver (DSM2) was made successfully on 29 September, with a delta-V of 31.791 m/s and an overall burn duration of about 52 minutes. The maneuver was performed without ground contact – the high-gain antenna could not be kept pointing at the Earth during the burn because it coincided with the plume zone of the thrusters. The high-gain antenna was moved away shortly before the start of the maneuver, then pointed back after it was completed.

The instruments remained off as various spacecraft checks were completed, notably the fifth AOCS checkout that took place between 11 and 13 October.

Another trajectory correction was carried out on 13 November, a modest delta-V of 9.9 cm/s to improve the accuracy of the Mars approach path.

Meanwhile, preparations got underway for the fourth Active Payload Checkout (PC4). The checkout operations for most of the instruments – except ROSINA and OSIRIS – started on 23 November. The two NAVCAMs were also operated during this slot. All of the checkout operations were executed successfully, apart from several minor glitches that were analyzed and rectified. By December, during the last series of instrument checkouts, the OSIRIS-WAC was able to take long exposure images of the fast approaching planet against the background of stars in which the planet was overexposed and surrounded by a halo of scattered light.

On 3-4 January 2007, the new year's activities began with OSIRIS observations of Lutetia, the second of the mission's asteroid targets. During this 36 hour campaign, a series of images were taken from a range of 245 million km.

On 9 February, 16 days ahead of closest approach to Mars, another small trajectory correction maneuver (TCM-16d) was made which lasted 54 seconds and consumed 58.28 g of fuel. All was now ready for the fly-by of the Red Planet.

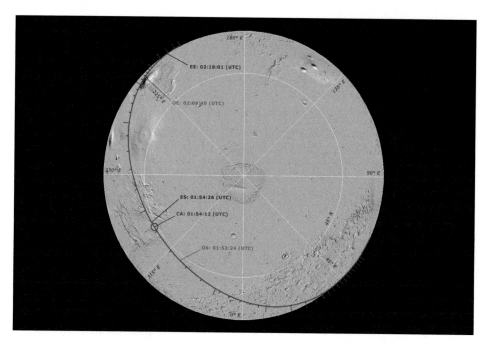

Fig. 7.10: A view of Mars's northern hemisphere showing the ground trace of Rosetta during the fly-by on 25 February 2007. The spacecraft's track around closest approach is projected onto Mars for the period between 23:49:29 UT on 24 February (lower right) and 02:30:48 UT on 25 February (upper left). Red ticks indicate 5 minute intervals, with the smaller dark ticks denoting intervals of 1 minute. Also indicated, with their nominal times are: OS (Occultation Start) – Rosetta behind Mars as seen from Earth; CA (Closest Approach) ~250 km above the surface. ES (Eclipse Start) – Rosetta enters Mars's shadow; OE (Occultation End) – Rosetta observable from Earth again; and EE (Eclipse End) – Rosetta exits the planet's shadow. The closest point of approach was above 43.5° N, 298.2° E. (ESA)

The X-band transmitter was switched off at 01:08 UT on 25 February, and telemetry was lost for almost 90 minutes. However, the S-band carrier signal was still received until 01:56 UT, when ground controllers lost contact for a 15 minute radio blackout as the spacecraft passed behind the planet with respect to ground stations on Earth.

At 01:57:59 UT, Rosetta made its closest approach, passing 250.6 km above 43.5° N, 298.2° E. The occultation (eclipse) started at 01:58 UT, with the spacecraft spending about 25 minutes in shadow. Tension was high in the mission control room until the S-band carrier signal returned on time, signifying Rosetta had powered up again successfully upon re-emerging into sunlight.

Rosetta passed over the planet's surface at a relative speed of 36,191 km/h. During the swing-by, the gravitational pull of Mars caused Rosetta to change direction, placing it on course for its second encounter with Earth in November 2007. The spacecraft was decelerated by an estimated 7,887 km/h relative to the Sun, putting it in the desired 0.78 × 1.59 AU orbit inclined 1.9 degrees to the ecliptic.

During the approach to Mars, SREM, the standard radiation monitor, observed the radiation environment for 48 hours and its RPC plasma suite spent 48 hours measuring the properties of the charged particles and the magnetic field along the trajectory.

The OSIRIS camera system captured a series of images of Mars and the small moon Phobos as it emerged from behind the planet and transited the Martian disk. These were later assembled into a movie.

Fig. 7.11: An OSIRIS-NAC true color view of Mars taken at 18:28 UT on 24 February at a distance of 240,000 km, about 7.5 hours before closest approach to the planet. It combines images obtained with orange (red), green, and blue filters. The spatial resolution is about 5 km per pixel. The greenish areas are clouds. The south polar ice cap is clearly visible. The image is centered on the red desert of Elysium Planitia. (ESA & MPS for OSIRIS Team MPS/UPD/LAM/IAA/RSSD/INTA/UPM/DASP/IDA, CC BY-SA 3.0 IGO)

Some magnificent full disk images of Mars were returned, showing the south polar cap, ice clouds in the northern polar region, and sparse morning clouds. Above one limb, the images provided an edge-on view of clouds some 50-60 km above the surface. Stereoscopic images were also obtained.

The fly-by was during the southern spring, a time of the Martian year when a large amount of carbon dioxide and water ice is vaporizing from the southern polar cap, prior to migrating to the northern polar cap during northern winter.

VIRTIS also studied Mars at infrared wavelengths and, together with ALICE, performed a spectroscopic analysis of the atmosphere. NAVCAM images were also obtained.

Because the upper atmosphere of Mars is similar to the ultraviolet spectrum of a comet (in that it is dominated by carbon monoxide, hydrogen, carbon, and oxygen, as well as their ions) this was the first opportunity for ALICE, the ultraviolet imaging spectrometer, to function in something that resembled a comet's environment.

It saw the sunlit upper atmosphere emitting ultraviolet light, thus providing the first in-situ observation of this 'dayglow' phenomenon on Mars. Such observations can give important information on how carbon dioxide, the principal gas in the planet's atmosphere, behaves at high altitude. ALICE also mapped the night side spectrum, studying the far ultraviolet 'nightglow' and seeking evidence of auroras. In addition, OSIRIS and ALICE searched for a dust ring above the equator.

Rosetta switched off its solar-powered instruments prior to executing the fly-by, because its path would take it through the shadow of Mars. Without sunlight, the instruments would have had to draw power from the batteries – which had to be husbanded for exploring Comet 67P. The fly-by trajectory also imposed a loss of communication during an occultation with Earth, starting 2 minutes prior to closest approach and lasted about 15 minutes.

Passing through the planet's shadow was an unfortunate consequence of re-targeting Rosetta at 67P, rather than 46P/Wirtanen, with different launch and fly-by timings. The mission team improvised a means of obtaining data anyway, by turning on two instruments on the battery-powered Philae lander – which operated independently of Rosetta for a total of three hours, including the eclipse. The ÇIVA camera took images of Rosetta in silhouette against the red-brown surface of Mars (see Fig. 7.12), and the Rosetta Magnetometer and Plasma monitor (ROMAP) observed the patchy magnetic field of the planet and the turbulence caused as the solar wind crossed the 'bow shock' upstream of it (see Fig. 7.13).

Fig. 7.12: This image was taken by the ÇIVA imager on the Philae lander during the Mars swing-by on 25 February 2007, four minutes before closest approach at a distance of about 1,000 km. It shows part of the main spacecraft and one of its solar arrays, with Mars in the background. The lander was attached to the side of the main spacecraft. (ÇIVA/Philae/ESA Rosetta)

The downlink of the science data acquired during the fly-by was completed on 9 March. The Mars Swing-by Phase was formally wrapped up by the third Deep Space Maneuver (DSM3) on 26 April, a minor adjustment that consumed ~6.96 kg of fuel and altered the spacecraft's velocity by 6.526 m/s. The departure trajectory took the spacecraft through the magnetotail of the planet.

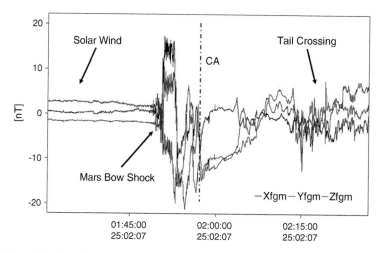

Fig. 7.13: This data from the ROMAP instrument on the Philae lander shows the magnetic environment around Mars. The incoming supersonic solar wind is undisturbed (left) until it encounters the boundary 'bow shock' of the magnetosphere, after which it is decelerated to subsonic speed and becomes turbulent. The horizontal axis shows time and the vertical axis shows the intensity of the magnetic field. (ROMAP/Philae/ESA Rosetta)

7.8   EARTH FLY-BY #2

Having left Mars behind, Rosetta turned its instruments toward Jupiter, the largest planet in the Solar System. At extremely long range, ALICE watched as NASA's New Horizons flew down the Jovian magnetotail, bound for distant Pluto. The instrument also observed Jupiter's auroras and the plasma torus created by its volcanically active moon Io. These observations were completed on 9 May.

Although Jupiter was just a pinprick of light, ALICE was able to separate its emitted light into the three components that made up its spectrum. These included sunlight reflected from Jupiter's cloud tops, ultraviolet emission given off by particles ejected in eruptions from the volcanoes on Io, and light from auroras caused by particles striking the planet's atmosphere – some from the Sun and some ejected from Io.

Rosetta passed the aphelion of its third orbit around the Sun on 20 May 2007, when it reached a heliocentric distance of 237.14 million km (1.58518 AU). After its fifth passive payload checkout, the spacecraft was again placed into Near Sun Hibernation Mode on 5 June, a quiet phase that was punctuated by weekly ground contacts. In early September, the spacecraft was reconfigured back to Active Cruise Mode, followed by a month of active payload checkout 6, starting on 17 September.

As Rosetta was rapidly approaching Earth for the second time, controllers chose to refine its path with a trajectory correction maneuver on 18 October that lasted for 46 seconds and gave a delta-V of 3.4 cm/s.

At 20:57:23 UT on 13 November 2007, the mission made its second Earth swing-by, with the closest point of approach 5,295 km above the Pacific Ocean (63.76° S, 74.58° W), south-west of Chile, at a relative velocity of 45,000 km/h (12.5 km/s).

On this occasion, the highest priority was given to spacecraft operations because the gravity assist was crucial for the success of the overall mission. In any case, on both the inbound and outbound tracks the solar illumination and temperature conditions were rather unfavorable. As a result, there were very limited time slots available for the instruments to be used safely.

Nevertheless, some instruments were activated for calibration, scientific measurements and imaging. These observations were scheduled during and around the time of closest approach, from 7 to 20 November.

Observations were made by the navigation cameras, the ALICE ultraviolet spectrometer, the OSIRIS cameras, and the VIRTIS infrared instrument. ALICE was used to obtain spectra of the illuminated Earth and Moon. VIRTIS studied fluorescent emission of carbon dioxide and molecular oxygen by scanning the atmosphere whilst pointing at the limb on the sunset side. VIRTIS also obtained spectra of the Moon.

The MIRO microwave instrument was also activated several times during the fly-by phase in preparation for the observations which were planned for asteroids Steins and Lutetia. And the ROMAP magnetometer and plasma monitor on the Philae lander was operative for about two weeks.

The SREM radiation monitor made measurements of Earth's radiation belts, and the Radio Science Experiment provided tracking data to investigate possible anomalous accelerations during swing-bys.

During its approach, OSIRIS captured images of Earth's night side, showing large cities that appeared as points of light, along with a narrow crescent of light above the South Pole. After closest approach, the instrument's NAC was able to image both Earth and Moon. The results included color views of Earth's daylight hemisphere, featuring Australia and south-east Asia, and the Earth-facing side of the Moon.

Rosetta first pointed toward Earth in order to perform observations of the atmosphere and the magnetosphere – including a search for shooting stars from space. It imaged urban regions in Asia, Africa and Europe, then aimed at the Moon and obtained spectra of its illuminated face. After closest approach to Earth, the spacecraft looked back and its NAVCAM took a series of black-and-white images that included Antarctica and a crescent Moon.

Fig. 7.14: During its second approach to Earth, Rosetta observed the planet's night side. This OSIRIS-WAC image was taken with a red filter at 19:30 UT on 13 November 2007. Its false colors highlight city lights in Europe, north Africa and the Middle East. (ESA ©2005 MPS for OSIRIS Team MPS/UPD/LAM/IAA/RSSD/INTA/UPM/DASP/IDA)

Rosetta observed the Moon for 11 hours shortly after the Earth fly-by, and then, as it receded, it performed observations of the Earth-Moon system on 15, 16, 18 and 20 November.

Additional passes using ground stations of the NASA's Deep Space Network were used to download the science data from the fly-by as rapidly as possible. The Santiago Station was used to cover the periods of closest approach that could not be covered using ESA's ground stations, in order to obtain a full set of data that might record any indications of a possible anomalous acceleration that had been reported at previous close Earth gravity assist manoeuvres (see Box 7.2: The Mystery of the Fly-by Anomaly). No effect was detected this time.

The second Earth fly-by helped Rosetta to gain sufficient energy to be catapulted towards the main asteroid belt, where it was to investigate the small asteroid, Steins, in September 2008. By this stage, Europe's comet chaser had flown a little over 3 billion km of its 7.1 billion km journey toward 67/P Churyumov-Gerasimenko.

Fig. 7.15: After its second Earth gravity assist, Rosetta looked back and took a number of images. This OSIRIS-NAC image was taken at 02:30 UT on 15 November 2007, and is a composition of orange, green and blue filters. The continent of Australia is evident at the bottom. (ESA ©2007 MPS for OSIRIS Team MPS/UPD/LAM/IAA/RSSD/INTA/UPM/ DASP/IDA)

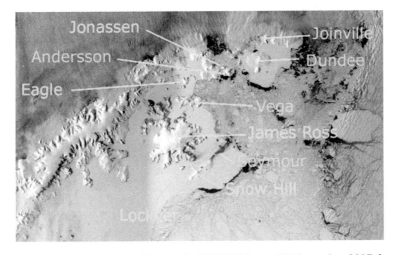

Fig. 7.16: An image by one of Rosetta's NAVCAMs on 13 November 2007 from an altitude of 5,350 km. Graham Land is the portion of the Antarctic peninsula that lies closest to South America. Various nearby islands are labeled. (ESA)

7.9   STEINS

After swinging by Earth on 13 November, Rosetta's operations focused on down-linking the science data, spacecraft reconfiguration and checkout activities, and, based on tracking during daily passes with the New Norcia ground station, correcting its trajectory.

The first of two planned TCMs to put Rosetta on course for Steins took place at 00:54 UT on 23 November. The maneuver was executed using onboard accelerometers, and lasted 538 seconds. The velocity change (delta-V) was 1.526107 m/s and the estimated total fuel consumed was 1.6 kg.

At perihelion on 17 December at about 136 million km (0.91077 AU) from the Sun, Rosetta was largely shut down. It started its fourth orbit around the Sun with a passive payload checkout (PC7) in early January 2008, and observations were made of the portion of sky that would be seen during the approach to asteroid Steins.

On 21 February, the second planned TCM prior to the Steins encounter was completed, with a tiny velocity change of 248.3 mm/s.

The next major activity was a test of the asteroid fly-by attitude dynamics on 24 March. This validated the spacecraft's behavior under the extremely dynamic conditions imposed by the fly-by scenario chosen for Steins. It was also an opportunity to test the behavior of the AOCS software in Asteroid Fly-by Mode (AFM), the influence of stray sunlight on the star trackers of the orbiter, and the cameras of the orbiter and lander.

Three days later, Rosetta adopted Near Sun Hibernation Mode attitude to cruise passively with its gyros switched off and attitude control relying on thrusters. It was not reactivated until 1 July, after which it was reconfigured once more to Active Cruise Mode.

The eighth active payload checkout (PC8) involved conducting a series of commissioning activities and software updates for the instruments over a period of four consecutive weeks. In parallel, the eighth attitude and orbit control system checkout was conducted.

The Steins navigation campaign got underway in early August, with long range imaging on 4, 7, 11, 14, 18, 21, and 25-31 August and 1-4 September. From mid-August the spacecraft was pointing permanently at the asteroid, apart from occasional re-alignments for instrument calibrations. This was the first time in the history of ESA spacecraft operations that such a technique had been used. Of the total of 324 observations by NAVCAM and OSIRIS-NAC, 75 were taken by OSIRIS.

The accuracy of the earlier measurements exceeded expectations, enabling them to be used for the execution of the first trajectory correction maneuver on 14 August. Designated CA-3 (Closest Approach Minus 3 Weeks), the 113 second maneuver changed Rosetta's velocity by 12.8 cm/s and adjusted its trajectory to open its closest approach to Steins from 554.2 km out to 792.4 km, very near the 800 km which would provide the best fly-by conditions consistent with spacecraft performance.

The CA-8d slot for trajectory correction on 28 August, eight days before closest approach, was not used. However, it was necessary to refine the orbit with an additional TCM on 4 September, around 36 hours before closest approach, in order to achieve the desired flyby conditions.

The CA-36h maneuver lasted for 103.5 seconds and consumed 127.1 g of fuel. The resultant velocity change was 11.8 cm/s, ensuring that Rosetta's estimated closest approach would take place at an ideal altitude of 800.7 km.

Scientific activities associated with the fly-by began on 20 August, when the OSIRIS camera completed another observation of the asteroid's light curve.

The fly-by was to occur on the sunlit side of Steins to facilitate continuous observation of the asteroid before, during, and after closest approach – while passing through a phase angle of zero (with the Sun directly overhead). To keep the small object in the field of view of the scientific instruments in the closest approach phase, the spacecraft would have to be aimed with an accuracy of better than 2 km.

It was also programmed to perform an attitude flip maneuver that would last for 20 minutes, starting 40 minutes prior to closest approach (see Fig. 7.17). Rosetta had to change its orientation rapidly, pushing the system to its design limits during the high speed fly past.

During this autonomous tracking of the asteroid, the attitude of the spacecraft was automatically driven by NAVCAM A in order to keep the small asteroid continuously in the field of view of the imaging instruments. A rehearsal in March 2008 had not gone quite to plan.

As operations manager Sylvain Lodiot explained:

The NAVCAMs were not behaving as expected and the 12 hours leading up to the Steins fly-by were spent trying to find proper settings for them. For both cameras the tracking data delivered was not satisfactory for the enabling of the autonomous tracking mode. The background noise was disturbing the measurements and delivering a distorted set of data.

The control team spent many hours finding a set of parameters that could allow the camera to deliver proper data. The final configuration was reached some tens of minutes before the fly-by itself. Those hours were pretty stressful. Even then, some spacecraft off-pointing was observed during the AFM mode, with Steins sometimes drifting out of view. Significant work had to be done to improve the technique before the fly-by of Lutetia.

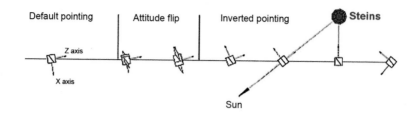

Fig. 7.17: A schematic illustrating Rosetta's attitude flip and autonomous pointing during the fly-by of Steins, with time running from left to right and the spacecraft's attitude indicated by its +Z axis (toward asteroid Steins, red) and +X axis (blue). (ESA)

As it closed in, Rosetta turned its high-gain antenna away from Earth, breaking the telemetry signal, and initiated its scientific observations.

The closest approach to Steins was at 18:38:19 UT on 5 September 2008, some 2.14 AU from the Sun and 2.41 AU from Earth. The spacecraft swept past the asteroid at a relative velocity of 8.62 km/s (about 31,000 km/h).

Fourteen instruments were switched on for the fly-by, providing detailed, multi-wavelength analyses of the asteroid and in-situ measurements of its dust, plasma, magnetic and radiation environment.

The remote sensing instruments – ALICE, OSIRIS, VIRTIS and MIRO – provided imaging and/or spectrometry at ultraviolet, visible, infrared, and submillimeter wavelengths. Both the OSIRIS cameras were used. The NAC was expected to obtain the highest resolution images, but unfortunately it ceased its programmed observations at a distance of 5,200 km from the asteroid and entered safe mode about 10 minutes prior to closest approach. Consequently, the highest resolution images, at about 80 meters per pixel, were taken by the WAC from a range of about 800 km around the time of closest approach (see Fig. 7.18).

The five instruments of the Rosetta Plasma Consortium – IES, ICA, LAP, MIP, MAG – as well as ROMAP and SREM, sensed the charged particle, magnetic and radiation environment near the asteroid. The ROSINA and GIADA instruments searched for gas and dust around it. Instruments on the lander were also operating, as was the Radio Science Experiment.

The Rosetta control room at the European Space Operations Center, Darmstadt, received the first radio signal via NASA's Goldstone antenna at 20:14 UT, confirming that the spacecraft was performing is programmed operations. Activities continued for several days after closest approach, including gravitational microlensing observations by OSIRIS. All the science data acquired during the asteroid fly-by were downlinked by the beginning of October, interleaved with OSIRIS science observations.

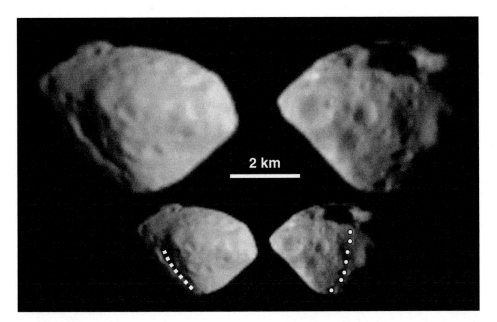

Fig. 7.18: Images of Steins taken by the Wide Angle Camera and Narrow Angle Camera of OSIRIS. The WAC image (top right), taken around closest approach, shows the surface at a resolution of 80 meters per pixel. The NAC image (top left) taken 10 minutes before closest approach, has a resolution of 100 meters per pixel. The scale is given by the 2 km bar. The difference in line of sight between the images is 91 degrees and they give two views of the asteroid. Steins's rotation is retrograde (east to west) and therefore, in accordance with IAU rules, its north pole points towards celestial south. As a result the large, 2 km diameter crater is close to the south pole. The positions of the seven pits in the linear catena (bottom right image) and the large fault on the opposite side (bottom left image) are indicated. (ESA © 2008 MPS for OSIRIS Team MPS/UPD/LAM/IAA/ RSSD/INTA/UPM/DASP/IDA)

Up to that time, seven asteroids had been encountered by spacecraft: five S-type (stony), one V-type (Vestoid) and one C-type (carbonaceous) (see Chapter 1). Asteroid (2867) Steins was the first of the rare E-type to be visited by a robotic explorer, so scientists were eager to learn more about its characteristics to improve their knowledge of the variety of asteroids, and how the Solar System formed and evolved. The surface composition of E-type asteroids resembles that of enstatite achondrite meteorites (having a high content of magnesium silicate) but their origin is uncertain. E-types are the dominant asteroids in the Hungaria group in the main belt between Mars and Jupiter.

Rosetta confirmed that the rotation period of Steins is roughly 6 hours, and found that it spins in a retrograde (east to west) direction, so its north pole faced toward celestial south.[1] The fact that the north-south axis of rotation was roughly perpendicular to the orbital plane meant the spacecraft flew right over its equator. OSIRIS imaged about 60% of the surface, revealing the asteroid to be shaped like a diamond with dimensions 6.67 × 5.81 × 4.47 km, widest at its equator.

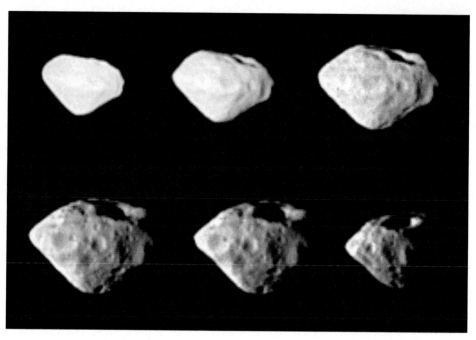

Fig. 7.19: OSIRIS-WAC images of Steins taken during the 5 September 2008 fly-by. *(ESA ©2008 MPS for OSIRIS Team MPS/UPD/LAM/IAA/RSSD/INTA/UPM/DASP/IDA)*

Compared with other asteroids imaged at close quarters by spacecraft, Steins has a low crater density, and almost all of them are shallow indentations. The overall crater shape and depth-to-diameter ratio are consistent with degradation (or infill) by material ejected from impacts and disturbance of loose regolith (surface material) by seismic shaking during large impacts.

Nevertheless, the ancient surface has at least 40 small, shallow impact craters and two larger ones (see Fig. 7.20). The largest, just over 2 km in diameter and nearly 300 meters deep, is at the south pole. It was named Diamond, in reference to the asteroid's overall shape. A linear chain of similarly sized hollows, called a catena,

---

[1] Many objects in the Solar System, such as Earth, rotate from west to east, so that their north pole points toward celestial north. However, some objects, such as Steins, spin in a backward or retrograde direction (east to west) and, in accordance with the rules of the International Astronomical Union (IAU), their north pole points toward celestial south.

extends north from it and crosses the equator (see Fig. 7.18). It may have been created by the excavation of Diamond opening a subsurface fracture, into which loose material later settled, creating hollows. On the opposite side of the asteroid is a large groove (fracture or fault) that stretches from the north pole to the equator.

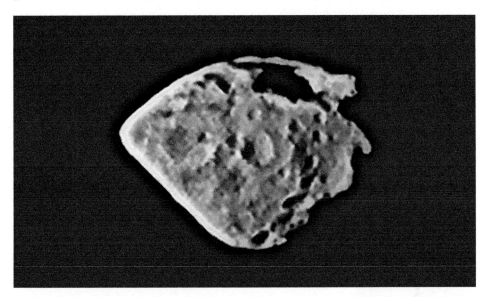

Fig. 7.20: Steins imaged by Rosetta's OSIRIS camera on 5 September 2008. Image stacking and processing by amateur astrophotographer Ted Stryk enhanced the shadows to emphasize the difference between bright crater rims and their shadowed floors. Over 40 craters can be discerned on the surface. The largest shown at the top of this frame is the 2 km wide crater named Diamond, which is actually at the south pole. (ESA ©2008 MPS for OSIRIS Team MPS/UPD/LAM/IAA/RSSD/INTA/UPM/DASP/IDA; processing by T. Stryk)

The explosive creation of Diamond was probably one of the major contributors to the general appearance of the asteroid. The fracturing would have given rise to a loosely bound 'rubble pile' which had a low overall density. Furthermore, debris from the impact would have been scattered across the surface, masking older craters.

Scientists believe that Steins was originally part of a larger, differentiated object which broke up long ago. It was then battered by impacts, most significantly by the creation of Diamond. Its unconsolidated nature suggests that a major collision in the future could easily shatter it into fragments.

Analysis of the data by the OSIRIS principal investigator Horst Uwe Keller, and colleagues, attributed the striking conical appearance of Steins and the relatively low number of craters with diameters of less than 0.5 km to surface reshaping as a result of the Yarkovsky-O'Keefe-Radzievskii-Paddack (YORP) effect. This was the first time that this effect had been seen in a main belt asteroid.

The YORP effect is a phenomenon that occurs when light photons from the Sun are absorbed by a small body and re-radiated as infrared emission. This carries away momentum as well as heat. Because more heat escapes from one side of the object than the other, its rate of rotation (and possibly its spin axis) will change.

If the rate of spin slowly increases, loose material may slide toward the equator. Landslides, or gradual movement of granular material and small boulders, could have erased many of the smaller craters and made the distinctive conical shape. However, if this was once the case, the current rotation rate of Steins is too slow for the YORP effect to further modify it. Seismic tremors generated by small objects striking the surface would have further disturbed the loose material.

Fig. 7.21: Asteroid Steins in 3D. This image, which gives a visual impression of the highs and lows of the surface, is based on data collected by Rosetta at a distance of about 800 km. It is best viewed using stereoscopic glasses with red-green or red-blue filters. (ESA ©2008 MPS for OSIRIS Team MPS/UPD/LAM/IAA/RSSD/INTA/UPM/DASP/IDA)

Detailed analysis of the OSIRIS images allowed Keller and his colleagues to confirm that the fairly high albedo (reflectivity) and spectral characteristics of Steins are consistent with its classification as an E-type asteroid. The low iron content and lack of measurable variation in surface color were evidence of a homogeneous composition.

Box 7.1: Naming Features on Steins

Following analysis of the images returned by Rosetta's OSIRIS cameras, a number of distinct surface features were identified and named on asteroid Steins. Because its shape bears some resemblance to a cut diamond, the International Astronomical Union (IAU), which decides on the official place names given to all worlds in the Solar System, accepted a proposal from the OSIRIS imaging team that craters on Steins be named after gemstones.

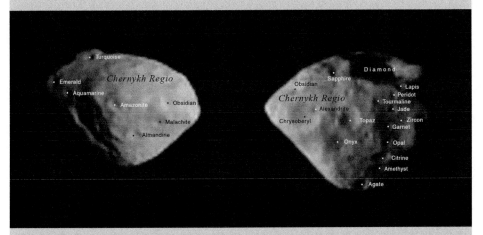

Fig. 7.22: Named features on asteroid (2867) Steins. (©ESA 2012 MPS for OSIRIS Team MPS/UPD/LAM/IAA/RSSD/INTA/UPM/DASP/IDA)

The process of naming major features on Steins was initiated by Sebastien Besse, a PhD student working with the OSIRIS imaging team in France, who went on to become a Research Fellow at ESA-ESTEC in the Netherlands.

During his studies of the craters on Steins, he began to think about a suitable nomenclature for them. Since Steins looks very much like a diamond, he decided to research the names of different gemstones, with a view to linking them to particular craters. One of the factors influencing his deliberations was a desire to select names that would be familiar or appealing to the general public, and relatively easy to pronounce.

After discussions with the rest of the team, it was agreed that the names of 23 gemstones or precious minerals would be recommended to the IAU. At the same time, it was decided to follow tradition by naming a large region after the asteroid's discoverer, Nikolai Chernykh, who first detected Steins in 1969. Although the name 'Ruby' was rejected as unsuitable and replaced by 'Diamond', most of the names were officially accepted in May 2012.

The name Diamond was assigned to the dominant landform, a 2.1 km diameter impact feature near the south pole. A circular crater about 650 meters wide and 80 meters deep that was imaged in the center of the asteroid during closest approach was called Topaz.

Another notable feature is a catena, a chain of pits (or crater-like hollows) that stretches from the asteroid's north pole all the way to Diamond crater. At least some of the craters in the chain may have been formed by loose material sinking into a subsurface fracture. These were named after familiar gemstones such as Agate, Opal and Jade. However, Besse also decided to include Citrine, since it sounds like the French word for lemon: citron.

Another elongated feature (a groove or fault), surrounded by small pits and craters, is visible on the other side of Steins, approximately opposite to the catena. This feature has yet to be named.

7.10   BACK TO EARTH

As tiny Steins was left behind, Rosetta continued its passage through the inner asteroid belt. With almost all of the instruments shut down, the top priority of the two weeks that followed the encounter was to downlink the science data. All of the measurements acquired during the fly-by were successfully received on the ground.

For four weeks following the fly-by, the only noteworthy scientific operations were periodic observations of gravitational microlensing events in the Galactic Bulge using the OSIRIS Narrow Angle Camera. This was one of several serendipitous science research programs that were added to the main scientific program during Rosetta's cruise phase.

Gravitational microlensing occurs when a massive object passes in front of a fainter object. The gravity of the foreground object serves as a lens, bending the light from the more distant star or planet, giving rise to two or more distorted, unresolved images which are noticeably magnified.

In practice, because the required alignment is so precise and difficult to predict, microlensing is very rare; hence the decision to view the densely populated center of our Galaxy. Starting on 7 September and continuing until 4 October, a series of seven observations were made. It was expected that observations of the dense star fields would include about 50 microlensing events at any one time. When compared with simultaneous observations taken with ground-based telescopes, it was expected the observations would provide accurate masses for dozens of stars and other faint objects, in particular 'failed stars' known as brown dwarfs and perhaps a couple of exoplanets.

For five days in October and November when mission activity was low, engineering teams investigated an issue that had occurred with the COSIMA instrument during the Steins fly-by. After settling on the instrument's Target Manipulation Unit (TMU), it was decided to upload software updates as part of the active payload checkout scheduled for September 2009.

During the latest active cruise phase, all instruments were once again switched off, except for the SREM radiation monitor that ran in the background.

On 17 December, Rosetta passed the aphelion of its current orbit around the Sun, reaching a record distance of 338.63 million km (2.26 AU). Eight days later, its distance from Earth set a new record for the mission at 485.12 million km (3.24 AU).

Meanwhile, between 17 December 2008 and 6 January 2009, Rosetta was within 3 degrees of the Sun as viewed from Earth, with the minimum separation on 27 December.

As the new year dawned, the spacecraft was given its ninth passive payload checkout with the instruments being activated and checked one at a time. In addition, various maintenance tasks were carried out.

On 16-18 February, while still near aphelion, a thermal characterization exercise was carried out to verify the thermal performance of the spacecraft and its instruments in preparation for the fly-by of asteroid Lutetia in July 2010, which would occur at 2.7 AU. It was conducted by exposing the spacecraft to the Sun with an elongation angle of first +175 degrees from the +Z axis (positive towards the +X axis) and later +192 degrees. Each attitude was maintained for 24 hours, sufficient to attain thermal stabilization. There were no problems in sustaining these attitudes.

However, some problems were noted with the ROSINA and the VIRTIS instruments, which reported non-ideal conditions when going to 192 degrees, exposing the rear of the spacecraft to the Sun. ROSINA noted outgassing that could disturb its asteroid measurements, obliging the ground team to favor early exposure of the back side of the spacecraft for the instrument. VIRTIS recorded high sensor temperatures that were likely to impair its science results. This could be obviated by limiting the exposure of the rear of the spacecraft to very short times. These factors would be taken into account when the mission team eventually defined the final operational requirements for the Lutetia fly-by.

Several recovery activities, test activities, and software updates were also performed, with the troublesome Target Manipulation Unit on COSIMA being fully recovered.

7.11   EARTH FLY-BY #3

The fourth Deep Space Maneuver (DSM4) was successfully performed on 18 March, slightly modifying the trajectory in preparation for Rosetta's third swing-by of Earth on 13 November 2009.

After a short tracking campaign to define the new orbit, the active cruise phase was replaced by passive cruise phase #6 on 2 April. With all of the instruments switched off apart from the SREM, the spacecraft was once again placed in Near Sun Hibernation Mode, where it would remain until 8 September. Weekly monitoring was conducted via ESA's New Norcia ground station.

On 9 September, the telemetry, tracking & commanding was reconfigured from the medium-gain antenna to the high-gain antenna, and the next day the tenth spacecraft checkout (SC10) started. This activity was briefly interrupted on 16 September by the spacecraft entering safe mode, but it was soon recovered and the checks were successfully completed. The tenth payload checkout (PC10) started on 17 September and finished on 8 October. It involved the staggered activation, testing and calibration of the instruments, as well as software updates.

In the first half of October, the mission team checked out the readiness of ESA's Kourou and Maspalomas ground stations, as well as the NASA Deep Space Network stations at Canberra, Madrid and Goldstone, using X-band and S-band radio frequencies. The navigation campaign leading to the Earth fly-by started with data about Rosetta's inbound trajectory provided by its telemetry, and from Doppler and ranging data received from the ESA and NASA ground stations. Based on this data, the flight dynamics team was able to calculate any changes to the spacecraft's velocity that might be required to refine the approach to Earth.

The primary trajectory correction was conducted on 22 October, three weeks prior to closest approach. The four axial thrusters fired for 86 seconds to achieve a delta-V of 8.8 cm/s. The accuracy of this maneuver meant that no further corrections would be necessary.

A safe mode was triggered on 5 November, as Sylvain Lodiot's operations team was trying to send commands to Rosetta to retrieve lost data from its memory. Fortunately, the spacecraft was soon returned to normal. Nevertheless, Lodiot carried out an analysis of the unexpected shut down, and realized there had been many similar close calls during previous data dumps. From then on, therefore, the operations team made sure never to attempt to redump lost data.

On 6 November, ALICE was switched on as part of the scientific operations planned during the Earth swing-by. This enabled researchers to test ALICE's performance by viewing Earth in ultraviolet light. The spectral data confirmed that the instrument was in focus, and showed the main ultraviolet spectral emission of our planet. MIRO, OSIRIS, RPC and VIRTIS were activated during the next few days. The encounter program would span more than a week on either side of closest approach.

Between 8 and 18 November, when not conflicting with the swing-by mission operations, the spacecraft slewed as necessary to make observations of Earth and the Moon. In addition, the various science teams used the opportunity to

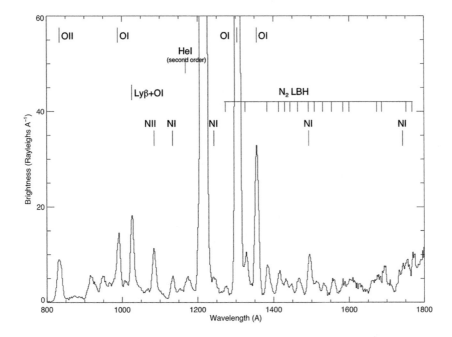

Fig. 7.23: A graph showing one of the spectra obtained by the ALICE ultraviolet instrument during the approach to Earth. Some of the oxygen (O) and nitrogen (N) emission lines have been labeled. (NASA/JPL/SwRI)

characterize the performance and calibrate their instruments in preparation for the program intended for the encounter with asteroid Lutetia in 2010 and, ultimately, with Comet 67P in 2014.

Rosetta's final visit to Earth was achieved at 07:45:40 UT on 13 November, with the closest point of approach 2,479.523 km being above the Indian Ocean, south of the Indonesian island of Java. The gravity assist meant that the spacecraft received a boost of 3.6 km/s, accelerating it into a new orbit with an aphelion at about 5.33 AU and a perihelion still bound to 1 AU.

Mission operations were primarily conducted by ESA's New Norcia ground station, with the support of tracking passes by Kourou in French Guiana and Maspalomas in Spain, as well as by NASA's DSN stations at Goldstone (DSS-24), Madrid (DSS-54) and Canberra (DSS-34). The success of the swing-by was confirmed when contact was established via Maspalomas at 08:05 UT.

All the science observations during the swing-by were successfully performed. In addition to imaging Earth and the Moon, OSIRIS sought auroras, lunar sodium and a hydroxyl (OH) tail during lunar stray light calibration.

Fig. 7.24: This image was obtained as Rosetta approached Earth for the third swing-by. It was acquired with the OSIRIS-NAC from 633,000 km at 12:28 UT on 12 November 2009, with a resolution of 12 km per pixel. The crescent is roughly centered on the South Pole. The outline of Antarctica is visible under the clouds which form the south polar vortex. Very bright spots are strong reflections from pack ice along the coastline. (ESA ©2009 MPS for OSIRIS Team MPS/UPD/LAM/IAA/RSSD/INTA/UPM/DASP/IDA)

Fig. 7.25: A high-resolution view of the South Pacific by the OSIRIS-NAC on 13 November. The image shows cloud structures associated with an anticyclone. This false-color composite was generated using the orange, green and blue filters. (ESA ©2009 MPS for OSIRIS Team MPS/UPD/LAM/IAA/RSSD/INTA/UPM/DASP/IDA)

Fig. 7.26: This OSIRIS image of the USA at night from a distance of 60,000 km was taken with a 10 second exposure on 13 November 2009 at 04:44 UT, three hours prior to closest approach. The bright spots show the cities on the east coast and the Gulf of Mexico. Some cities are clearly visible, but others like New York are covered by clouds, making the light diffuse. (ESA © 2009 MPS for OSIRIS-Team MPS/UPD/LAM/IAA/ RSSD/INTA/UPM/ DASP/IDA)

During its approach to Earth, Rosetta took a sequence of full-disk images of the night side using the VIRTIS Mapper channel (VIRTIS-M). This hyperspectral imager collected image data simultaneously in 864 narrow, adjacent spectral bands (colors) from distances ranging between 240,000 and 228,000 km. When the spacecraft was less than 55,000 km from Earth, it carried out limb scans of the night side, starting from 150 km above the surface. Between 14:00 and 14:25 UT on 13 November, the instrument looked back at Earth from 230,000 km and studied the composition of the observed areas. Combining the images with spectral data enabled two-dimensional compositional maps to be created (see Fig. 7.27).

The results confirmed the ability of VIRTIS to isolate desired features of the observed object, using the spectral data. They included color composites of Earth composed by using different spectral bands in the visual wavelength range. One false-color image revealed the distribution of chlorophyll, a key signature of vegetation, in the South American rainforest. Other images showed Earth in the infrared.

The SREM instrument was used to determine the spatial distribution of Earth's radiation belts along the spacecraft's trajectory. The data included count rates for electrons and protons, the strength of the magnetic field, and the derived electron fluxes in different energy ranges.

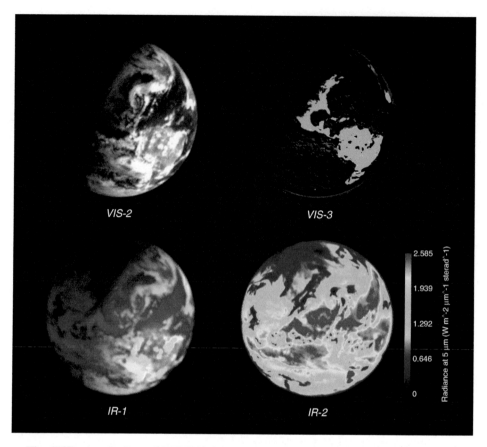

Fig. 7.27: A selection of VIRTIS imagery from the final Earth fly-by. VIS-2 was obtained over the Gulf of Mexico and the Americas at 14:00 UT on 13 November, in the visible and near-infrared spectral range with a spatial resolution of about 50 km. The western coast of Africa can also be seen at the top right of the image. The blue, green and red channels were selected to provide maximum contrast to the ground, clouds and sea: 0.474 micrometers for sea water, 0.785 micrometers and 1.0 micrometers for land. The VIS-3 image was obtained several hours after closest approach, about 230,000 km from Earth. The composite image shows the distribution of the chlorophyll present in vegetation, with maximum chlorophyll abundance shown in green. Black patches over parts of South America are caused by dense clouds which completely masked the ground. IR-1 is an infrared view of Earth using colors with the R, G and B channels at 4.92 micrometers, 2.25 micrometers and 1.20 micrometers respectively. Being sensitive to reflected solar radiation and emitted thermal radiation, the VIRTIS-M data allowed the night side to be seen. The cyan areas are related to high altitude clouds, which are particularly bright in the near-infrared on the day side. Landforms appear pink. IR-2 was obtained using thermal emission at 5.0 micrometers. The color scale (right) shows the measured radiance. The warmest areas (orange-red) are on the day side and in the equatorial region. The coldest areas (violet) are the tops of the clouds and the region (upper left) that includes Canada and the Arctic. (INAF-IFSI/INAF-IASF/ASI)

Starting at 20:00 UT on 9 November, the RPC suite of five instruments made complementary measurements of Earth's plasma environment and magnetospheric studies for a week around closest approach. On 10 November, its Langmuir Probe (LAP) detected clear signals from the spacecraft firing its thrusters.

A clean-up maneuver with a delta-V of 58 cm/s was performed on 23 November to eliminate an inaccuracy in the departure trajectory. Rosetta was now on course to re-enter the asteroid belt, but still five years away from its historic rendezvous with Comet 67P.

Box 7.2: The Mystery of the Fly-by Anomaly

Since NASA's Galileo spacecraft experienced a speed increase of 3.9 mm/s greater than expected when it swung past Earth in December 1990, mission controllers and scientists at ESA and NASA have noticed that some spacecraft experience unexpected changes in speed during Earth fly-bys. This appears to be due to a strange variation in the amount of orbital energy they exchange with Earth.

The changes in speed are extremely small and may occur as either an increase or a decrease in velocity. However, attempts to explain and predict the variation by means of fundamental physics have all failed.

The largest unexpected variation – a boost of 13.0 mm/s – was observed when NASA's NEAR spacecraft made its Earth swing-by in January 1998.

Scientists were hoping to gain more insight into the anomaly when Rosetta swung by Earth on three separate occasions. Curiously, the spacecraft had an unexplained velocity increase of 1.8 mm/s during the first fly-by in 2005, but no evidence for the mysterious variation was detected in either 2007 or 2009.

"It's a mystery as to what is happening with these gravity events," said Trevor Morley, lead flight dynamics specialist working on Rosetta. "Some studies have looked for answers in new interpretations of current physics. If this proves correct, it would be absolutely ground-breaking news."

Attempted explanations range from tidal effects of the near-Earth environment, atmospheric drag, or the pressure of radiation emitted or reflected by the Earth, to much more extreme possibilities, such as dark matter, dark energy, or previously unseen variations in General Relativity.

One American research team, led by ex-NASA scientist John Anderson, has even looked at the possibility that Earth's rotation may be distorting spacetime – the fundamental fabric of our Universe – more than theory predicts, and affecting spacecraft, but there is, as yet, no explanation for this mechanism, and why it should occur during some fly-bys and not others.

Before even considering such exotic explanations, all the usual causes of spacecraft speed errors have been thoroughly eliminated by numerous investigations at both ESA and NASA. Software bugs, calculation errors, tracking uncertainties and other, much more mundane causes have all been systematically eliminated or accounted for, leaving the speed anomaly maddeningly unexplained. The mystery remains!

7.12   BACK TO THE ASTEROID BELT

On 23 November, Rosetta passed through the perihelion of its current orbit, at 146.6 million km (0.98 AU) from the Sun. It would never again be within the radius of Earth's orbit around the Sun.

Apart from the SREM radiation monitor, which continued to run in the background, all of the instruments were off.

The new year, 2010, began inauspiciously, with Rosetta entering an unplanned safe mode on 1 January, in the middle of a week that was between ground communication passes. It proved to be an issue with one of the remote terminal units. After further investigation on 4 January, the spacecraft was restored to normal operations over the next three passes.

On 11 January, a test was successfully completed to characterize the telecommand link performance on the low-gain antenna, using both the ESA New Norcia ground station and a NASA DSN ground station at Goldstone. This cleared the way for a test of the Deep Space Hibernation Mode (DSHM), the never-before-tried operating mode which the spacecraft was to employ between June 2011 and January 2014, during its passive cruise to Comet 67P.

The aim of the test was to validate the spacecraft behavior and the operational scenario for the hibernation phase, including:

- The spacecraft configuration for DSHM entry.
- The spacecraft spin-up maneuver.
- Detection and processing of Rosetta's strobing signal on the ground.
- Commanding hibernation entry.
- The spacecraft in DSHM configuration for 1 week.
- The spacecraft's autonomous DSHM exit.
- The spacecraft spin-down.
- Recovery of the spacecraft after DSHM exit.

The DSHM test was successfully completed between 20 and 27 January. During this period, the largely dormant spacecraft was monitored daily for about 8 hours.

Rosetta was spin-stabilized on 20 January (rather than the normal three-axis stabilization) and commanded to enter hibernation mode the next morning. After the spin-up, the downlink was alternated several times between modulated telemetry using the medium-gain antenna and an unmodulated carrier signal using the high-gain antenna to monitor the systems and to test the signal acquisition of the 'strobing' pulses from the spinning spacecraft, respectively.

After DSHM began, the spacecraft switched off the subsystems for attitude and orbit control (AOCS) and for telemetry, tracking and command (TTC). The TX-2 transmitter was switched on again later for monitoring purposes during the test.

At the start of the week-long test, the spin rate was commanded to 4 degrees per second (or two-thirds of a revolution per minute). Data showed that it decreased very slowly over time, at a rate of about 0.0002 degrees per second per day, presumably because of solar radiation pressure acting on the large solar arrays and the high-gain antenna dish. The principal inertial axis, or spin axis, which had been estimated beforehand to be at about 7 degrees in the +X/+Z quadrant of the spacecraft's reference frame, shifted to 10.4 degrees.

Throughout the DSHM test, spacecraft temperatures remained within limits, and it was not necessary for the mission operations team to intervene.

The hibernation was completed on 27 January as planned, although an autonomous action of the computer caused Rosetta to enter a safe mode. By mid-afternoon, the reaction wheels had resumed attitude control and the spacecraft was once again three-axis stabilized. Another safe mode was triggered during the post-DSHM recovery. However, the spacecraft was promptly restored to its normal operating mode.

The solid-state mass memory, which had been powered down on 21 January and was off for the entire test, was fully re-configured as part of the nominal recovery activities after exiting DSHM.

With Rosetta returned to normal cruise mode, early February saw re-lubrication of reaction wheel B. When the so-called run-in phase was completed on 11 February, it was returned to the control loop of the attitude and orbit control system to enable the spacecraft to use all four reaction wheels. Several payload tests and software updates were also conducted.

Between 14:33 UT and 15:15 UT on 24 February, the solar generator on the Philae lander was tested. Isolated from the Rosetta orbiter in terms of power, the lander drew current from its own battery and solar arrays while being exposed to the Sun.

By now, thoughts were turning to the forthcoming fly-by of asteroid Lutetia. In a rehearsal on 14-15 March, the spacecraft repeated the attitude flip made at Steins and then autonomously tracked a virtual point which corresponded to what would be Lutetia's position.

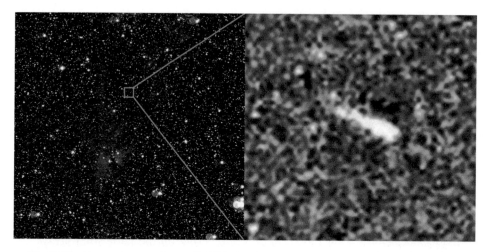

Fig. 7.28: Rosetta image of asteroid P/2010-A2. The OSIRIS-WAC image (left) spans 2.2 × 2.2 degrees; the NAC image (right), spanning 200 × 200 seconds of arc, shows the asteroid and its trail. The halo-like features around the bright stars are internal reflections within the camera. (ESA – OSIRIS-Team; MPS/UPD/LAM/IAA/RSSD/INTA/UPM/DASP/IDA)

On 16 March, the OSIRIS-NAC made observations of the comet-like object P/2010-A2 (see Fig. 7.28) whose position in the inner portion of the asteroid belt, small nucleus and long dust tail suggested it was a member of a group of rare 'main belt comets' which were suspected of being icy asteroids undergoing episodes of cometary activity.

Despite its great distance from P/2010-A2, Rosetta's unique vantage point enabled scientists to see the strange object from a unique perspective. By combining data from ground-based observatories and the Hubble Space Telescope, they were able to reconstruct the three-dimensional shape of P/2010-A2's diffuse trail in great detail. This confirmed that the object and its apparent tail were actually the remains of an asteroid collision that happened one and a half years earlier.

Three weeks of active payload checkout 12 (PC12) started on 23 April, when most of the instruments were activated for their final check and update prior to the Lutetia fly-by. These activities were successfully completed on 17 May.

Meanwhile, on 14 May, Rosetta flew beyond 2.26 AU, its previous farthest distance from the Sun. By late May, the spacecraft was being tracked regularly by the ESA New Norcia ground station and by NASA DSN stations at Goldstone (DSS-15, DSS-24 and -25), Canberra (DSS-34 and -45) and Madrid (DSS-54, -55, -63 and -65) for radio navigation purposes inbound to Lutetia.

In addition, Rosetta's cameras began tracking the asteroid for optical navigation. The first set of observations was completed on 31 May, when the spacecraft was 53 million km from its objective. These observations were initially made twice a week, but as the range reduced the rate was increased to daily. They utilized the two navigation cameras (NAVCAM-A and -B) and OSIRIS-NAC. However, OSIRIS remained off on 31 May, owing to low temperatures on the instrument's electronics box.

The orientation of the spacecraft was adjusted to allow those components to warm up. The camera was activated on 2 June, for use during the remainder of the navigation campaign. On 18 June, the temperature of the CCDs in the NAVCAMs was lowered from –25°C to –30°C in order to reduce the number of 'warm pixels'.

On 11 June, Rosetta's attitude and orbit control system (AOCS) and the telemetry, tracking and commanding (TT&C) subsystem were configured for the fly-by. On 14 June, NASA's 70 meter dish antenna in Madrid (DSS-63) was used to rehearse the fly-by operations, including a change of telemetry bit rate during the pass, with data acquisition in X- and S-band for the RSI experiment.

At about 06:28 UT on 18 June, a minus-3-weeks trajectory correction maneuver (TCM) with a nominal delta-V of 27.5 cm/s was performed to refine the trajectory for Lutetia. By 27 June, the range had reduced to 16.5 million km, and Rosetta was closing in rapidly. One week later, the distance was 7.33 million km and the spacecraft was following a perfect path for its close fly-by.

Box 7.3: Enigmatic Lutetia

As described in Chapter 1, asteroid discoveries were few and far between during the 19th century. Only the largest and brightest of these small, rocky objects could be detected by large ground-based telescopes – in most cases by chance.

Rosetta's main asteroid target, Lutetia, was only the 21st such object to be numbered since the discovery of the first asteroid, Ceres, in 1801. Discovered by Hermann Goldschmidt in Paris in November 1852, it was given the Roman name for the French capital city: Lutetia.

The asteroid follows a fairly normal orbit in the inner asteroid belt, at a mean distance of approximately 2.4 AU, or 364 million km, from the Sun. The orbit lies almost in the plane of all the planets – the ecliptic – and is moderately eccentric (elliptical). Its orbital period is 3.8 years.

Nevertheless, prior to Rosetta's fly-by, Lutetia posed a riddle for planetary scientists. In particular, there was uncertainty about its size, mass, density, composition and origin.

After Lutetia was announced as a target for Rosetta in 2004, it was subjected to intense ground-based scrutiny. These observations enabled astronomers to construct its shape and spin rate from the changing light curve. Most asteroids tend to be irregularly shaped, so that different amounts of light are reflected as they rotate. The ratio between the three major axes of an asteroid, and its rotational properties, can be determined from measuring how this reflected light changes with time. Assuming a certain reflectivity (albedo), the dimensions of the asteroid can also be estimated.

The results indicated Lutetia to be an irregular, cratered body with dimensions of 132 × 101 × 76 km. In addition, the rotational axis of the asteroid is highly inclined and almost in its orbital plane, so that it resembles a spinning top knocked on its side – much like the planet Uranus. This led to the prediction that Rosetta would see only the northern hemisphere, and that the southern hemisphere would be dark and cold during the fly-by.

It was also estimated that Lutetia rotates with a period close to 8.17 hours – knowledge that was of great help in planning the scientific measurements during Rosetta's fly-by.

Initial observations recorded a high albedo, suggesting a high metallic content, and this led to the body being classified as an M-type asteroid. However, the lack of clear features in its light spectrum led some researchers to suggest it more closely resembled a C-type, a form of primitive, carbon-rich object similar in composition to meteorites classified as carbonaceous chondrites.

When Lutetia was at opposition in 2008-2009, the opportunity was taken to test this theory further. A team of researchers used the giant VLT, Keck and Gemini telescopes to estimate that Lutetia has a high bulk density – in the range 3.98 to 5.00 g/cc, depending on the model that was used. This range of density would support a carbonaceous composition. Meanwhile, visual spectroscopic studies noted variations with rotation phase, possibly representing local differences in the mineralogical or chemical content of the surface.

Bearing this in mind, scientists were looking forward to solving the following questions about Lutetia through analysis of Rosetta's fly-by data:

- *What is the asteroid's true composition? Is it a C-type or M-type asteroid?*
- *What are its precise mass and density?*
- *What is its surface topography and geology? What processes have shaped its surface?*
- *When and where did Lutetia originate? As a primitive planetary building block, it could provide important clues about how the inner Solar System formed.*
- *Is it surrounded by an exosphere – an extremely sparse envelope of gases? If so, what is its composition?*

In addition, the fly-by would provide ground-truth to improve the calibration of observations obtained by ground-based telescopes. It would also be a rare opportunity to test the scientific instruments on board Rosetta before it reached its final destination.

7.13   LUTETIA

Rosetta passed by asteroid (21) Lutetia at 15:44:57 UT on 10 July 2010, at a relative speed of 15 km/s and a range of 3,162 km. The asteroid was 2.72 AU (406.9 million km) from the Sun and 3.05 AU (456.2 million km) from Earth, giving a signal travel time of about 20 minutes.

The closest approach phase was followed using the 70 meter dish (DSS-63) at NASA's DSN ground station in Madrid. This allowed the spacecraft to transmit telemetry at the maximum bit rate of 91 kbps. The large dish was also used to amplify the signal as much as possible, in support of the two-way RSI radio link for tracking purposes.

Fig. 7.29: A view of Lutetia provided by OSIRIS during closest approach. The large ridged crater on the left shows numerous dark boulders and landslides. (ESA 2010 MPS for OSIRIS Team MPS/UPD/LAM/IAA/RSSD/INTA/UPM/DASP/IDA)

The fly-by strategy allowed continuous observations before, during, and for 30 minutes after closest approach, at which time the spacecraft had reached the minimum permitted angle with respect to the Sun and had to slew away from the asteroid.

About four hours prior to closest approach, Rosetta performed a flip maneuver to acquire the correct attitude, preparatory to entering Asteroid Fly-by Mode. In this mode, the attitude of the spacecraft was automatically driven by one of the navigation cameras so that Lutetia was continuously in the field of view of the imaging instruments. The flyby was conducted with the spacecraft autonomously tracking the asteroid, as practiced during the March rehearsal.

Seventeen instruments were operated during the encounter, including remote sensing and in-situ measurements and some of the payload of the Philae lander (the main exceptions being CONSERT and GIADA). Together, they looked for evidence of a highly tenuous atmosphere (exosphere) and magnetic effects, as well as studying the asteroid's surface composition and density. They also attempted to capture any dust grains that were floating near the asteroid for onboard analysis. Analysis of the fly-by data provided no significant evidence for it having either an exosphere or an internal magnetic field.

The fact that Lutetia rotated on its side meant that only about half of the asteroid was imaged by OSIRIS – almost the entire northern hemisphere and portions of the southern hemisphere, revealing a cratered, irregular object. At approximately 121 × 101 × 75 km, it was the largest asteroid yet encountered by a spacecraft. Scientists suspect that Lutetia was once more or less spherical, and a battering by incoming objects blasted large amounts of rock from its surface.

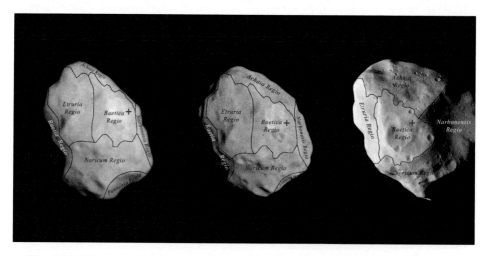

Fig. 7.30: The named regions on asteroid Lutetia. The three images were taken (left to right) 60, 30 and 3 minutes prior to Rosetta's closest approach on 10 July 2010. They were taken at ranges of 53,000, 27,000 and 3,500 km respectively, with spatial resolutions of about 1,000, 500 and 60 meters per pixel. The Rosetta science team identified seven distinct topographic and geological regions, named for provinces of the Roman Empire. The unobserved area of the southern hemisphere was named after Hermann Goldschmidt, who discovered Lutetia in 1852. The blue cross indicates the north pole. (ESA 2010 MPS for OSIRIS Team MPS/UPD/LAM/IAA/RSSD/INTA/UPM/DASP/IDA)

As Rosetta drew close, a giant bowl-shaped depression stretching across much of the asteroid rotated into view. Images showed that some portions of Lutetia are heavily cratered, implying considerable age. Based on crater counts, parts of its surface appear to be about 3.6 billion years old. Areas with few craters suggest a surface age of 50-80 million years old, which is regarded as young by astronomical standards.[2]

The detailed images enabled scientists to identify seven distinct topographic and geological regions, which the IAU named for provinces of the ancient Roman Empire: Baetica, Achaia, Etruria, Narbonensis, Noricum, Pannonia and Raetia. The unobserved region of the southern hemisphere was named after Hermann Goldschmidt, who discovered the asteroid in 1852.

The Baetica region is situated around the north pole, and includes a cluster of superimposed craters named the North Polar Crater Cluster (three of which exceed 10 km across), as well as their impact deposits. This is the youngest surface unit on the asteroid.

Baetica is covered by a smooth ejecta blanket approximately 600 meters thick that partially buried older craters. Other features include landslides, gravitational rock slides, and blocks of ejecta up to 300 meters in size. The landslides and corresponding rock outcrops are generally brighter than their surroundings.

The two oldest regions – Achaia and Noricum – are both heavily cratered. The Narbonensis region coincides with the largest impact crater on Lutetia, 56 km wide Massilia, which was modified by pit chains and grooves at a later date. Pannonia and Raetia are probably impact craters. Noricum is transected by a prominent groove 10 km long and about 100 meters deep.

Numerical simulations have shown that the impact that produced the largest crater on Lutetia seriously fractured, but did not shatter, the asteroid. So Lutetia has likely survived intact from the beginning of the Solar System. The presence of linear fractures and the morphology of the impact craters also indicate the interior has considerable strength and therefore is not a rubble pile like many smaller asteroids.

Studies of the fracture patterns indicate that an impact crater up to 45 km across may exist in the unseen southern hemisphere (see Fig. 7.33). This was given the nickname Suspicio crater. Lineaments imply a location for this crater in the southern hemisphere consistent with that of hydrated minerals detected by ground-based telescopes.

---

[2] Generally, an ancient surface displays many more impact craters than a younger surface, because more time has passed for the object to experience bombardment by meteoroids and comets. There are models for the size distribution of potential impactors over time.

Fig. 7.31: Named features on Lutetia. The most notable impact craters were named after cities which existed around the same time as the city of Lutetia (the Roman name for Paris, France). The largest, 56 km in diameter, is Massilia, the Roman name for Marseille. Other features were named after rivers of the Roman Empire and adjacent parts of ancient Europe. (ESA 2010 MPS for OSIRIS Team MPS/UPD/LAM/IAA/RSSD/INTA/UPM/DASP/IDA)

Apart from 19 craters named after cities in the ancient Roman Empire, with the largest, Massilia, named after the ancient city now known as Marseille, most of the other features on Lutetia were named after rivers in Europe around the Roman era.

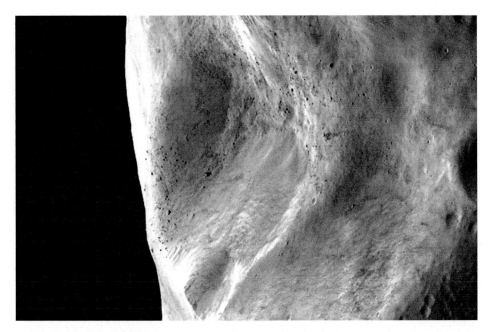

Fig. 7.32: This close-up view of a crater in the Baetica region shows boulders and land-slides. Several hundred dark boulders up to 400 meters across are scattered around the crater's sides. These are some of the largest rocks found thus far on small bodies in the Solar System. The landslides probably occurred when impacts elsewhere on the asteroid caused vibrations that dislodged loose regolith. (ESA 2010 MPS for OSIRIS Team MPS/UPD/LAM/IAA/RSSD/INTA/UPM/DASP/IDA)

Some parts of Lutetia are covered in a thick layer of regolith, or unconsolidated material that comprises loosely aggregated dust particles around 50-100 microm-eters in size. Over billions of years the surface has probably been pulverized by impacts. Much of the ejecta would have escaped to space, but a lot of it would have fallen back, producing a layer of loose material up to 600 meters thick. In the low gravity, the vibrations from nearby impacts have created giant landslides.

OSIRIS also imaged numerous dark boulders strewn across the surface (see Fig. 7.32), some of them 300-400 meters across, or about half the size of Ayers Rock (Uluru) in Australia.

The VIRTIS spectrometer recorded a maximum surface temperature of –19°C on the sunward hemisphere and a minimum of –103°C on the night hemisphere (see Fig. 7.35). There was no evidence of ancient chemical processes involving water, or space weathering.[3]

---

[3] Angioletta Coradini, the VIRTIS principal investigator, passed away shortly after Rosetta's fly-by, during the preparation of a key paper for the journal Science that was eventually pub-lished on 28 October 2011.

Fig. 7.33: An OSIRIS image of Lutetia centered on the North Pole Crater Cluster (the purple outline), with Massilia crater at the lower left (red outline). Red and purple lines indicate the concentric grooves or 'lineaments' associated with these craters. Those colored blue suggest there is a large crater (nicknamed Suspicio) on the unseen hemisphere. Yellow denotes other linear features. (ESA/Rosetta/MPS for OSIRIS Team MPS/UPD/ LAM/IAA/SSO/INTA/ UPM/DASP/IDA)

So what is Lutetia made of? VIRTIS obtained hyperspectral images with a spatial resolution varying from 12 km to less than 1 km at closest approach. These indicated the composition of the surface to be remarkably uniform and mainly primordial chondritic (silicate) material, but this could merely represent a bland surface veneer.

Additional observations by OSIRIS, MIRO and ALICE aboard Rosetta, visible studies by the New Technology Telescope of the European Southern Observatory, Chile, and near-infrared and mid-infrared data from NASA's Infrared Telescope Facility in Hawaii and by the Spitzer Space Telescope were combined to study Lutetia across a very wide range of wavelengths in order to deduce its surface composition. This was the most complete spectrum of an asteroid ever assembled. It was compared against spectra of meteorites found on Earth that have been extensively analyzed in laboratories. Only the enstatite chondrite type of meteorite

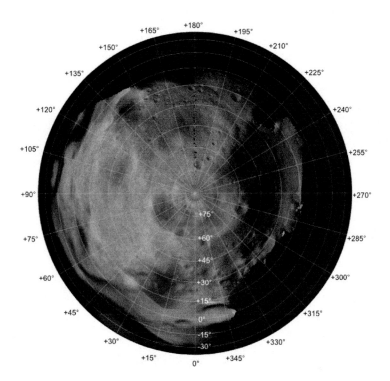

Fig. 7.34: A map of Lutetia centered on its north pole showing the area viewed by Rosetta. It covers most of the northern hemisphere and part of the southern hemisphere. The number of craters in different regions indicates the age of the surface, ranging from 3.6 billion years old to a relatively youthful 50-80 million years old. (ESA 2011 MPS for OSIRIS Team MPS/UPD/LAM/IAA/RSSD/INTA/UPM/DASP/IDA)

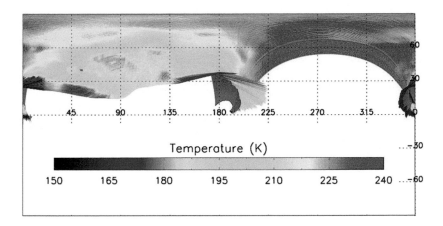

Fig. 7.35: A false-color temperature map of Lutetia obtained by the VIRTIS instrument. The highest temperature (red) was on the daylight hemisphere, where the Sun was overhead. The minimum temperature measured (purple) was on the night side. Notice the low temperatures in some of the smaller craters (left). (LESIA)

proved to have properties that matched Lutetia's over the full range of colors,[4] but other characteristics suggested a surface composition compatible with primitive, carbon-rich meteorites known as carbonaceous chondrites.

Fig. 7.36: This 3D image of Lutetia can be viewed using stereoscopic glasses with red-green or red-blue filters. It was created by combining two separate images obtained several minutes prior to Rosetta's closest approach to the asteroid. (ESA/H. Sierks MPS)

Based on images taken by OSIRIS, Rosetta scientists modeled the asteroid's shape. They also studied deflections in the spacecraft's trajectory caused by the weak gravity, inferred from the radio signals received at Earth. This indicated a mass of 1.7 million billion tonnes.

"The mass was lower than expected. Ground-based observations had suggested much higher values," reported Martin Pätzold from the University of Cologne, Germany, the leader of the radio science team.

---

[4] Enstatite chondrites (a.k.a. E chondrites) account for only about 2% of recovered meteorite falls. The fact that they are the only chondrites to have the same isotopic composition as Earth and the Moon strongly implies that enstatite chondrite formed at about 1 AU from the Sun, and that our planet was formed by the accretion of this material.

Having a mass and a volume, it was possible to calculate the bulk density. At 3.4 g/cc, which is comparable to our Moon, Lutetia proved to have one of the highest asteroid densities thus far measured. This would be fairly easy to explain if Lutetia were completely solid, free of voids or cracks, but researchers studying the impact craters found huge fractures throughout, suggesting it is fairly porous. This would lower its bulk density considerably. To explain the combination of a high density and a porous interior, it has been argued that the asteroid must contain significant quantities of iron.

Like the rocky terrestrial planets, many of the larger asteroids, such as Lutetia, are thought to have grown large enough to differentiate internally, melting with heat released by radioactive isotopes in the rocks. The denser elements, such as iron, would then sink to the center and the lighter rocky material would float to the top.

In the case of Lutetia, it would seem that it was subjected to some internal heating early in its history, but did not melt completely and so did not end up with a well-defined iron core.

If this is the case, Lutetia, as the first asteroid known to be partially differentiated, provides a rare snapshot of the early development of planetary bodies. Perhaps it achieved sufficient size to develop a core that was at least partially molten, whilst avoiding the larger collisions which accelerated planet formation.

This implies that Lutetia is more like a planetesimal, or a planet precursor, than a fragment of a larger parent asteroid or a rubble pile that broke apart, as most known asteroids seem to be. Most of the bodies of the inner Solar System disappeared after several million years, as they were incorporated into the newly accreting planets. Some of the largest, such as Lutetia, were ejected to safer orbits farther from the Sun.

What caused this ejection? One possibility is that its orbit was dramatically altered by a close encounter with one of the rocky planets. An encounter with the young Jupiter during the giant planet's migration to its current orbit could also explain the change in orbit that left Lutetia in what became the main asteroid belt.[5]

Lutetia's birthplace in the inner Solar System, and subsequent relocation, would explain why its color and surface properties differ so much from most of the asteroid population. Similar asteroids represent less than 1% of the population of the main belt. The new findings explain why Lutetia is different: it is a very rare survivor of the material from which the rocky planets were formed.

---

[5] Many astronomers believe that Jupiter orbited closer to the Sun in the early Solar System, before migrating out to its current position. Its huge gravitational attraction would have disrupted the orbits of the other objects in the inner Solar System at that time.

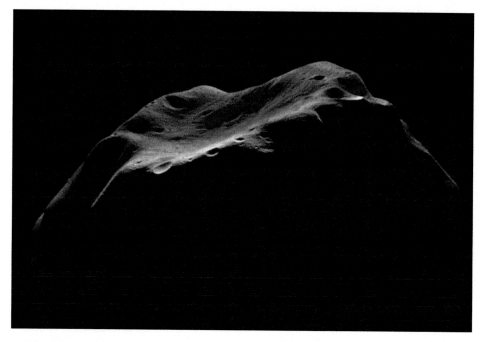

Fig. 7.37: An OSIRIS view of Lutetia as Rosetta left the asteroid behind. (ESA/MPS/ UPD/LAM/IAA/RSSD/INTA/UPM/DASP/IDA/Daniel Machacek)

7.14   DEEP SPACE HIBERNATION

Payload operations associated with the Lutetia fly-by were completed on 11 July, with all of the instruments apart from SREM shut down. The next three weeks were mainly dedicated to downlinking the large volume of science data acquired.

On 12 July, when Rosetta was more than 406 million km from the Sun, it captured the record for the farthest distance from the Sun to be flown by a solar-powered spacecraft, surpassing the distance record previously held by NASA's Stardust comet mission. As it pushed on, the energy in sunlight became weaker and the temperatures continued to fall. On 22 July, Rosetta exceeded its own previous maximum distance from Earth. During August, Rosetta remained in active cruise mode. On 20 August, there were tests of the high-gain antenna and low-gain antennas. These verified that it was possible to command the spacecraft from the New Norcia and DSS-43 ground stations across 3 AU using the low-gain antenna at a data rate of 7.8 bps. Both stations also detected a regular variation in the signal from the high-gain antenna that was achieved by commanding the spacecraft's dish to simulate a strobing signal by sweeping its beam across Earth several times.

On 3 September, the performance of the solar arrays was tested by tilting them relative to the Sun direction in order to simulate deep space conditions.

Meanwhile, there was concern about the condition of the spacecraft's four reaction wheels. High friction in wheel B had been noted after the Steins fly-by, with the likelihood it would cease to operate within several months. A similar problem arose with another reaction wheel the following year.

Rosetta began using three reaction wheels for attitude control on 15 July 2010, when wheel B was switched off because of two failed relubrications.[6] There were no plans to restart it until after the end of the long period of deep space hibernation. Then, on 27 August, wheel C started to exhibit a very noisy frictional torque. With its behavior degrading, the ground team decided to undertake a relubrication, and to operate wheel C at lower speeds (less than 1,000 rpm) except for slew maneuvers. It was also decided not to use the reaction wheels in safe mode from 13 November onwards. Fortunately, on 24 November, a second relubrication resulted in some improvement in wheel C's performance.

However, the condition of the wheels continued to cause considerable concern. In case the spacecraft woke up after hibernation with only two operational wheels, the operations team developed a new attitude and orbit control system (AOCS) mode with industry, based on the use of two active wheels and the thrusters. This software was fully ready and tested for uplink in case it was ever needed, but it was never used.

"Meanwhile we fully changed how we operated the wheels," explained Sylvain Lodiot. "The user manuals told us we had to operate them at high speed, but the evidence indicated the opposite. So we embarked on a year-long test with a real spare wheel on a full scale model of Rosetta at ESOC. This also involved sending the wheel back to the manufacturer for opening and bearing checking. After all the tests, the wheel was still fine, so at wake up, we changed all the settings on board to allow operating around zero speed, and that worked!"

Another unwelcome complication arose on 9 September, when a test procedure to switch to the redundant reaction control system (RCS) pressure regulator was completed. This revealed that there was a leak of pressurant in a section of piping which was common between the two regulators.

The first indication of such a problem had arisen in September 2006, when a pressure reading in the RCS had dropped unexpectedly to zero at a time when Rosetta was not in contact with Earth. The reason for this sudden drop in pressure was uncertain, but a leak was regarded as a possibility.

Once the leak was confirmed, the faulty section of piping was isolated, forcing the mission team to modify planned operations and to cancel the planned repressurization of the RCS in January 2011. The system would have to be operated using the existing pressure, in so-called 'blow-down' mode.

An investigation was then launched to determine the implications of the new procedures. Ground tests on one thruster were undertaken in an effort to characterize the behavior of Rosetta's 24 thrusters at low pressure, whilst other work

---

[6] Each wheel had its own internal oil reservoir.

ensured optimal thermal control of the propellant tanks in future rendezvous maneuvers. Although the new procedure increased operational uncertainties, the mission team eventually decided they still had plenty of safety margin.

The development of a new rendezvous maneuver strategy was greatly assisted by the fact that fuel allocated for dealing with uncertainties on maneuvers in the early part of the mission was still available. The new strategy meant delaying the start of the post-hibernation rendezvous maneuver at Comet 67P by about a week, but this was within the uncertainties involved in the overall timeline for comet operations. The rest of the mission maneuvers would be performed with the propulsion system at low pressure, delivering lower (but still acceptable) efficiency. There should be no impact on the comet science operations, and the date for lander delivery would remain within the planned window.

During 2-31 October, Rosetta was in superior conjunction, while passing behind the Sun as viewed from Earth. It operated in active cruise mode, but with severely limited ground station communication because of the influence of the Sun on the radio transmission. NASA's Deep Space Network provided additional support to the radio science investigations. Downloading of the mass memory was not attempted between 10-23 October, when the telemetry link was particularly poor.

Meanwhile, preparations continued for the forthcoming entry into Deep Space Hibernation Mode (DSHM), the prolonged period when the spacecraft would be largely shut off in order to conserve power and propellant.

On 19 November, the team carried out a similar test to the one on 20 August, to ensure that the ESA and NASA ground stations would be able to command Rosetta when using its low-gain antenna and detect the strobing signal from the high-gain antenna as the spacecraft was spinning. Undertaken with continuous low-gain antenna coverage, this test validated the radio communication activities and performance to be expected after entering DSHM in July 2011, when Rosetta would be at about 4.4 AU from Earth.

The final payload checkout prior to hibernation (PC13) was successfully carried out in early December, with no major problems arising.

The first key task in January 2011 was a large trajectory correction, known as the first rendezvous maneuver. This was intended to include five burns of the propulsion system, with an optional extra firing in the event that a final tweak was needed.

The first burn (RDVM 1A) proceeded as planned on 17 January, with the 393 minute event altering Rosetta's velocity by 300 m/s.

On 18 January, during the second planned burn, there was a large performance anomaly in the reaction control system (RCS), evidently caused by an issue with thruster 9A. This led to an attitude pointing error that prompted the spacecraft to enter a level two safe mode. RDVM 1B was to have delivered a delta-V of 274 m/s, but it was terminated after only about 30 m/s.

After recovering the spacecraft, the mission control team at ESOC revised the RCS operating mode to preclude a repeat of the problem. In addition, the replanned maneuver sequence was to be performed using the redundant set of thrusters.

The new sequence comprised the RDVM1 Test Burn (35 m/s) on 21 January, followed over the next three days by RDVM 2A (160 m/s), RDVM 2B (200 m/s) and RDVM 2C (45 m/s). Overall, this achieved 98% of the required trajectory change. The residual (17.3 m/s) was left to a final trim maneuver scheduled for 10 February.

Unfortunately, this maneuver was aborted after several seconds because Rosetta entered safe mode in response to an attitude pointing error that exceeded the failure detection isolation and recovery (FDIR) system's limit of four degrees for the first 100 seconds of the maneuver. The pointing error was due to an operational oversight, in which a command that was to adjust the spacecraft's attitude prior to the burn was not uplinked.

The good news was that, during the few seconds that the burn lasted, RCS branch A had operated as expected, using the commands issued by the attitude and orbit control system. The safe mode was promptly recovered and the spacecraft was prepared once again to carry out the trim maneuver.

The spacecraft trim maneuver, rescheduled for 17 February, successfully provided a delta-V of 17.29 m/s. The burn, which lasted about 29 minutes, was completed about 30 seconds earlier than predicted, indicating a slight over-performance of the thrusters relative to expectations.

Despite the off-nominal performance of the RCS during the trim maneuver and while making the short, sporadic pulses to offload the reaction wheel, the performance of RCS-A confirmed that no permanent failure was present in this branch. The 'problem' appeared to lie with the modeled predictions the engineers were using for cold-starting the thrusters at low pressures.

Precise orbit determination following the RDVM 1 burns showed this maneuver was almost perfect, so plans for a possible further correction were shelved. The outcome was that Rosetta was now on a fly-by trajectory for Comet 67/P with a perfectly acceptable 'miss distance' of about 50,000 km.

After the completion of the RDVM 1 activities, the spacecraft was changed back to its normal configuration on 4 March, before completing pre-hibernation activities. One key activity was a check of the performance of the solar arrays at 4 AU (600 million km from the Sun), giving satisfactory results.

During March, Rosetta continued in active cruise mode, with most scientific instruments off. The main activities were the preparation of the instruments for deep space hibernation, using OSIRIS and the NAVCAMs to image Comet 67P and the area of sky that would be observed during the approach to the comet, as well as characterizing the spacecraft's inertia properties and the Sun acquisition sensors.

During the long range imaging on 26 March, an anomalous acceleration of the spacecraft was observed. The most likely explanation was thought to be outgassing of products accumulated during the rendezvous maneuver in January.

On 15 May, Rosetta was 654 million km (4.37 AU) from the Sun. It was also 508 million km (3.39 AU) from Earth, so the one-way signal travel time was 28 minutes 15 seconds.

Deep Space Hibernation Begins

Starting in mid-May, the mission team concentrated on their final preparations for the Deep Space Hibernation Mode. A final performance test of the solar arrays was conducted, and all the commands for entering into hibernation were uploaded in a disabled state. By early June, the spacecraft was configured in readiness for the onset of the hibernation.

Rosetta was successfully spun-up and commanded into DSHM on 8 June. The maneuver was executed without any problems. The spacecraft's signal was lost on the ground at about 08:00:35 UT when the spin-up maneuver was triggered on board. (The slow spin would impart stability while the spacecraft was asleep.)

The 70 meter diameter antenna of NASA's DSS-43 in Canberra detected an extremely weak strobing signal. This confirmed that the spacecraft had successfully executed the most critical phase of the maneuver. About 50 minutes later, after estimating the real spin axis, Rosetta's onboard software refined the high-gain antenna direction. The strength of the received signal increased sharply to the same level as when the spacecraft was not spinning.

Once the ground checks of the strobing period and signal strength were completed, the telecommand link was verified by switching the spacecraft's telemetry modulation on and off twice and monitoring the change in signal strength. The final telecommand to authorize entry into hibernation was released at 12:57:40 UT. Confirmation of success came at 14:12:00 UT, when the final pulse from the spacecraft was received by the ground stations.

To ensure that all had proceeded as planned, a passive monitoring phase of six days was then undertaken.

Rosetta now began the coldest, most distant leg of its journey as it raced away from the Sun, out toward the orbit of Jupiter. It was oriented so that its solar panels faced the Sun to collect as much of the diminishing sunlight as possible.

For the next two and a half years, the spacecraft was on its own, leaving the mission team to look forward in eager expectation to the long-awaited rendezvous with Comet 67P.

During its hibernation, Rosetta would achieve the following record distances: 792 million km (5.29 AU) from the Sun on 3 October 2012, then 937 million km (6.26 AU) from Earth on 1 December 2012.

Rosetta's awakening was set for 10:00 UT on 20 January 2014, at which time it would be 672 million km (4.49 AU) from the Sun and 807 million km (5.39 AU) from Earth.

REFERENCES

Rosetta Begins its 10-year Journey to the Origins of the Solar System, 2 March 2004: http://www.esa.int/Our_Activities/Space_Science/Rosetta/Rosetta_begins_its_10-year_journey_to_the_origins_of_the_Solar_System

Rosetta On Its Way, 2 March 2004: http://sci.esa.int/rosetta/34784-02-march-2004-launch/

Ariane 5 Sends Rosetta on its way to Meet a Comet, Arianespace, 2 March 2019: http://www.arianespace.com/mission-update/ariane-5-sends-rosetta-on-its-way-to-meet-a-comet/

Ariane 5 V158 press kit: http://www.arianespace.com/wp-content/uploads/2017/06/04_feb_26-en.pdf

Rosetta – Rendezvous With A Comet, ESA BR-318, 23 July 2014: https://sci.esa.int/web/rosetta/-/54814-esa-brochure-318-rosetta-rendezvous-with-a-comet

Rosetta – Living With A Comet, ESA BR-321, 1 August 2015: https://sci.esa.int/web/rosetta/-/56351-esa-br-321-rosetta-living-with-a-comet

DLR Rosetta website: https://www.dlr.de/dlr/en/desktopdefault.aspx/tabid-10394/

The Second Rosetta Launch campaign: http://sci.esa.int/rosetta/34152-2nd-rosetta-launch-campaign/

Rosetta Status Report Archive: https://sci.esa.int/web/rosetta/status-report-archive

Where Is Rosetta? Animation: https://sci.esa.int/where_is_rosetta/

ESA PR 15-2004: Two Asteroid Flybys For Rosetta, 11 March 2004

Barucci, M. A., et al, Asteroid target selection for the new Rosetta mission baseline: 21 Lutetia and 2867 Steins, Astronomy & Astrophysics, Vol. 430, pp 313-317, 1 January 2005

ESA PR 29-2004, Rosetta's scientific 'first' – observation of Comet Linear, 26 May 2004 http://www.esa.int/Our_Activities/Space_Science/Rosetta/Rosetta_s_scientific_first_-_observation_of_Comet_Linear

Earth and Moon through Rosetta's eyes, 3 May 2005: https://www.esa.int/Our_Activities/Space_Science/Rosetta/Earth_and_Moon_through_Rosetta_s_eyes/(print)

In-flight Commissioning of the Near Sun Hibernation Mode, 17 May 2005: http://sci.esa.int/rosetta/37178-rosetta-status-report-no-41/

Rosetta: ESA's Comet Chaser Already Making Its Mark, ESA Bulletin, Vol. 123, pp 62-66, 15 August 2005 http://www.esa.int/esapub/bulletin/bulletin123/bul123i_ferri.pdf

OSIRIS camera on Rosetta obtains 'light curve' of asteroid Steins, 20 March 2007: http://www.esa.int/Our_Activities/Space_Science/Rosetta/OSIRIS_camera_on_Rosetta_obtains_light_curve_of_asteroid_Steins/(print)

Evidence for more dust than ice in comets, 12 October 2005: http://www.esa.int/Our_Activities/Space_Science/Evidence_for_more_dust_than_ice_in_comets

A large dust/ice ratio in the nucleus of comet 9P/Tempel 1, Michael Küppers et al, Nature, Vol. 437, p. 987-990, 13 October 2005: https://www.nature.com/articles/437958a

Rosetta Mars Flyby: http://sci.esa.int/rosetta/40697-rosetta-mars-swing-by/

Beautiful new images from Rosetta's approach to Mars: OSIRIS UPDATE, 25 February 2007: http://www.esa.int/Our_Activities/Space_Science/Rosetta/Beautiful_new_images_from_Rosetta_s_approach_to_Mars_OSIRIS_UPDATE

Rosetta delivers Phobos transit animation and 'sees' Mars in stereo, 27 February 2007: http://www.esa.int/Science_Exploration/Space_Science/Rosetta/Rosetta_delivers_Phobos_transit_animation_and_sees_Mars_in_stereo

Martian atmosphere as observed by VIRTIS-M on Rosetta spacecraft, A. Coradini, Journal of Geophysical Research, Planets, 2 April 2010 https://doi.org/10.1029/2009JE003345

At last: Rosetta's Mars flyby photos have been released!, Emily Lakdawalla, Planetary Society, 24 January 2012: http://www.planetary.org/blogs/emily-lakdawalla/2012/3340.html

Asteroid Steins: http://sci.esa.int/rosetta/43356-2867-steins/

E-Type Asteroid (2867) Steins as Imaged by OSIRIS on Board Rosetta, H. U. Keller et al, Science, 8 January 2010, Vol. 327, Issue 5962, pp. 190-193. DOI: 10.1126/science.1179559

Gemstones on diamond-like Steins: http://sci.esa.int/rosetta/54380-gemstones-on-diamond-like-steins/

The first European asteroid flyby, ESA Bulletin 137, February 2009: http://sci.esa.int/rosetta/47482-esa-bulletin-137/

Rosetta's gravitational microlensing programme, 9 October 2008: http://sci.esa.int/rosetta/43545-rosetta-s-gravitational-microlensing-programme/

Rosetta's third Earth swingby, 30 October 2009: http://sci.esa.int/rosetta/45841-rosetta-s-third-earth-swingby/

Rosetta's observations during the third Earth swingby, 10 November 2009: http://sci.esa.int/rosetta/45865-rosetta-s-observations-during-the-third-earth-swingby/

ESA spacecraft may help unravel cosmic mystery, 12 November 2009: http://www.esa.int/Our_Activities/Space_Science/Rosetta/ESA_spacecraft_may_help_unravel_cosmic_mystery

Rosetta sees a living planet, 13 November 2009: http://www.esa.int/Our_Activities/Space_Science/Rosetta/Rosetta_sees_a_living_planet

Preliminary science results from Rosetta's third Earth swingby: http://sci.esa.int/rosetta/46261-preliminary-science-results-from-the-3rd-earth-swingby/

VIRTIS spectral images of the Earth, 26 November 2009: http://sci.esa.int/rosetta/45972-virtis-spectral-images-of-the-earth/

Terrestrial OH nightglow measurements during the Rosetta flyby, A. Migliorini et al, Geophysical Research Letters, 26 June 2015 https://doi.org/10.1002/2015GL064485

Hubble and Rosetta unmask nature of recent asteroid wreck, 13 October 2010: http://sci.esa.int/rosetta/47830-hubble-and-rosetta-unmask-nature-of-recent-asteroid-wreck/

Asteroid (21) Lutetia – pre Rosetta flyby: http://sci.esa.int/rosetta/47389-21-lutetia/

Ground-based images of asteroid Lutetia complement spacecraft flyby, Southwest Research Institute, 7 October 2010: https://phys.org/news/2010-10-ground-based-images-asteroid-lutetia-complement.html

Rosetta triumphs at asteroid Lutetia, 10 July 2010: http://www.esa.int/Our_Activities/Operations/Rosetta_triumphs_at_asteroid_Lutetia

Lutetia: A Rare Survivor From The Birth Of The Earth, ESO, 11 November 2011: http://www.eso.org/public/news/eso1144/

Asteroid Lutetia: Postcard From The Past, 27 October 2011: http://www.esa.int/esaSC/SEMG93HURTG_index_0.html

Rosetta reveals mysterious Lutetia, 27 October 2011: http://sci.esa.int/rosetta/49543-rosetta-reveals-mysterious-lutetia/

Images of Asteroid 21 Lutetia: A Remnant Planetesimal from the Early Solar System, H. Sierks et al, Science, Vol. 334, Issue 6055, pp. 487-490, 28 Oct 2011. DOI: 10.1126/science.1207325

Asteroid 21 Lutetia: Low Mass, High Density, M. Pätzold et al, Science, Vol. 334, Issue 6055, pp. 491-492, 28 Oct 2011. DOI: 10.1126/science.1209389

The Surface Composition and Temperature of Asteroid 21 Lutetia as Observed by Rosetta/VIRTIS, A. Coradini et al, Science, Vol. 334, Issue 6055, pp. 492-494, 28 Oct 2011. DOI: 10.1126/science.1204062

Rosetta flyby uncovers the complex history of asteroid Lutetia, 29 May 2012: http://sci.esa.int/rosetta/50394-rosetta-flyby-uncovers-the-complex-history-of-lutetia/

Rosetta Fly-by at Asteroid (21) Lutetia. Special issue of Planetary and Space Science, Vol. 66, Issue 1, pp. 1-212, June 2012: https://www.sciencedirect.com/journal/planetary-and-space-science/vol/66

Roman Empire remembered on maps of asteroid (21) Lutetia, 11 October 2013: http://sci.esa.int/rosetta/49523-map-of-lutetia/

Asteroid (21) Lutetia as a remnant of Earth's precursor planetesimals, Icarus. Preprint: http://www.eso.org/public/archives/releases/sciencepapers/eso1144/eso1144.pdf

Lutetia's dark side hosts hidden crater, 8 October 2014: http://sci.esa.int/rosetta/54741-lutetias-dark-side-hosts-hidden-crater/

Lutetia's lineaments, S. Besse et al, Planetary and Space Science, Vol. 101, pp. 186-195, 15 October 2014. https://doi.org/10.1016/j.pss.2014.07.007

Rosetta operations strategy modified following a RCS test, 23 September 2010: http://sci.esa.int/rosetta/47733-rosetta-operations-strategy-modified-after-rcs-test/

8

Unveiling A Cosmic Iceberg

8.1 THE AWAKENING

Like the heroine of some fairy tale Rosetta awakened and called home on 20 January 2014, after a record 957 days in hibernation. At the speed of light, the signal took 44 minutes and 53 seconds to cross the 807 million km gulf to Earth.

During its long outward cruise toward the orbit of Jupiter, the intrepid explorer had traveled almost 800 million km from the Sun. Now, as Rosetta's orbit brought it back to a heliocentric distance of 'only' 673 million km, its giant solar wings began once again to receive sufficient energy to restore the spacecraft's systems to life.

Although it was still about 9 million km from its target, Comet 67P, Rosetta was awoken by its pre-programmed internal alarm clock at 10:00 UT. After warming up its star trackers, then activating its thrusters to cancel the slow spin that had stabilized it while inert, the spacecraft entered a planned safe mode. Over a period of about 6 hours, this resulted in Rosetta aiming its high-gain radio antenna at Earth. After switching on its transmitter, it sent the long-awaited signal which informed mission operators that it had survived what was perhaps the most hazardous part of its long trek.

The signal was received at 18:18 UT by the 70 meter diameter antenna (DSS-14) at NASA's Goldstone ground station in California, during the spacecraft's first window of opportunity to communicate with Earth.[1] As soon as the renewal of

---

[1] NASA's DSN network continued to provide routine tracking and telecommand support for several weeks after Rosetta's awakening on 20 January, until the Earth distance had decreased to enable ESA's 35 meter stations at New Norcia, Cebreros and Malargüe to take over. Tracking passes lasting 7 hours were scheduled every day for the first week, then daily 5 hour passes over DSN and ESA stations.

© Springer Nature Switzerland AG 2020
P. Bond, *Rosetta: The Remarkable Story of Europe's Comet Explorer*,
Springer Praxis Books, https://doi.org/10.1007/978-3-030-60720-3_8

contact was confirmed by ESA's Space Operations Center (ESOC) in Darmstadt, Germany, the message "Hello, world!" was posted on the @ESA_Rosetta twitter account.

Acquisition of signal (AOS) came 18 minutes later than hoped, but well within expectations. This delay was due to the onboard computer automatically rebooting itself at the beginning of the sequence to exit hibernation, but this was not considered to be an issue.[2] In safe mode, it transmitted a simple radio tone using the S-band transmitter and waited for instructions from Earth.

Within several hours, the Flight Control Team had established full control, and switched on the more powerful X-band transmitter to facilitate a faster download rate of about 9 kbps. The incoming stream of high-rate housekeeping data (telemetry) gave a detailed look at the health and status of crucial propulsion, attitude-keeping and power systems, among others.

The propellant tank temperatures were now running at 7-9°C, slightly colder than the 10-15°C expected but well within predictions. The solar arrays appeared to have suffered little, if any, degradation during the 31 months of spacecraft silence, and their power levels were similar to those prior to hibernation.

As Andrea Accomazzo, the ESA spacecraft operations manager, explained, "We were most concerned about power, and seeing if the solar arrays were generating sufficient electricity to support the planned recommissioning activities. But even though we were still 673 million km from the Sun, we were getting enough power and the arrays appear to have come through hibernation with no degradation."

"We have our comet chaser back," said Alvaro Giménez, the ESA Director of Science and Robotic Exploration.

"This was one alarm clock *not* to hit snooze on," said Fred Jansen, the ESA Rosetta mission manager, "and after a tense day we are absolutely delighted to have our spacecraft awake and back online."

Over the next several days, detailed health checks verified that the rest of the systems were as expected, and Rosetta was declared fully functional.

"We're now recording tracking station data, and in a few days will be able to conduct the first full orbit determination since wake up," announced Frank Dreger, the head of flight dynamics at ESOC.

Three of the four reaction wheels – gyroscopes used to control the spacecraft's orientation – were reactivated flawlessly, although they were initially operated at low speed (around 250 rpm). The fourth, wheel B, which had caused concern earlier (see Chapter 7), was activated on 3 February, and, once controllers were happy with its behavior, they established four-wheel control by the attitude and orbit

---

[2] A similar computer reboot had occurred midway through the long hibernation, due to a 'memory leak' caused by background software processes that continued to run, filling up the buffers.

control system (AOCS) on 7 February. The next few weeks were spent testing and configuring onboard systems, including the solid-state mass memory that would store science and operations data until it could be downloaded.

The next phase, from mid-March to late April, saw the gradual recommissioning of Rosetta's 11 science instruments. The Rosetta Mission Operations Center at ESOC coordinated these tests to ensure that each instrument would be ready to perform on arrival at Comet 67P. The first instrument to be switched on was the OSIRIS imaging system. Brief initial checkouts for ALICE and RPC were also conducted.

Rosetta got its first sighting of the tiny, distant comet since awakening when two 'first light' images were taken by OSIRIS on 20-21 March. However, at a range of some 5 million km, the comet covered less than one pixel. OSIRIS was then tasked to take navigation images that could be used for fine-tuning a series of comet rendezvous maneuvers, scheduled for May. In addition, it started taking light curve measurements to enable the rotation period of the comet to be determined. By 4 May, the distance to 67P had closed to 2 million km and the imagery showed the comet developing a coma that soon extended some 1,300 km into space.

Box 8.1: Mars Express Helps to Arouse Rosetta

Capturing Rosetta's wake-up signal was not without its challenges, although ESOC was able to predict the spacecraft's location in the sky to within 2,000 km at a distance of 807 million km, equivalent to a tiny fraction the diameter of a full Moon.

In order to ensure that the communications procedures to be carried out were correct, a test campaign was carried out using ESA's Mars Express, which carried a similar radio system to that of Rosetta. The spacecraft, in orbit around the Red Planet, 'pretended' to be Rosetta transmitting to the NASA ground stations that were to be used to listen for the signal, DSS-14 in Canberra and DSS-63 in Goldstone, to reduce the possibility of any problems in receiving the comet chaser's wake up call.

The test called for ESOC to command Mars Express to use its S-band transponders (normally only used for radio science or during emergency communications) to transmit at a very low bit rate, just as Rosetta was programmed to do.

This involved a lot of behind-the-scenes work from both ESA's Mars Express team and their colleagues at NASA's DSN (including having them come in to work on weekends and on the U.S. Thanksgiving holiday). But it paid off: a series of five test passes demonstrated that the 70 meter antennas and the teams manning them were ready for Rosetta's wake up.

By the end of March, eight instruments were undergoing reactivation: ALICE, CONSERT, COSIMA, GIADA, MIDAS, ROSINA, RPC and RSI. Some, such as MIDAS and COSIMA, were given software uploads to improve their performance. In the case of the COSIMA dust analyzer, the update was uploaded in early April and, after being switched off and on again, it was up and running with a fresh memory.

Meanwhile, the Philae lander was also switched on for the first time since hibernation. This wake-up command from Earth (via Rosetta) had been uploaded the previous week and it was executed at 06:00 UT on 28 March. A confirmation signal was received when the spacecraft next communicated with Earth, at 11:35 UT. The first image from ÇIVA on the lander, taken on 15 April, showed both of Rosetta's solar arrays.

8.2 COMMISSIONING PROBLEMS

Not all of the instrument reawakening and commissioning procedures went to plan, however. For example, the Ion Composition Analyzer (ICA) suffered some problems during its initial commissioning, and six weeks were to pass before it was finally given a clean bill of health. In order to avoid another 'latch up', in which a circuit in the instrument would start to draw excessive current and become too hot, it was decided to limit the operations of the ICA until Rosetta was close to the comet.

Similarly, all three of the ROSINA spectrometers were affected either by low temperature or software issues, producing error event messages when they were first turned on in April. One issue was due to the temperature of the detector in the Double Focusing Mass Spectrometer (DFMS) being 0.8°C below its operating limit of −30°C. Also, one of the Reflectron Time-of-Flight parameters was wrongly set and the Comet Pressure Sensor was too cold, producing erroneous offset values. These issues were solved three weeks later by a software patch and a spacecraft turn to warm the DFMS. However, it was not until May that the instrument team expressed their satisfaction with its performance.

In the case of the VIRTIS instrument, principal investigator Fabrizio Capaccioni recalled:

> We were quite tense and also a bit rusty in the instrument handling; in fact, we ended up uploading the wrong telecommand sequences!

> The onboard software returned a flurry of unpleasant comments directed to us and to some of our relatives – including deceased ones! We did not expect to have made mistakes, so we got really scared when the instrument responded as it did. The worst thoughts came to our minds. Was it a broken sensor? A faulty component? All those years of expectations and planning, for what?

> We had to quickly verify the received housekeeping data and the uploaded sequences as we didn't have much time: we had an interactive session of only few hours and we had many checks to do.

An accurate verification of the parts in small print of the user manual got us back on our feet. We indeed had reversed the order of some telecommanding calls, and we had to go back to the sequences we uploaded in the cruise phase and redesign the initialization procedure in a few minutes, hoping not to have made any further mistakes. Thanks to all the people involved, it all went well and the onboard software was finally satisfied. We remained on good terms with it ever since.

The post-hibernation checkout of the MIRO microwave instrument also gave an unpleasant surprise, even though the instrument team was very cautious, turning on one component at a time and monitoring its performance carefully.

Early in the turn-on sequence, and near the end of a tracking pass, one of MIRO's sensors reported that the spectrometer's temperature shot up from around 20°C to 45°C for several seconds. Was that an indication of a short circuit? But everything else appeared normal, and staff in the control facility in Darmstadt said they saw nothing unusual.

MIRO's original principal investigator, Sam Gulkis, had just a few minutes to decide what to do, before radio contact with the spacecraft was lost for several hours. Should he shut down MIRO to prevent further damage in case something was short-circuiting, or keep it on to avoid the risk of something having broken that would prevent it being turned back on in the future?

Based in Darmstadt, Gulkis decided to leave MIRO on while Rosetta was out of contact. It was after midnight in Darmstadt when this happened, and it had been a long day, so he went to sleep while team members at JPL (where it was still daytime) continued investigating. Sam later told them he fell asleep thinking the instrument was lost before Rosetta had even arrived at the comet.

At JPL, someone asked, "Why didn't the control center in Darmstadt, who monitor all of the instruments, warn us that they'd seen a temperature spike?" Instead, Darmstadt had reported everything was normal. He was surprise to find that Darmstadt had not been asked to monitor the particular spectrometer temperature that spiked.

At the same time, a second person contacted the lead engineer for MIRO and described the situation to her, saying one of the temperature sensors was known to be unreliable – even on the ground prior to launch, it would sometimes give crazy readings.

Looking back through the instrument's documents, the team confirmed that the temperature sensor that spiked was the unreliable one.

"Because we knew it was unreliable when we launched it in 2004, we'd told ESOC *not* to monitor it," recalled *Mark Hofstadter*, who succeeded Gulkis. "But by 2014 we had forgotten the problem, and at JPL we were looking at all of our temperature sensors and saw the (false) spike. When Sam woke up, we were able to tell him that he made the right choice in leaving MIRO turned on, and when Rosetta came back in contact with the Earth, we saw that MIRO was operating normally with all temperatures and currents as expected."

There were also nerve-wracking moments during the post-hibernation commissioning of the Rosetta Plasma Consortium (RPC), which comprised five sensors tasked with investigating the magnetic, electric and plasma environment of Comet 67P.

"A critical moment for RPC was the loss of the main power supply of our plasma unit," said team member *Ingo Richter.* "A capacitor died and got a short circuit, causing the complete failure of our main power supply. Fortunately, the plasma unit is equipped with a completely redundant power supply, which was immediately started in order to provide the power. The back-up system was working fine after ten years in space!"

Meanwhile, the RPC's ICA sensor suffered overheating, and was automatically switched off. This was a problem that had occurred several times during the cruise phase, but the team had hoped to see less of it when the instrument temperature was lower.

Another team member, *Hans Nilsson,* recalled, "We discovered that a loss of data that we had sometimes seen was due to a systematic problem in the instrument. It was causing the loss of up to half our data. We ran the instrument for only brief periods of time, fearing that the next overheating event might kill the instrument. The overheating appeared to get more common, and when we had two within a few days, we feared that the end was near for our instrument. But the opposite happened. Suddenly the overheating events became very rare, and we could start to operate the instrument continuously. After some more time, we discovered that, using an alternative mode, we also got rid of the data loss. Seldom has an instrument improved so much during a mission!"

After all the alarms and tension, the payload was finally ready to scrutinize Comet 67P. On 13 May, the Rosetta team held a commissioning 'close out' review, with each of the orbiter and lander instruments being given a formal 'Go' for routine science operations.

8.3   NAVIGATING TOWARDS A COMET

At the time of coming out of hibernation, Rosetta was still about 9 million km from the comet, and traveling at 800 m/s relative to 67P. In order to rendezvous with the comet, and fly alongside, it would be necessary to close to within 100 km and then slow their relative speed almost to zero.

The first major milestone occurred on 7 May, when Rosetta made the first of 10 planned orbit correction maneuvers (OCM) designed to put it onto the proper trajectory to arrive at 67P on 6 August. The first thruster burn, which was a test to check that all the systems were working properly, began at 17:30 UT. This reduced the vehicle's speed relative to the comet from about 775 m/s to 755 m/s.

Fig. 8.1: Paolo Ferri, ESA's head of operations (left), Rosetta mission manager Fred Jansen (right) and Rosetta spacecraft operations manager Sylvain Lodiot (seated) monitoring the 21 May orbit correction maneuver. (ESA)

The first of three large OCMs in the Near Comet Drift (NCD) set, took place on 21 May. The thrusters fired for 7 hours 16 minutes and reduced Rosetta's relative speed by 289.9 m/s. This was one of the longest burns in ESA spaceflight history. The burn started at 15:23 UT (spacecraft time) and was carried out autonomously, using commands that had been uploaded two days earlier. The burn consumed about 218 kg of fuel.

The second major OCM, on 4 June, consumed another 190 kg of propellant and delivered a delta-V of 269.5 m/s. This burn began at 14:21:58 UT and ran for 6 hours 39 minutes, which was about two minutes less than expected, but the outcome was well within margins. The third major OCM followed at 13:17 UT on 18 June, lasted for 136 minutes 41 seconds, and yielded a delta-V of 88.7 m/s.

These four burns had provided 667.8 m/s of the roughly 775 m/s needed to reduce the relative velocity to less than 1 m/s when Rosetta completed its rendezvous. The specification for each OCM was based on the performance of the previous burn and refinement of 67P's position from OSIRIS and the navigation camera images. The commands were uploaded several days in advance, and timed for when the spacecraft would be visible to a ground station, so that the mission control team at ESOC could receive telemetry and monitor the maneuver in near-real time.

The next phase involved four Far Approach Trajectory (FAT) burns at weekly intervals in the period 2-23 July. The first began at 12:05:57 UT on 2 July and was scheduled to continue for 1 hour 33 minutes 13 seconds. It ended about a minute earlier than expected, but achieved the desired change in velocity of 58.7 m/s.

The next three FAT burns were made on 9, 16 and 23 July. Each was shorter in duration and designed to generate a smaller delta-V than the previous burn. All were successful. The burn on 9 July lasted 46 minutes 2 seconds and achieved a delta-V of 25.7 m/s. The third burn was started at 11:36 UT on 16 July. It lasted about 26 minutes and reduced Rosetta's speed by 11 m/s relative to the comet.

The final FAT burn was made when Rosetta had closed in to about 4,500 km of its target. It began at 10:38 UT on 23 July (spacecraft time), lasted 16 minutes 35 seconds, and was designed to reduce the speed by about 4.82 m/s, after which the spacecraft's relative speed would be a modest 3.5 m/s.

Fig. 8.2: The rugged, bi-lobed configuration of Comet 67P is clearly visible in this OSIRIS image taken on 3 August 2014, three days prior to Rosetta's arrival. (ESA/Rosetta/MPS for OSIRIS Team MPS/UPD/LAM/IAA/SSO/INTA/UPM/DASP/IDA)

The CATP (Close Approach Trajectory, Pre-Insertion) burn started at 09:00 UT on 3 August and continued for about 13 minutes 12 seconds. It bent the spacecraft's trajectory towards the comet to trim the 'miss distance' from 200 km to 70 km. It also trimmed some 3.2 m/s off the relative rate, so that the final approach would occur at a walking pace of about 1 m/s.

8.4 ARRIVAL!

The moment that everyone involved in the Rosetta mission had been awaiting since launch a decade ago, arrived on 6 August, with the CATI (Close Approach Trajectory, Insertion) burn sending the spacecraft looping around its final target.

Table 8.1: Planned Timetable for Rosetta's Arrival at Comet 67P on 6 August 2014. (Times are UT)

08:00 – BoT New Norcia tracking station. AoS telemetry data flow. Rosetta slews into position for thruster burn.
09:00:01 – Start of CATI thruster burn. Start of orbit entry maneuver. One-way light time delay for signal confirmation on ground.
09:06:27 – End of CATI thruster burn. Rosetta now on first leg of comet orbit. Rosetta slews back to comet-pointing mode.
09:22:30 – Start of thruster burn confirmed on ground.
09:28:56 – End of thruster burn confirmed on ground.
19:43 – EoT New Norcia.
19:48 – BoT Malargüe tracking station.

Key: BoT: Beginning of track. EoT: End of track. AoS: Acquisition of signal. LoS: Loss of signal. One-way signal time was 22 minutes 29 seconds. The planned duration of the CATI thruster burn was 6 minutes 26 seconds.

Although the thruster firing of only 6 minutes 26 seconds was termed the final insertion burn, CATI did not actually put Rosetta into orbit around the comet – the spacecraft was still too far away to be captured by its feeble gravity. Instead, the delta-V of about 1 m/s achieved a speed that was approximately equal to that of the comet at a stand-off distance of about 100 km.

ESA Director General Jean-Jacques Dordain reported, "After ten years, five months and four days traveling towards our destination, looping around the Sun five times and clocking up 6.4 billion kilometers, we are delighted to announce finally – 'We are here.' Europe's Rosetta is now the first spacecraft in history to rendezvous with a comet, a major highlight in exploring our origins."

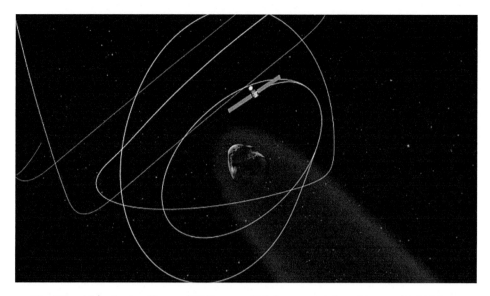

Fig. 8.3: After chasing Comet 67P, Rosetta slightly overtook it and 'entered orbit' from the 'front' of the comet, as the spacecraft and the comet traveled through space. The first three orbits were actually triangular arcs on the sunward side. Over the following weeks, Rosetta executed a complex series of maneuvers that reduced the separation from around 100 km to 25-30 km, in the process circularizing its orbit. This image is not to scale: Rosetta's solar arrays spanned 32 meters and the nucleus of the comet was approximately 4 km wide. (ESA – C. Carreau)

With its speed and direction of flight now matching those of the comet, the spacecraft began to follow a series of triangular (tetrahedral) arcs, each about 100 km long, around the nucleus. A small CAT thruster firing was performed at the apex of each triangle – which was initially every Wednesday and Sunday – to move Rosetta onto the next arc and enable it to remain near the comet. Each burn also lowered the fly-by distance (altitude). As time went by, these arcs above the surface were to be progressively lowered until 67P's gravity finally captured the spacecraft. As the altitude dropped below 30 km, the orbit would become circular.

The burns took place as follows:

- 10 August: CAT Change 1 lasted 6 minutes 25 seconds and the delta-V of 0.88 m/s put Rosetta onto the next arc at approximately 100 km fly-by altitude.

- 13 August: CAT Change 2 lasted 6 minutes 22 seconds and the delta-V of 0.87 m/s put Rosetta onto the next arc, which was also at about 100 km fly-by altitude.

- 17 August: CAT Change 3 lasted 6 minutes 19 seconds and the delta-V of 0.85 m/s started the descent toward the next triangular orbit ('Little CAT') which reached an altitude of about 80 km on 19 August.

- 20 August: CAT Change 4 lowered orbit to about 60 km on 22 August.
- 24 August: CAT Change 5 started the first arc of the 'Little CAT' triangle, when the altitude was about 72 km. Closest approach between 24 and 25 August was about 50 km.
- 27 August: CAT Change 6 burn.
- 31 August: Rosetta initiated the third and final arc of 'Little CAT', followed by the transition to the two maneuvers of the Transfer to Global Mapping (TGM).

An example of the incredible precision of the flight dynamics calculations in support of these intricate maneuvers, was the one made after the 13 August burn. It showed that the thrusters had over-performed by about 0.2%, which equated to approximately +2 mm/s.

8.5 GLOBAL MAPPING

The Global Mapping Phase (GMP) started on 10 September, and was scheduled to last until 7 October. Its aim was to gather high-resolution imagery and other science data to characterize the potential landing sites for the Philae lander, while also continuing to monitor how Rosetta responded to the environment of an active comet, prior to moving closer in. Five candidate landing sites for Philae had already been identified on 23-24 August 2014 (see Chapter 9).

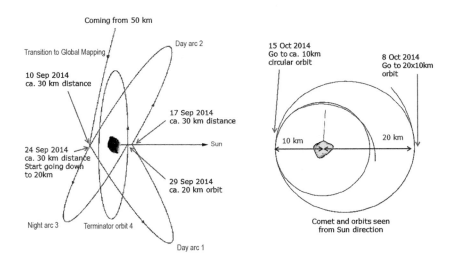

Fig. 8.4: Rosetta's Global Mapping Phase Orbital Changes. (ESA)

Rather than flying one complete orbit around the nucleus, Rosetta conducted two, seven-day-long, half orbits at about 30 km, in different planes. The transition to the GMP began at 09:00 UT on 10 September, when the spacecraft was traveling above the terminator – the boundary between night and day on the comet's surface. It performed a 19 cm/s burn to adopt a roughly 30 km circular orbit. At the same time, the plane of the orbit was orientated 60 degrees to the Sun direction, so that it would orbit over areas of the nucleus in their 'morning hours'.

Table 8.2: Global Mapping Phase Burns During September 2014

Burn	Date	Delta-V (m/s)	Duration (min: sec)
TGM1	03/09	0.56	04:55
TGM2	07/09	0.45	04:18
GMP1	10/09	0.193	02:19
GMP slot 1	14/09	0.025	00:32
GMP2A	17/09	0.085	01:23
GMP2B	17/09	0.087	01:25
GMP slot 2	21/09	0.018	00:25
GMP3	24/09	0.016	00:24
GMP4	29/09	0.106	01:37

Key: TGM: Transfer to Global Mapping. GMP: Global Mapping Phase.

Seven days later, when the spacecraft was again above the terminator, it performed a burn to change the plane of its orbit in order to fly over 'afternoon' areas. From 18 September, it was in a 28 × 29 km orbit with a period of 13 days 14 hours 59 minutes.

When Rosetta's minimum altitude dropped to just 29 km, it was low enough to be captured by the comet's feeble gravity, and for its orbit to become circular.

On 24 September, at the end of the second 30 km arc, the orbit was shifted above the comet's night side, just before dawn. Flying 30 degrees ahead of the terminator plane, the instruments could determine the thermal characteristics of the night side of the nucleus.

Shortly prior to entering the night arc, Rosetta completed a small maneuver to lower its orbit, so that when it crossed the terminator plane five days later, it would be at an altitude of about 20 km.

On 29 September, another maneuver circularized the orbit at 20 km in the terminator plane. After a week there, the mission team decided that it was safe to go closer.

A series of maneuvers reduced Rosetta's distance from an 18.6 km orbit with a period of 7 days to an intermediate orbit of approximately 18.6 × 9.8 km with a period of about 5 days.

The orbit was circularized at 9.8 km, with a period of approximately 66 hours, on 15 October, and the mission began its Close Observation Phase (COP). This remarkably low altitude was designed to provide higher resolution images of the

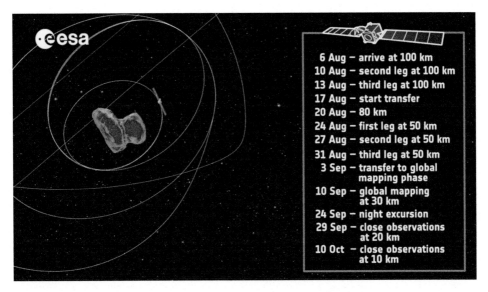

Fig. 8.5: The landmark events in Rosetta's move to a 10 km orbit above the comet. (ESA)

primary landing site, in the region named Agilkia. The new orbit also enabled Rosetta's instruments to collect dust and measure the composition of gases close to the nucleus.

8.6 LANDER RELEASE

Rosetta stayed in the 10 km COP orbit until 28 October, when a brief thruster burn placed it on a very elliptical trajectory that rapidly moved the spacecraft out of the 10 km circular orbit in the terminator plane. The thruster burn lasted 82 seconds and the delta-V of 0.081 m/s began the transfer into a slightly elliptical 'lander pre-delivery' orbit approximately 30 km from the center of the comet. (See Chapter 9 for more details on Philae's landing and outcome)

Three days later, on 31 October, a 90 second thruster burn provided a delta-V of 9.3 cm/s that moved Rosetta into the planned lander delivery orbit, at a height of about 30 km. This orbit was to be maintained until the pre-delivery maneuver planned to occur 2 hours prior to the release of the lander on the morning of 12 November. Some minor 'touch up' burns were added in order to maintain Rosetta on this orbit.

Despite several hiccups in preparing the lander for release and descent, the first 'Go/No Go' decision was made at 19:00 UT on 11 November. This confirmed Rosetta was in the correct orbit for delivering Philae to the surface at the required time. At midnight, the commands to control the separation and delivery were ready for uploading. The orbiter was also verified to be ready for the upcoming descent operations.

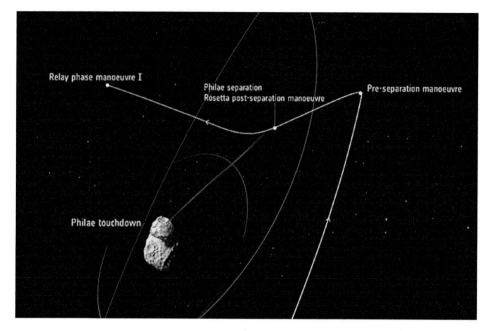

Fig. 8.6: Rosetta's maneuvers on 12 November, the day that the Philae lander was released. The descent trajectory is shown in red. (ESA, modified by Peter Bond)

The final pre-delivery maneuver by Rosetta was conducted at 07:35 UT on 12 November, enabling the orbiter to achieve the planned release point about 22.5 km from the comet's center. Next, was the final 'Go/No Go' decision which verified that the two spacecraft, the orbit, the ground stations, the ground systems and the flight teams were all ready for landing. The release of Philae occurred at 08:35 UT (spacecraft time).

Forty minutes later, Rosetta fired its thrusters to depart the elliptical delivery orbit. This shifted Rosetta to a safe distance from the comet, whilst helping to guarantee visibility of Philae at touchdown.

At 10:23 UT (spacecraft time), OSIRIS-NAC took a sequence of images that showed details of the lander, confirming the deployment of its three legs and antennas.

After a descent lasting seven hours, Philae touched down at 15:34 UT. The teams at ESOC in Darmstadt and the DLR Lander Control Center in Cologne had to wait 28 minutes 20 seconds until the signal, relayed via the orbiter, was picked up by both the ESA station in Malargüe, Argentina, and the NASA DSN station in Madrid, Spain (see Chapter 9).

A clear strong signal was received, with some breaks, but the lander's telemetry stabilized at about 17:32 UT and communication was maintained until Rosetta dropped below the lander's horizon at 17:59 UT, breaking its line-of-sight link. The fact that this occurred about an hour earlier than expected for the target landing site at Agilkia was explained away as being due to local horizon interference.

However, it soon became clear that although the descent had gone well, something had gone seriously wrong with the landing.

Later on 12 November, having analyzed the lander's telemetry, the Lander Control Center in Cologne and the Philae Science, Operations and Navigation Center in Toulouse reported that there were touchdowns at 15:34, 17:25 and 17:32 UT. Philae had bounced twice. Its precise whereabouts were unknown, but at least it was on the surface.

Fortunately, Philae's instruments were operating and delivering images and data (see Chapter 9). All went well during the second lander-orbiter communication slot between 06:01 UT and 09:58 UT on 13 November.

"We had a perfect pass," said Andrea Accomazzo, the Rosetta flight director. "The radio link was extremely stable and we could download everything according to the nominal plan."

Further communications sessions occurred over the next two days. The last time that Rosetta achieved contact was at 22:19 UT on 14 November. The signal was initially intermittent, but quickly stabilized and remained good until 00:36 UT on 15 November. But Philae's power was rapidly depleting. It completed its primary science mission after nearly 57 hours on the comet's surface.

From now on, no contact with Philae would be possible unless sufficient sunlight reached its solar panels to generate the power needed to awaken it. To increase the odds of this occurring when the comet was nearer the Sun, controllers sent the lander commands to raise its body by about 4 cm and rotate it through 35 degrees.

Meanwhile, whenever the opportunity arose, the orbiter's OSIRIS instrument would take high-resolution images to enable scientists to search for the tiny lander in the rugged region on the small lobe of 67P, where it was thought to have come to rest.

8.7   ESCORTING A COMET

With Philae in hibernation, a new phase of operations began for the Rosetta team.

"With lander delivery complete, Rosetta will resume routine science observations and we will transition to the Comet Escort Phase," said flight director Andrea Accomazzo. "This science-gathering phase will take us into next year as we go with the comet towards the Sun, passing perihelion, or closest approach, on 13 August, at 186 million kilometers from our star."

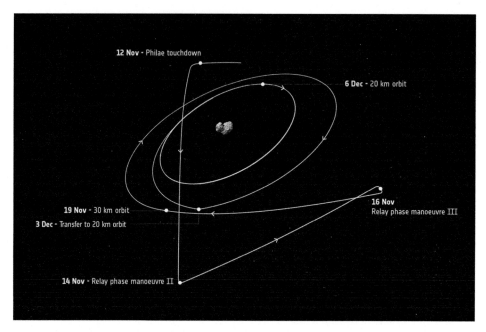

Fig. 8.7:   Rosetta's main maneuvers after Philae's release on 12 November 2014. (ESA)

On 16 November, the flight control team moved from the large main control room at ESA's Space Operations Center in Darmstadt, where key operations during landing were performed, to the smaller dedicated control room that was usually employed to fly the spacecraft.

Rosetta carried out thruster burns on 14, 16, 19, 22 and 26 November to temporarily reinstate its circular orbit at about 30 km. All future trajectories would be designed purely to optimize the use of the scientific instruments.

Two thruster burns on 3 and 6 December caused Rosetta to drop down to a 20 × 20 km circular orbit again. The spacecraft was then orbiting above the terminator, the day-night boundary on the comet's surface. This orbit would be used to map large parts of the nucleus at high resolution and to conduct gas, dust and plasma measurements as surface outgassing and jet activity continued to increase.

Rosetta remained in this orbit until 20 December, when the first of two burns began its climb back to a 30 km circular orbit. After the second burn on 24 December, the spacecraft stayed in this orbit for about 6 weeks before initiating a new phase involving a series of close fly-bys of the nucleus.

8.8 FLY-BY OPERATIONS

Near-comet operations during 2014 had involved frequent maneuvers that required rapid adjustments and intense activity by mission planners and the flight control team.

After Philae had been delivered onto the nucleus, the mission entered its main science phase, a very different type of operation with its own unpredictable challenges. With little time to take a breath and relax, the off-duty ESOC flight control team remained on call around the clock.

During the main science phase, the orbiter's trajectory was set by the Rosetta Science Ground Segment (RSGS) at ESAC in Spain, where 16-week planning calendars were drawn up by the long term planning (LTP) process. Sixteen weeks prior to the start of an LTP cycle, the RSGS proposed a trajectory for that cycle, and the flight dynamics team at ESOC checked it against current spacecraft and mission constraints. Pursuing the main science phase, the flight control team adopted a regular weekly planning cycle that had two very short term plans (VSTPs), one drawn up on Monday for Wednesday to Saturday, and the other on Thursday for Saturday to Wednesday.

The procedure was as follows:

- On planning days, the most recent navigation inputs, i.e. NAVCAM images and radiometric data from the ground stations, arrived early in the morning. The flight dynamics team at ESOC took optical images with the navigation camera five times per day. These images were then used to reconstruct the position and trajectory of Rosetta with respect to 67P.
- Production of all flight dynamics commands needed for the upcoming VSTP and covering (among others) trajectory and pointing strategy for that period.
- Merging of flight dynamics products with instrument commands and spacecraft platform commands.
- Ensuring that this merger was conflict free by reconciling instrument, spacecraft and mission constraints.
- Generating commands to be uploaded to the spacecraft, as well all data files needed on the ground. These included, for example, the ground tracking station instructions for the ESTRACK and NASA DSN stations for that VSTP, and the instructions for the automation tool in the ground control system.

Mission planners maintained two parallel trajectory plans, described as 'preferred' and 'high activity'. The desire was always to fly the preferred path, operating as close to the nucleus as conditions would permit, based upon specific assumptions about the level of comet activity. The second plan was a fallback in case the preferred orbit became unsafe. The flight control team always had to be ready to respond to an increase in the level of activity by moving the spacecraft farther from the nucleus.

As Laurence O'Rourke of the Rosetta Science Operations Center in Spain, emphasized, "The desire is to place the spacecraft as close as feasible to the comet before the activity becomes too high to maintain closed orbits."

Also, if the optical navigation process were to fail, the operations teams, with the approval of the mission manager, could command Rosetta to move clear.

8.9   THE COMET COMES ALIVE

Starting on 26 January 2015, Rosetta entered conjunction season – a four-week period when it was more or less on the opposite side of the Sun from Earth, when the Sun interfered with line-of-sight radio transmission. The consequent reduction in data rates restricted the amount of data that could be downloaded during a given ground station pass.

"Now, using ESA's 35 meter ESTRACK stations we get 14 kilobits per second (kbps), and this goes up to 45 kilobits per second when we use NASA's 70 meter stations," said Sylvain Lodiot, the spacecraft operations manager.

By June, when the orbital geometry had improved and the spacecraft was closer to Earth, data rates via the ESTRACK stations recovered to the maximum rate of 91 kbps.

The first opportunity for in-depth study of the comet's evolving nucleus and coma began with a thruster burn on 4 February that placed Rosetta on an elliptical orbit that took it more than 140 km from the comet on 7 February, and then headed inward again.

Table 8.3:   Thruster Burns Associated With The 6 km Fly-by

4 Feb: Depart from 26 km terminator orbit.
7 Feb: Reach 142 km from comet, then turn back.
11 Feb: Arc back down to 101 km.
14 Feb: Reach 50 km stand-off distance; turn and burn for the closest fly-by arc.
14 Feb: 6 km flyby at 12:40:50 UT

At about 12:41 UT on 14 February, Rosetta made its closest ever fly-by of 67P by swooping over the Imhotep region on the larger lobe at an altitude of only 6 km.

Fig. 8.8: A mosaic of four images taken on 14 February 2015 during Rosetta's first dedicated close fly-by. It features the Imhotep region on the larger lobe of the comet. The images were taken at 14:15 UT from a distance of 8.9 km. The scale is 0.76 meters per pixel, and the area measures 1.35 × 1.37 km. The closest point of approach, at about 6 km from the surface, was at 12:41 UT. (ESA/Rosetta/NAVCAM – CC BY-SA IGO 3.0)

This fly-by took Rosetta over the most active regions of the comet, allowing its instruments to obtain images and spectra of the surface with unprecedented resolution and over a range of wavelengths.

It also provided a unique opportunity to directly sample the innermost regions of the coma, where material escaping from the nucleus is fed into the coma and the tail. In particular, the instruments were looking for zones where the outflowing gas and dust were accelerated away from the surface, to determine how these constituents evolved as a function of distance from the comet.

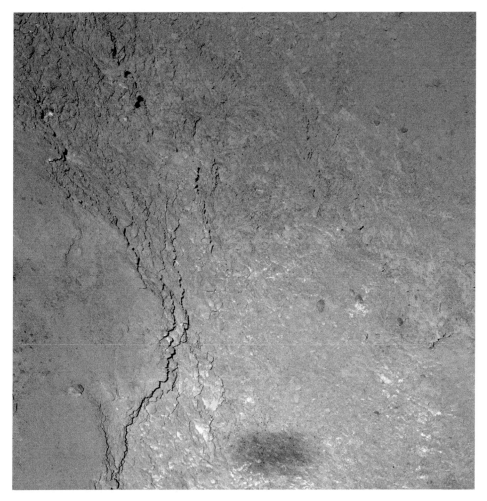

Fig. 8.9: A close-up view of the steep slopes and smoother terrain in the Imhotep region by the OSIRIS-NAC at an altitude of 6 km during Rosetta's fly-by on 14 February 2015. The area covers 228 × 228 meters and the spatial resolution is 11 cm per pixel. The penumbral shadow of the spacecraft can be seen at the bottom. (ESA/Rosetta/MPS for OSIRIS Team MPS/UPD/LAM/IAA/SSO/INTA/UPM/DASP/IDA)

The Sun was directly behind the spacecraft during the fly-by, enabling shadow-free images to be taken. The one exception was Rosetta's shadow on the surface, which appeared in OSIRIS images as a dark, roughly rectangular, penumbra (see Fig. 8.9). The geometry meant that, in a departure from the norm, the flat face of the solar panels was oriented more or less parallel to the comet's surface.

By studying the reflectivity of the surface as it varied with the angle at which the sunlight fell on it, scientists hoped to gain a more insight into the dust grains on the nucleus. The surface was already known to be very dark, reflecting a mere 6% of the sunlight that it receives.

Box 8.2: The OSIRIS Controversy

In November 2014, scientists attending a conference in Tucson, Arizona, presented some "staggering images" of Comet 67P, including the first color pictures of the nucleus, that had been taken by the OSIRIS high-resolution camera. However, none of these images had been released to the public by Rosetta's operator, the European Space Agency (ESA), in contrast to the regular flow of less detailed images from the orbiter's navigation cameras.

The journal Science reported that Project Scientist Matt Taylor was "reduced to learning about the new results at the Arizona conference by thumbing through Twitter feeds on his phone." So what was going on?

Writing in Science, journalist Eric Hand wrote: "For the Rosetta mission, there is an explicit tension between satisfying the public with new discoveries and allowing scientists first crack at publishing papers based on their own hard-won data." He continued: "In particular, the camera team, led by principal investigator Holger Sierks, has come under special criticism for what some say is a stingy release policy."

Mark McCaughrean, an ESA senior science adviser at ESTEC in the Netherlands, was quoted as saying: "It's a family that's fighting, and Holger is in the middle of it, because he holds the crown jewels."

Media and others used to rapid dissemination of images and other data from NASA missions argued that a similar policy should be followed by ESA for the flagship Rosetta mission. However, as James Green, the director of NASA's planetary science division in Washington, D.C., pointed out, ESA relies much more on contributions from Member States, whereas NASA pays for most instrument development directly.

"It's easier for [NASA] to negotiate [data release] because we're paying the bills," said Green, whereas ESA has to do it "by influence".

On Rosetta, for example, ESA contributed very little toward the cost of developing the €100 million OSIRIS camera, and the agency therefore had less control over how its images were disseminated.

Furthermore, prior to Rosetta's launch in 2004, a data embargo of 6 months was set for all the instrument teams. However, McCaughrean pointed out that mission documents also stipulated that instrument teams must provide "adequate support" to ESA management in its communication efforts.

"I believe that [the OSIRIS camera team's support] has by no means been adequate, and they believe it has," McCaughrean was quoted as saying. "But they hold the images, and it's a completely asymmetric relationship."

Sierks, based at the Max-Planck-Institute for Solar System Research in Göttingen, Germany, argued that the OSIRIS team had already been providing a fair amount of data to the public – about one image every week. He also said that the priority should go to researchers who wanted to use the images in their papers for scientific journals, some of which enforced strict embargoes.

The OSIRIS data releases during the mission typically lagged far behind the actual date they were taken. It was not until December 2015 that ESA announced a new website on which the OSIRIS team would be releasing images on a regular basis – at least one per week. In May 2016, ESA released a large quantity of science data to the public which covered the mission through to 19 December 2014. However, it was not until April 2019 that the entire catalogue of more than 70,000 OSIRIS images of Comet 67P was published on a new website called the OSIRIS Image Viewer.

In a recent personal communication, Holger Sierks revisited the OSIRIS image release policy:

> *I think we were all overwhelmed by the public interest, ESA and us. The public interest was stimulated by ESA, and pushed all of us to increase the efforts on PR, focusing primarily on imagery, understandably, and so the tension on releases came up.*

> *We were in good agreement with ESA on the release of images in approach to the comet, as we were for the fly-bys at Earth, Mars, Steins, and Lutetia. We had a plan for the comet. With the increased demand on images, and the first papers hitting editors from outside the team based on released public images, tension increased. There is no agreement with the outside world not using the data other than for following the mission.*

> *We were providing ESA with images required for all operational aspects, including regular navigation and the landing of Philae and the orbiter, in real time. We reported to the ESA project appropriately, and showed all detail of our imagery at the Science Working Team meetings.*

> *A full access of ESA to our image archive, as ESA may have liked, was not possible, as the images had to get calibrated, and interpreted by the team to a level. With the huge investments by a number of key European research institutes, and space agencies, we also had to find a balance in return.*

> *The signed agreement on the proprietary period reads 12 months; we had discussions shortening it to 6 months as due for NASA missions and our NASA Rosetta partners, but shortening could not be agreed on with the groups that delivered their hardware 10 years ago on the basis of the agreed 12 months. And it is a true statement, the instruments are not*

owned by ESA, ESA is providing and flying the bus, as ESA'S director of science once stated.

All OSIRIS images were delivered to the PSA (Planetary Science Archive) of ESA in due time, raw and calibrated, following the 12 months proprietary phase, and the ESA archiving time line. They are publicly available ever since ESA put them on the archive.

For the fun of sharing the images, and enjoying them browsing for specific topics, we ran a student's project and published the full archive [on website] https://rosetta-osiris.eu/ . It is far more handsome than the science archives by the space agencies. We see more than 500 unique visitors per day still accessing the server.

8.10 SWAMPED BY FALSE STARS

Apart from an increase in drag on the spacecraft and its giant solar wings as a result of flying through denser regions of outflowing gas and dust, Rosetta almost lost its way.

Some two hours prior to closest approach, the star tracker began to lose the ability to identify enough stars to navigate correctly, and occasionally locked onto 'false stars'. This was due to the reflective particles in the coma. The spacecraft switched to some back-up units, including the back-up star tracker, but this also experienced tracking problems owing to the number of dust particles in the field-of-view during the fly-by.

"With a lot of luck, the spacecraft did not end up in safe mode," pointed out Sylvain Lodiot, the operations manager. "Although in this case we could have recovered the spacecraft and resumed operations as planned, the science instruments would have automatically switched off in the meantime. By the time they were switched back on, we would have been relatively far away from the comet again."

After this close fly-by, the mission entered its next phase, in which Rosetta was to execute a series of fly-bys at distances that varied from about 15 km to 100 km. This shift from 'bound orbits' to fly-bys had always been planned for this phase of the mission, based on predictions of increasing cometary activity and the safety of the spacecraft.

The range of fly-by distances also balanced the needs of the instruments, to optimize their respective scientific returns. For example, during some of the close fly-bys, Rosetta would move over the comet almost in step with its rotation to allow the instruments to monitor a single point on the surface. In contrast, the more distant fly-bys provided a broader context with wide-angle views of the nucleus and its growing coma.

Immediately after Rosetta's extremely close fly-by on 14 February, it traveled away from the comet, reaching a distance of 253 km from its center three days later.

The next pass by the nucleus involved a new trajectory arc which took the spacecraft within about 55 km of the surface on 23 February, as part of the effort to sample the material which was flowing away from the comet at various distances.

During 12-20 March, the orbiter transmitted continuously to the Philae lander in the hope of a response, but to no avail. The most likely time to establish contact occurred during the 11 fly-bys in which the orbiter's path placed it in a particularly favorable position with respect to the lander in local daytime, when Philae's solar panels would be receiving sunlight and charging its battery.

The smooth progress of the near-nucleus operations was interrupted on 28 March, during a fly-by that passed within about 14 km of the large lobe. Navigational issues similar to those of 14 February recurred. Again, the primary star tracker experienced difficulties in locking on to its guide stars while approaching the active comet. Attempts to regain tracking capabilities were hindered by the amount of background noise, which produced hundreds of 'false stars', and it was almost 24 hours before tracking was properly re-established.

In the meantime, a spacecraft attitude error had built up, resulting in the high-gain antenna pointing away from Earth. As a result, a significant drop in the strength of the radio signal received by ground stations was registered.

After recovery of the star tracking system, the high-gain's error was corrected automatically, returning the received signal from the spacecraft to full strength.

But the 'false stars' continued. When comparisons with other navigation mechanisms showed inconsistencies with the star trackers, some onboard reconfigurations were begun. While this was underway, the same error occurred again, this time resulting in an automatic safe mode that turned off many key systems, including the science instruments. Nevertheless, some data were returned, and the NAVCAMs took images during the inward leg and shortly after closest approach.

Although the operations team toiled successfully to recover the spacecraft from safe mode to normal status over the next 24 hours, it took much longer to reactivate all of the science instruments.

Meanwhile, Rosetta moved onto a pre-planned escape trajectory which took it some 400 km from 67P. An orbital correction maneuver was performed on 1 April to start to bring it back. After a second maneuver on 4 April, the desired distance of 140 km was achieved on 8 April.

Nevertheless, the continuing navigational difficulties associated with 'false stars' necessitated a new strategy for near-comet operations. The desire to keep Rosetta safe whilst maximizing its scientific return required mission planners to modify the existing flight plans as the comet spewed out increasingly large amounts of dust.

"This has ultimately meant a complete replanning of the upcoming fly-by trajectories," said Sylvain Lodiot. "We're first moving to a terminator orbit at a distance of 140 km and then we are targeting 100 km. Then we will adopt a similar strategy to when we first approached the comet, in August last year. That is, we'll fly 'pyramid' trajectories, starting at about 100 km on 11 April, and we'll monitor how the spacecraft reacts before moving closer."

During May and June, the operations team flew Rosetta at the safer distance of roughly 200 km from the comet in order to avoid further navigation issues. It was also maneuvered into a terminator trajectory, in the hope this would reduce the degree to which the dust confused the star trackers.

As the weeks went by, the plan was to slowly edge Rosetta nearer to the comet whilst closely monitoring the performance of the star trackers in 'continuous tracking mode'. By late June, its closest approach was down to 165 km.

Meanwhile, from March 2015 onward, Philae's environmental conditions were expected to improve, with higher surface temperatures and better illumination. When the orbital geometry was suitable, Rosetta periodically switched on its receiver to listen for signals from the dormant lander.

These efforts paid off on the evening of 13 June, when communications with Philae were re-established after 211 days of silence. The orbiter maintained a weak, but steady, radio link for some 85 seconds. The relayed signals were received by ESOC in Darmstadt at 20:28 UT (see Chapter 9).

A second burst of lander data was received at about 21:26 UT on 14 June, lasting just a few seconds. These data were confirmed to be giving the lander's current status, indicating that its internal temperature had already risen to –5°C.

Philae issued short bursts of housekeeping telemetry over the following days, including data from its thermal, power, and computer subsystems.

On 15 June, teams at ESA, DLR and CNES agreed to a new trajectory that would optimize the opportunities for lander-to-orbiter communication. This planning required a great deal of coordination between the flight dynamics experts at ESOC in Darmstadt, the Rosetta Science Ground Segment (RSGS/ESAC), the Philae Lander Control Center team (DLR Cologne), and the Science Operations Navigation Center (SONC) at CNES/Toulouse.

Their efforts resulted in Rosetta making two 'dog-leg' burns, beginning on 16 June, to put it into a modified orbit designed to establish more frequent and longer contacts with Philae. The new orbit included an already planned lowering from 200 km to 180 km, and 'nadir pointing' (continuously pointing Rosetta's communications unit at the comet) in the latitudes where the lander was believed to be located.

These modifications enabled the lander to make a further six intermittent contacts on 19, 20, 21, 23 and 24 June, and 9 July. Housekeeping data were transferred from Philae to Rosetta on all of these dates, apart from 23 June, but the communications links were too short and unstable to enable any scientific measurements to be commanded.

During the first weeks of July, Rosetta flew along the terminator plane at distances between 180 km and 153 km, and at latitudes between 0 and 54 degrees, seeking the best location for communicating with Philae. However, by this time the increased levels of outflowing gas and dust meant the star trackers were partially swamped over the weekend of 10-11 July, and the spacecraft had to retreat to safer distances of 170-190 km.

Despite Philae's temperature rising above 0°C as the comet approached perihelion in August, no further contacts were made with the lander. Perhaps it was transmitting, but the orbiter, at its safe altitude, was too far away to pick up the weak signal.

With activity subsiding after perihelion, Rosetta was able to safely approach the nucleus once again. However, despite repeated passes, no signals were detected, and attempts to send commands 'in the blind' to trigger a response from it also failed to produce a result.

8.11   PERIHELION AND BEYOND

On 23 June 2015, ESA confirmed that the Rosetta mission would be extended to the end of September 2016. The nominal mission was funded to the end of December 2015, but ESA's Science Program Committee issued formal approval to continue it for a further nine months.

This extension meant that the orbiter would continue to accompany Comet 67P after it made its perihelion passage on 13 August 2015 and headed back into deep space.[3] By continuing its study of the comet as it receded from the Sun, Rosetta would provide scientists with a more complete picture of how the activity of a comet increases and decreases along its orbit.

The additional observations by Rosetta would also provide improved in-situ data to compare with complementary Earth-based observations. (Professional and amateur astronomers across the globe had observed the comet as it neared perihelion, and observations using professional telescopes were planned every night around closest approach.)

As expected, the amount of material being ejected by the comet's jets increased dramatically up to, and after, the closest approach to the Sun. Scientists estimated that, at peak activity, the nucleus was shedding 1,000 kg of dust per second, producing perilous conditions for Rosetta.

In the days leading up to perihelion, the orbiter had to retreat to altitudes of between 325 km and 340 km to prevent its star trackers being confused by 'false stars'.

The first year of exploration at 67P had been largely focused on near-comet studies, but this changed on 23 September, when Rosetta was inserted into a new trajectory that carried it up to 1,500 km from the nucleus in the direction of the Sun.

---

[3] Perihelion occurred at 02:03 UT on 13 August, 186 million km (1.24 AU) from the Sun, between the orbits of Earth and Mars. At that time, the one-way travel time for a signal was 14 minutes 44 seconds.

During this three-week excursion, the orbiter retreated much farther from the nucleus than at any time since its arrival in August 2014 (see Appendix 3). The maximum distance was achieved by the end of September, with the spacecraft heading back toward the nucleus by mid-October.

The main science goal driving this course of action was to make a broader study of the coma while the comet's activity remained high, post-perihelion. Almost all of the instruments were operating, but this phase was of particular interest to the Rosetta Plasma Consortium, who were keen to investigate the bow shock, the boundary between the comet's magnetosphere and the ambient solar wind environment.

"Previous measurements performed during fly-bys [by other missions] only provided limited data points about the bow shocks of a handful of comets," explained Claire Vallat, a Rosetta scientist at the European Space Astronomy Center (ESAC). "Rosetta, instead, will take data over several days, monitoring the evolution of the plasma environment of 67P shortly after its perihelion."

Meanwhile, another significant scientific opportunity opened in May 2015, as the southern hemisphere of the nucleus started to be bathed in sunlight after more than five years of darkness. Previously, only the MIRO microwave instrument had been able to study this frigid region. Now Rosetta was able to undertake the first multi-instrument observations of the comet's south polar regions.

A number of passes over the southern hemisphere were achieved before the brief southern summer ended in early 2016 – significantly, both before and after perihelion. In late October, after the three-week trip into the coma, the mission team scheduled a number of orbits that would compare the northern and southern hemispheres, making slower passes in the south in order to maximize the observational opportunities.

In March-April 2016, Rosetta went on another far excursion, this time on the night side, to study the wider coma, tail, and plasma environment. It reached a distance of about 1,000 km on 30 March and then headed back in, flying past the nucleus at an altitude of 30 km about a fortnight later.

The operations plan for the following weeks included a close fly-by and orbits dedicated to a range of science observations. Unfortunately, despite a decline in activity after perihelion, the environment was still dusty enough to cause navigational errors.

8.12   ANOTHER DRAMATIC SAFE MODE

The continuing false detections by the star trackers led to Rosetta entering safe mode on the weekend of 28-29 May, when it was just 5 km above the nucleus. After the star trackers were confused by dust, they locked on to a 'false star' which led to spacecraft pointing errors that would trigger an automatic shift to safe mode. Unknown to controllers on Earth, the lack of star tracker data prevented the spacecraft's computer from entering Safe Hold Mode. Instead, it became locked in an intermediate Sun-keeping Mode. As a result, contact with Rosetta was lost for nearly 24 hours.

"We had no signal from the spacecraft, so we were 'in the blind', which is the worst situation possible," said Sylvain Lodiot, the spacecraft operations manager. "Also we had not received the telemetry telling us the spacecraft was going to safe mode. So we were not sure what had happened."

In an effort to recover the spacecraft, Lodiot persuaded the flight dynamics team and the mission manager to send commands in the blind to power cycle (i.e. switch on and off) the star trackers.

As Lodiot says:

Once I convinced them and got the 'Go', my team sent the command via the low-gain antenna. And it worked! The hung star tracker began to acquire stars, which allowed the transition from Sun-keeping Mode to Safe Hold Mode, and then we got the signal back from the spacecraft.

It was an extremely dramatic weekend. After we sent commands 'in the blind', which successfully tackled the hung star tracker issue, an additional false star detection almost sent the spacecraft back into safe mode again.

I remember that weekend very well. 29th May is the birthday of my daughter, so I was in France, but I spent the previous night and most of the 29th on the phone before driving back and going straight to work on Sunday afternoon. I was not very popular that day at home ... I even missed the (birthday) cake.

In accordance with normal practice during such an event, extra ground tracking station time was sought in order to provide additional support for recovering the spacecraft. The already scheduled Rosetta tracking slot with ESA's New Norcia station in Australia was extended by claiming time from Mars Express operations. The 'blind' command was sent by New Norcia and, later, ESA's Cebreros station in Spain assisted with the recovery.

Eventually, the spacecraft was returned to three-axis stabilized safe mode. However, it took some time for the ground team to receive the NAVCAM images needed to confirm its exact position relative to the comet. Despite the safe mode, controllers were able to resume science operations by 2 June, and to continue with the plan to maneuver Rosetta into a 30 km orbit.

Less than a month prior to the end of the mission, imagery from OSIRIS-NAC revealed the long-lost Philae lander, wedged in a deep crack in the Abydos region of the nucleus – as had been inferred. Taken on 2 September when the orbiter came within 2.7 km of the surface, the images clearly showed the 1 meter wide body of the lander, along with two of its three legs, lying on its side (see Chapter 9).

"With only a month left of the Rosetta mission, we are so happy to have finally imaged Philae, and to see it in such amazing detail," said OSIRIS team member Cecilia Tubiana, who was the first person to see the images when they were downlinked.

8.13   THE GRAND FINALE

As comet activity diminished post-perihelion, the operations team could move Rosetta much closer in order to undertake a detailed survey of the changes that had occurred on its surface and in its environment. There was also the opportunity to use the experience gained in operating Rosetta close to the nucleus in order to carry out new and slightly riskier investigations, including flights over the night side to observe plasma, dust, and gas interactions, and collection of samples of dust ejected close to the nucleus.

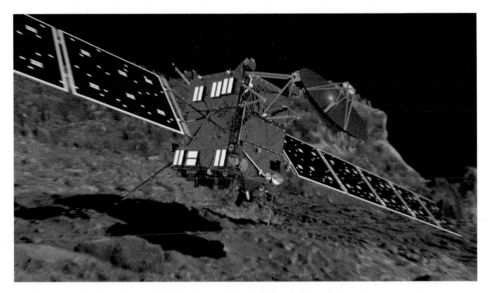

Fig. 8.10: An artist's impression of the Rosetta spacecraft shortly prior to its landing on 67P on 30 September 2016. (ESA)

When discussing the eventual termination of the mission, the mission team favoured setting the huge orbiter down on the surface of the nucleus. This would provide a new science bonanza as the instruments would return data right up until the end of the mission, gathering unique data at unprecedentedly close distances.

By the end of September 2016, Rosetta and 67P would have retreated to a distance at which there was not enough sunlight to power the scientific instruments and operate the spacecraft efficiently and safely. Also, communication would be restricted for several weeks because the comet would be behind the Sun from the vantage point of Earth, making it difficult to relay both scientific data and operational commands.

"This time, as we're riding along next to the comet, the most logical way to end the mission is to set Rosetta down on the surface," said Patrick Martin, Rosetta's mission manager. "But there is still a lot to do to confirm that this end-of-mission scenario is possible. We will first have to see what the status of the spacecraft is after perihelion and how well it is performing close to the comet, and later we will have to try and determine where on the surface we can have a touchdown."

On 21 July, ESA reported that the controlled impact on 67P would be made on 30 September 2016, in the Ma'at region of the small lobe. This target was chosen for its scientific potential, and for being consistent with key operational constraints imposed on the descent.

Starting on 9 August, Rosetta began to fly elliptical orbits that brought it progressively closer to 67P. Every three days, it made another fly-by, facilitating unprecedented close-up studies. At its closest, it was within 1.76 km of the rugged surface.

As Sylvain Lodiot said:

Although we've been flying Rosetta around the comet for two years now, our biggest challenge yet will be keeping it operating safely for the final weeks of the mission in the unpredictable environment of this comet and so far from the Sun and Earth.

We are already feeling the difference in gravitational pull of the comet as we fly closer and closer. It is increasing the spacecraft's orbital period, which has to be corrected by small maneuvers. But this is why we have these fly-overs, stepping down in small increments to be robust against these issues when we make the final approach.

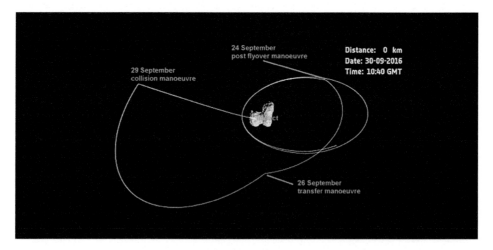

Fig. 8.11: A simplified overview of Rosetta's last week of maneuvers at 67P. After the final fly-over on 24 September, it transferred towards the 'initial point' of a 16 × 23 km orbit. A short series of maneuvers over the following days lined up Rosetta with the desired impact site. The collision course maneuver took place on 29 September, initiating the descent from an altitude of about 20 km on 30 September. (ESA)

On 24 September, after the final fly-over, flight controllers initiated a series of maneuvers to line up the spacecraft for its final descent. Two days later, Rosetta completed a short burn to put it on the 16 × 23 km trajectory around the comet from which it would eventually descend to the surface.

To prevent the spacecraft from entering a safe mode and losing contact prior to impact, its operators reprogrammed some of its tolerance to errors, and some star tracker checks were disabled for several weeks prior to the terminal descent.

Rosetta's final maneuver was a thruster firing which began at 20:48:11 UT on 29 September and lasted for 208 seconds. This nudged the spacecraft onto a collision course with the nucleus, initiating a leisurely 14 hour descent from an altitude of 19 km.

Shortly thereafter, its navigation cameras took a set of five images to confirm that it was following the planned trajectory and to refine the predicted impact time. After they were downloaded in the early hours of 30 September, the flight team used them to calculate the time of impact within a 4 minute window.

Initially the velocity was only 30 cm/s (about 1 km/h), but this gradually increased due to the gravity of the comet. At about 55 minutes prior to touchdown, some 2 km above the surface, Rosetta was closing in at 60 cm/s. The speed at impact was expected to be 90 cm/s, similar to that of Philae when it made its first touchdown.

During what was essentially a slow free-fall, six of the spacecraft's science instruments were providing unprecedented high-resolution images and data on the gas, dust and plasma in close to the comet. It was hoped they would continue to provide data until the vehicle was between 20 meters and 5 meters from the surface.[4]

The planned touchdown site was a flat area on the small lobe of the nucleus, close to a cluster of active pits in the Ma'at region. The intriguing pits were more than 100 meters in diameter and 50-60 meters deep. The specific target was a site adjacent to a 130 meter wide pit that the team had nicknamed Deir el-Medina, after a man-made pit in an ancient Egyptian town of the same name. Of particular interest were the walls of the pits, which exhibited intriguing lumpy structures referred to as 'goose bumps'. The speculation was that these structures, which were about a meter across, were related to the 'cometesimals' that accreted to produce the nucleus.

Early in the descent, OSIRIS imaged the regions of the large lobe the spacecraft was passing over, and then, as the small lobe loomed, it provided close-ups of the walls of the Ma'at pits and the intended landing site, nicknamed Sais.

---

[4] Rosetta was now so far from the Sun that it could not generate enough power from its solar panels to operate all of the instruments simultaneously. MIDAS, COSIMA and VIRTIS were shut down because their potential science return was considered to be lower.

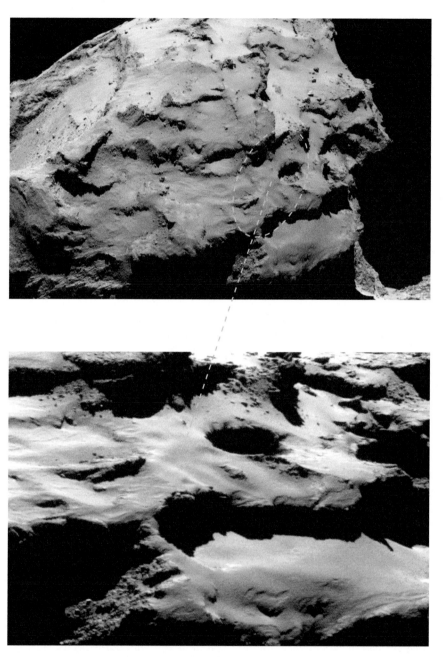

Fig. 8.12: The site selected for Rosetta's final touchdown was in the 500 × 700 meter yellow ellipse. The target was a flat area in the Ma'at region, close to some active pits on the small lobe of the nucleus. The image was taken with Rosetta's NAVCAM. (ESA)

In order to downlink the maximum number of images in the limited time available, especially in the final phase of the descent, the images were compressed by a factor of up to 20 times. In addition, instead of returning a small number with 2,048 × 2,048 pixels, a larger number of smaller ones were to be returned, with sizes from 1,000 × 1,000 down to 480 × 480 pixels.

Fig. 8.13: OSIRIS-NAC captured this image of Comet 67P on 30 September 2016 from an altitude of about 16 km above the surface, as the spacecraft made its controlled descent. It spans about 614 meters with a resolution of about 30 cm per pixel. (ESA/Rosetta/MPS for OSIRIS Team MPS/UPD/LAM/IAA/SSO/INTA/UPM/ DASP/IDA)

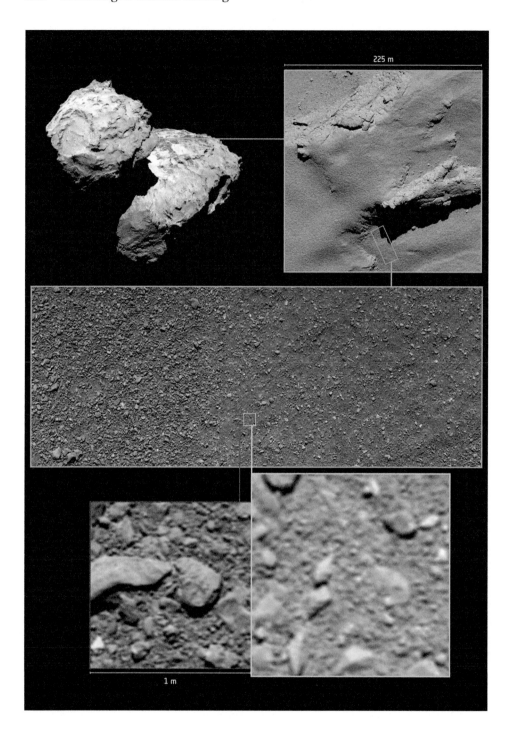

Below an altitude of 200-300 meters, the images were blurred because the cameras were not designed to image the comet from such close proximity: the NAC lost its focus below an altitude of around 1 km, and the WAC when the range fell below 200-300 meters.

What ESA and the OSIRIS team believed to be the final image from the spacecraft showed a slightly blurred view of a patch of stony ground, taken from an estimated altitude of 51 meters and thought to span an area about 2.4 meters across. Subsequent analysis revised the altitude to 24.7±1.5 meters. The uncertainty arose from the precise method of altitude calculation and the model used for the shape of the nucleus.

After the excitement of the historic landing had died down, the camera team found several unprocessed telemetry packets on their server and set out to investigate. As OSIRIS principal investigator Holger Sierks reflected, ""We thought, wow, that could be another image."

The transmission had been interrupted when Rosetta hit the ground, but three full packets of data had been received: 12,228 bytes in total, or just over half of a complete image. This was not recognized as an image by the automatic processing software, but engineers in Göttingen were able to make sense of these data fragments and process them into an image.

Released in May 2017, the reconstructed image was taken from a height of 19.5±1.5 meters and showed a patch of ground equivalent to a 1 × 1 meter square at a scale of 2 mm per pixel (see Fig. 8.14).

Meanwhile, ROSINA collected unique data on the density and composition of gas close to the comet. It was able to detect a tiny increase in the gas pressure around the nucleus as the spacecraft descended. MIRO complemented OSIRIS and ROSINA by measuring the surface temperature. GIADA measured the dust density

Fig. 8.14: An annotated image indicating the approximate locations of some of Rosetta's final images. Top left: a global view shows the area in which Rosetta touched down in the Ma'at region of the smaller lobe. This image was taken by the OSIRIS Narrow Angle Camera on 5 August 2014 from a distance of 123 km. Top right: an image taken by the same camera from an altitude of 5.7 km, during Rosetta's descent on 30 September 2016. The scale is about 11 cm per pixel, and the image measures about 225 meters across. The final touchdown point, named Sais, is seen in the bottom right of the image, within a shallow, ancient pit. Exposed, dust-free terrain is evident in the pit walls and cliff edges. The image is rotated 180 degrees with respect to the global context image at top left. Middle: an OSIRIS Wide Angle Camera image taken from an altitude of about 331 meters. The scale is about 33 mm per pixel and the image measures about 55 meters across. A mix of coarse and fine-grained material is visible. Bottom right: the last complete image returned by Rosetta, taken from a height of 24.7±1.5 meters. Bottom left: the final image (reconstructed after Rosetta's landing) was taken at an altitude of 19.5±1.5 meters. The scale is 2 mm per pixel and it measures about 1 meter wide. (ESA/Rosetta/ MPS for OSIRIS Team MPS/UPD/LAM/IAA/SSO/INTA/UPM/DASP/IDA)

and the acceleration of grains away from the nucleus. The RPC suite took a unique close-up look at the interaction between the solar wind and the surface of the comet and sampled levitating charged dust grains. ALICE obtained its highest resolution ultraviolet spectra of the surface, plus measurements which complemented some of the RPC data. RSI made the most accurate measurements of the gravity field.

During the 14 hour descent, the spacecraft sent an uninterrupted stream of data, until its final signal arrived at ESOC in Darmstadt at 11:19 UT (after the 40 minute time delay due to the spacecraft's distance from Earth), then there was silence as the spacecraft ceased transmitting upon impact.

Thousands of engineers and scientists who had worked on the mission, along with members of the international press, watched a live broadcast of the Rosetta Finale, which was streamed to venues in different countries.

The curtain had come down on Rosetta's remarkable comet chase.

The ROSINA principal investigator, Kathrin Altwegg, who had been involved in the project since 1994, spoke for everyone when she said, "I feel a little melancholic. But it's a colossal end!"

Rosetta operations manager Sylvain Lodiot said, "We've operated in the harsh environment of the comet for 786 days, made a number of dramatic fly-bys close to its surface, survived several unexpected outbursts from the comet, and recovered from multiple spacecraft safe modes. The operations in this final phase have challenged us more than ever before, but it's a fitting end to Rosetta's incredible adventure to follow its lander down to the comet."

Box 8.2: Shutting Down Rosetta

The International Telecommunication Union (ITU) regulations required that Rosetta's radio transmitter be switched off permanently at the end of its mission. Since the spacecraft wasn't designed to have its transmitter permanently off, the operations team had to upload a patch to modify the software.

The basic idea was to trigger a safe mode when the spacecraft touched the surface. Rosetta would experience a sudden shift in its attitude or motion that its software would register as being beyond the permitted limits. Specifically, the software patch would trigger an FDIR (Failure Detection, Isolation and Recovery) response which would result in a safe mode. On completion of the safe mode sequence, the spacecraft would be entered into a passive, non-reactive mode that was initially designed only for ground testing prior to launch. This would mean that all of the attitude and orbit control units were turned off, as well as the transmitter.

Three hours prior to the expected impact on 30 September, the so-called 'point of no return', the ground team fully activated the passivation instructions. The first safe mode to occur after that would permanently shut down the spacecraft.

REFERENCES

ESA Rosetta websites: http://rosetta.esa.int/ and https://sci.esa.int/web/rosetta/

Rosetta blog: http://blogs.esa.int/rosetta/

Rosetta Mission (Facebook): https://www.facebook.com/RosettaMission/

Rosetta on Twitter: https://twitter.com/ESA_Rosetta

Max-Planck-Institute for Solar System Research: https://www.mps.mpg.de/en/Rosetta

Rosetta Hibernation Wake Up FAQs, 20 January 2014: https://download.esa.int/esoc/shares/rosetta_wake-up_faq_20_jan_2014.pdf

ESA's 'Sleeping Beauty' Wakes Up From Deep Space Hibernation, 20 January 2014: http://www.esa.int/Our_Activities/Space_Science/Rosetta/ESA_s_sleeping_beauty_wakes_up_from_deep_space_hibernation

Rosetta Sets Sights On Destination Comet, 27 March 2014: http://www.esa.int/Science_Exploration/Space_Science/Rosetta/Rosetta_sets_sights_on_destination_comet

Thruster Burn Kicks Off Crucial Series of Manoeuvres, 7 May 2014: http://blogs.esa.int/rosetta/2014/05/07/thruster-burn-kicks-off-crucial-series-of-manoeuvres/

The Chase Is On, ESA Bulletin 158, p. 60-65, May 2014: https://sci.esa.int/documents/34878/36335/1567260105661-Rosetta_ESA_Bulletin_158.pdf

How Rosetta Arrives at a Comet, 1 August 2014: http://blogs.esa.int/rosetta/2014/08/01/how-rosetta-arrives-at-a-comet/

Access to Rosetta Data, 16 July 2014: http://blogs.esa.int/rosetta/2014/07/16/access-to-rosetta-data/

What's Happening in Rosetta Mission Control Today, 4 August 2014: http://blogs.esa.int/rosetta/2014/08/04/whats-happening-in-rosetta-mission-control-today/

Rosetta Arrives at Comet Destination, 6 August 2014: http://www.esa.int/Our_Activities/Space_Science/Rosetta/Rosetta_arrives_at_comet_destination

Rosetta Arrival at Comet 67P Churyumov-Gerasimenko press kit, August 2014: https://sci.esa.int/documents/34878/36335/1567260117961-Rosetta_media_kit_arrival_6August.pdf

Rosetta's Orbits Around the Comet, August 2014 – video, ESA-C. Carreau: http://blogs.esa.int/rosetta/2014/08/06/what-rosetta-does-now/

Living With A Comet, Rosetta in Pictures, August-November 2014, ESA Bulletin 160, November 2014: http://www.esa.int/About_Us/ESA_Publications/ESA_Publications_Bulletin/ESA_Bulletin_160_Nov_2014

Comet Landing – the Most Critical Moments, 7 November 2014: http://blogs.esa.int/rosetta/2014/11/07/landing-operations-the-most-critical-moments-you-should-watch-for/

Tensions Surround Release of New Rosetta Comet Data, Eric Hand, Science, 11 November 2014: https://www.sciencemag.org/news/2014/11/tensions-surround-release-new-rosetta-comet-data

Rosetta Swoops In For A Close Encounter, 4 February 2015: http://www.esa.int/Our_Activities/Space_Science/Rosetta/Rosetta_swoops_in_for_a_close_encounter

Rosetta's Big Day in the Sun, 13 August 2015: https://blogs.esa.int/rosetta/2015/08/13/rosettas-big-day-in-the-sun/

Rosetta Mission Extended, 23 June 2015: http://www.esa.int/Our_Activities/Space_Science/Rosetta/Rosetta_mission_extended

From One Comet Landing to Another: Planning Rosetta's Grand Finale, 12 November 2015: http://blogs.esa.int/rosetta/2015/11/12/from-one-comet-landing-to-another-planning-rosettas-grand-finale/

Rosetta's final hour (video), 30 September 2016: https://www.esa.int/ESA_Multimedia/Videos/2016/09/Rosetta_s_final_hour

Rosetta end of mission, grand finale – press kit, 25 September 2016: http://sci.esa.int/rosetta/58334-rosetta-end-of-mission-grand-finale-press-kit-september-2016/#

Science 'Til The Very End, 28 September 2016: https://blogs.esa.int/rosetta/2016/09/28/science-til-the-very-end/

Unexpected Surprise: A Final Image From Rosetta, 28 September 2017: http://sci.esa.int/rosetta/59618-unexpected-surprise-a-final-image-from-rosetta/

How Rosetta Gets Passivated, 29 September 2016: https://blogs.esa.int/rosetta/2016/09/29/how-rosetta-gets-passivated/

Rosetta Special, CAP Journal, Issue 19, March 2016: https://www.capjournal.org/issues/19/

9

Landing on a Comet

Once Rosetta was safely captured by Comet 67P's gravity, the orbiter and lander teams began to prepare for one of the most ambitious and daring phases of the entire mission: the historic landing of Philae on the icy nucleus.

As Rosetta gradually moved closer to its target, scientists examined the rugged, dual-lobed nucleus in search of a suitable touchdown site. Although time was short for identifying the optimum site, preparations for this moment had started some 15 years earlier, when the lander was being designed.

9.1 THE PLANNED LANDING SEQUENCE

The planned landing sequence had been carefully analyzed and prepared. Once a landing site was selected from mapping of the surface, the details of the ejection and descent of the lander were calculated.

Rosetta was placed on a trajectory that would bring it close to the planned landing site. Once the orbiter was aligned correctly, ground control would command Rosetta to eject Philae. The lander was then programmed to unfold its three legs, ready for a gentle touchdown at the end of its ballistic descent. Data from the lander would be relayed back to Earth via the orbiter.

The landing gear was designed to damp out most of the kinetic energy caused by the impact, in order to reduce the chance of a bounce, because the escape velocity of the comet was only about 1 m/s (3.6 km/h). The legs could also rotate, lift, or tilt to return the lander to an upright position if it were to topple over or land on its side.

The landing gear had to be modified after Rosetta missed its launch window to rendezvous with Comet 46P/Wirtanen (see Chapter 6). The new target, 67P, was

© Springer Nature Switzerland AG 2020
P. Bond, *Rosetta: The Remarkable Story of Europe's Comet Explorer*,
Springer Praxis Books, https://doi.org/10.1007/978-3-030-60720-3_9

more massive, so the lander's impact velocity was expected to be significantly higher than initially planned. In this case, its tilt capability would be restricted.

Prior to release, the flywheel on the lander was switched on in order to keep the orientation of the main body axis constant in space throughout the entire descent. The intention was for the orientation of the main axis of the lander to be perpendicular to the surface upon reaching the landing point.

The lander separation worked through a spindle drive system that could push the lander away from the orbiter at a speed selectable between 0.05 m/s and 0.51 m/s. In the event that this drive system was not performing properly, a spring eject mechanism could be used to propel the lander away at a fixed speed of 19 cm/s. Both of these mechanisms were to be activated automatically, within less than 30 seconds, to ensure the lander would escape from the orbiter within the correct time window.

The release procedure began by opening the so-called wax actuator that attached the lander to the orbiter. So as not to release the lander too early after being freed via the wax actuator, the spindle drive functioned in 'pull mode', softly holding the lander against its interface until release time.

The release procedure also pulled away the umbilical cable that linked the orbiter and lander. This cable had been used for power, command and data exchange between the two vehicles. A short time prior to this, the lander began to run on internal battery power with all of its key systems up and running.

The two elements of the CONSERT instrument, one on the orbiter, the other on the lander, were communicating with each other prior to release. They were to keep in contact for almost the entire 7 hour descent, until shortly before touchdown. CONSERT (see Chapter 5) was to collect science data during the descent that would also assist in determining the separation of the two spacecraft and the descent path of the lander.

Shortly after it ejected the lander, the orbiter slewed in space and moved to a safe distance in such a manner that it could maintain radio link contact throughout the descent.

Once the lander was a safe distance from Rosetta, it released its landing gear from the stowed position, unfolding the tripod legs and 'lifting' the touchdown drive to its maximum capacity.

The landing gear release was initiated by so-called 'thermal knives' that were heated so that they burned through the material that restrained the preloaded unfolding mechanism. Meanwhile, the CONSERT antenna unfolded and the ROMAP instrument's arm was put into its optimum position for measurements.

Two hours after separation, the orbiter and the lander were in position to resume a continuous data and command link. From that time onward, the ground control team ought to be able to follow closely how the lander performed during its descent – although with a delay of half an hour because of the travel time of the radio signal from the comet to Earth.

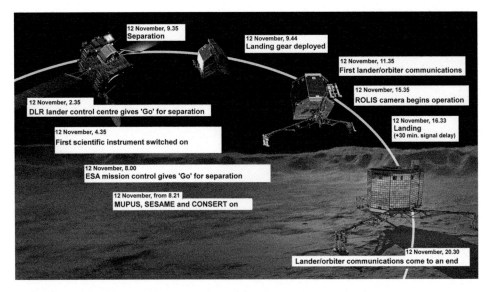

Fig. 9.1: The main activities during Philae's planned descent to the comet's surface. All times are given in 'comet time' and in Central European Time (CET, one hour ahead of GMT/UT). Confirmation at mission control was delayed because the signals from the lander took about 30 minutes to reach Earth via the Rosetta orbiter. (Peter Bond, DLR, ESA)

During the last hour of the descent, the lander prepared for touchdown, which was expected to occur at a velocity of about 1 m/s. To sense this moment, the lander was equipped to detect a sudden deceleration of its downward travel. The trigger was expected to come from upward motion of the central landing gear (now operating as a current generator), as that was pushed back into the lander body by the tripod making contact with the surface.

The landing gear also carried two accelerometers that were installed to sense the touchdown. However, during the long cruise to the comet it was discovered that these devices could not be used because the flywheel of the lander "made so much noise" that the accelerometers would trigger the touchdown sequence in flight. After detailed analysis, it was decided that, because nothing could be done, the accelerometers would have to be disabled for the descent.

As each foot touched the surface, the momentum would drive an ice screw into the ground. The screw would act as a drill, driven by the mechanical energy generated by the touchdown motion. Together, they would help to anchor the lander to the ground and also prevent lateral motion in coming to rest.

Due to its downward inertia, the body of the lander would continue to press further towards the surface after the touchdown. In response, the landing gear motor would move downward and convert the energy of motion into electrical energy. By this process, it was hoped that the touchdown would absorb most of the kinetic energy from the descent.

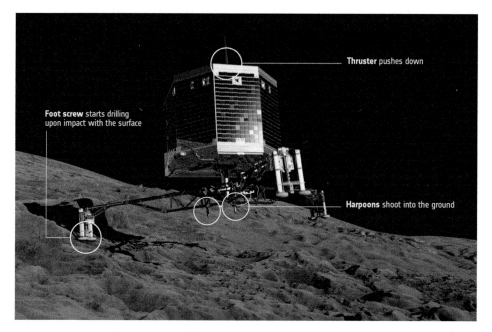

Fig. 9.2: The Rosetta lander was to be securely held to the comet's surface by a number of mechanisms. A cold gas thruster would fire upon touchdown to prevent recoil after impact with the surface. A screw on each foot would drill into the surface upon impact. Finally, a pair of harpoons would fire to anchor the lander to the ground. (ESA)

The touchdown signal from the landing gear was to trigger an active anchoring process. This was essential, because the comet's gravity was so weak that the lander might otherwise bounce back into space. Moreover, any future surface operation, such as drilling for subsurface samples, would lift the lander from the surface.

This process was to start with the firing of the active descent system, a cold gas thruster that was embedded in the top surface of the lander, to push the lander against the surface.

After several seconds, two pointed harpoons were to be ignited, one after the other, and fired into the ground at 70 m/s (250 km/h). After these had halted, the ropes that were connected to the harpoons were to be retracted to tightly bind the lander in position. The top thruster would remain firing throughout this operation to overcome the recoil from the explosive firing of the harpoons.

The anchoring process was expected to be achieved within a minute, opening the way for the historic in-situ investigation of the comet.

9.2   SITE SELECTION

The final decision on where Philae should touch down on the nucleus of the comet was made by the Landing Site Selection Group (LSSG), which included engineers and scientists from the Philae Science Operations and Navigation Center in France (SONC/CNES), the Lander Control Center in Germany (LCC/DLR), scientists of Philae's instrument teams and the ESA Rosetta team.

The time available for selecting the optimal landing site was very short. The rapidly changing conditions provided a brief window of opportunity, and a balance had to be agreed between competing needs. Because the comet's level of activity was driven by the inexorable increase in heating as it approached the Sun, Philae had to attempt its mission before conditions could jeopardize a safe landing.

On the other hand, the landing could not take place too early, since there had to be sufficient sunlight for Philae's solar cells to generate the power to operate the lander for several weeks after touchdown. Furthermore, the surface temperature had to be within the tolerances of the lander's design.

These factors dictated that the landing should take place in mid-November 2014, when the comet was about 3 AU (450 million km) from the Sun.

Clearly, the site selection process could only begin once Rosetta was close enough to be able to image and characterize the nucleus. The first priority was to determine its shape, rotation rate and orientation, gravity field, albedo (spectral reflectivity), surface features, and surface temperature. It was also important to assess the outgassing from the nucleus and quantify the density and velocity distribution of particles in the coma.

In July 2014, images from the OSIRIS cameras began to reveal the shape of 67P. By the time the spacecraft rendezvoused with the comet on 6 August, it was clear that the nucleus was a double-lobed structure with a 'head' and 'body' separated by a narrow 'neck'. From then on, the LSSG received increasingly detailed data about its shape, surface and environment.

Factors which had to be taken into consideration in drawing up a shortlist of potential landing sites included the physical nature of each site and the conditions at touchdown, in particular the slope of the surface, the number and size of boulders or crevasses, the speed at which the lander would strike the ground at the end of its free-fall, and its orientation.

During the early evaluations, the priority was to discover whether a landing at any particular site would be technically feasible, independent of its scientific interest. Factors that had to be taken into account included the duration of Philae's descent, the illumination conditions, and its ability to maintain clear, lengthy radio communications with Rosetta.

The landing site should also provide a balance between day and night conditions, in order to cater for the scientific needs of the instruments, as well as ensuring that the solar cells could recharge the battery to power the instruments, whilst not overheating the lander.

The return of data to Earth would require periodic communication between the lander and the orbiter to be maintained as much as possible during the lander's First Science Sequence (the preliminary intensive phase that was expected to last about three days) leading into the Long Term Science Sequence that might last until March 2015.

Finally, the sites must allow the CONSERT experiment to transmit radio signals through the solid nucleus between the orbiter and lander.

If the physical and environmental conditions could be met, the scientific value of each potential site was evaluated. To assist this assessment, essential observations were made by Rosetta's instruments, in particular OSIRIS, MIRO and VIRTIS, with some contributions from ALICE and ROSINA.

NAVCAM and OSIRIS images were used not only to model the shape and rotation of 67P's nucleus, but also to calculate the spacecraft's trajectory as a means of mapping the irregular gravitational field.

MIRO and VIRTIS measurements were used to predict the surface temperature of the comet for a landing three months in the future. Measurements of the pressure and density of the gas surrounding the nucleus by MIRO, VIRTIS, ROSINA and ALICE facilitated modeling of the environment in which the spacecraft would operate.

Measurements of the comet's surface and environment were also gathered using other orbiter instruments to assist in assessing the scientific merit of the various candidate sites.

A preliminary selection of 10 landing sites, referred to by the letters A to J, was drawn up by a subset of the LSSG at a meeting on 20 August, just two weeks after rendezvous. After the LCC/SONC had made a technical study of each site, the results were presented at the first full meeting of the LSSG on 23-24 August.

At this meeting, participants reviewed the results from the technical analyses and discussed the scientific merits of each candidate site. Then the shortlist was whittled down to five sites (A, B, C, I and J) for further detailed investigation.

Between 25 August and 13 September, the ESA Rosetta Mission Operations Center (RMOC) at ESOC carried out a comprehensive analysis to identify possible trajectories for delivering the lander and to confirm that the proposed landings could be achieved with the required accuracy.

During this period, more detailed measurements were taken from the orbiter, as its trajectory passed within 30 km of the comet's center.

At this stage, a shape model of the nucleus was generated. It had a resolution of 3.6 meters on the surface and 0.5 meter in elevation. The model was used to simulate a view of the horizon as seen from Philae's panoramic cameras, and this was also factored into the selection of the primary site.

At the second meeting of the LSSG, on 13-14 September, technical aspects, such as the flight dynamics of the orbiter and the lander, and recent scientific measurements from orbiter, were reviewed. The possible landing scenarios and their implications on the science program of the lander were discussed.

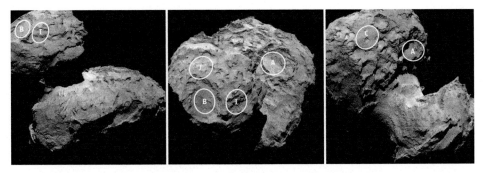

Fig. 9.3:   Five candidate landing sites for Philae were identified on 67P during the Landing Site Selection Group meeting held 23-24 August 2014. The approximate locations of these sites are shown on these OSIRIS Narrow Angle Camera images taken on 16 August from a distance of about 100 km. They were assigned a letter from an original pre-selection of ten possible sites labeled A through J. The lettering scheme did not signify any ranking. Three sites (B, I and J) were located on the smaller of the comet's two lobes, and two sites (A and C) were on the larger lobe. The comet nucleus is about 4 km across. (ESA/Rosetta/ MPS for OSIRIS Team MPS/UPD/LAM/IAA/SSO/INTA/UPM/DASP/IDA)

Finally, after the remaining candidate sites were ranked, the LSSG decided that the primary target would be site J, situated on the smaller lobe. Not only did it present the least hazardous terrain, it also had great scientific potential, with hints of surface activity nearby. The back-up would be site C, on the larger lobe. The entire selection process had taken only six weeks.

On 4 November, ESA announced that site J had now been named Agilkia, after an island in the Nile in southern Egypt. A complex of ancient Egyptian buildings, including the famous Temple of Isis, were relocated to Agilkia from the island of Philae when that was flooded by the building of the Aswan dams. The name was selected by a jury comprising members of the Philae Lander Steering Committee as part of a public competition run 16-22 October by ESA and the German, French and Italian space agencies.

Although the general location had been chosen, it was not possible to definitively pin-point a touchdown spot. Instead, it was envisaged that the landing would occur within an ellipse that measured hundreds of meters across. The precise dimensions of the actual landing area would depend on a number of factors, including the precision with which the location of the orbiter was known at the time that Philae was released, and where the lander was released along the planned delivery trajectory.

Detailed analyses and operational preparations for the landing were carried out between 16 September and 11 October. In that time, Rosetta was close enough to the comet for OSIRIS-NAC to map the distribution of boulders at both landing sites.

Fig. 9.4: The Agilkia region (site J) on Comet 67P was selected as the primary landing site. The circle indicates Philae's first touchdown point on 12 November 2014. The dashed line marks the equator. The large depression was named Hatmehit. This view is a composite of five OSIRIS-NAC images taken on 14-15 September 2014. (ESA/Rosetta/MPS for OSIRIS Team MPS/UPD/LAM/IAA/SSO/INTA/UPM/DASP/IDA)

On 12 October, the LSSG met to make the crucial 'Go/No Go' decision for Philae to land at the primary site. They used the latest high-resolution imagery from OSIRIS to calculate slight adjustments to the target coordinates within the landing ellipse, with the intention of reducing the risk of landing amongst boulders. Recent scientific measurements from other instruments were also used to support the risk analysis.

Two days later, the formal Lander Operations Readiness Review took place in order to give the official go-ahead for the landing. Between then and 3 November, the SONC and the LCC prepared the final operational sequences for both the Separation, Descent and Landing (SDL) and First Scientific Sequence (FSS) operations of Philae.

In parallel, the spacecraft operators at ESOC prepared the command sequences which would deploy Philae onto the comet's surface. Due to the extremely high precision required for this maneuver, the spacecraft's trajectory and commands were updated continually until, finally, with three days to go, the separation and landing sequence was uplinked on 9 November (see Table 9.1).

Table 9.1: Key Dates In Philae Landing Site Selection

19 August 2014	Due date for all data needed for the Landing Site Selection Group (LSSG) meeting on 20 August.
20 August	LSSG meeting to pre-select up to 10 candidate sites: Lander Control Center (LCC) and Science Operations and Navigation Center (SONC) carry out technical analysis.
22-24 August	Meeting of LSSG to select a maximum of five of the candidate landing sites; Coordinates of sites are sent to Rosetta Mission Operations Center (RMOC).
25 August	Delivery of OSIRIS Digital Terrain Models of the candidate sites.
5 September	RMOC experts report on first analysis of landing sites.
13-14 September	Meeting of LSSG to select primary and back-up landing sites; Landing Site coordinates are sent to RMOC.
23 September	Delivery of OSIRIS high-resolution images and boulder distribution file for primary and back-up landing sites.
26 September	RMOC report on final analyzes and operational scenario of primary landing site.
12 October	Meeting of LSSG for final 'Go/No Go' decision about primary landing site.
14 October	Lander Operations Readiness Review.
7 November	Final landing sequences are sent from LCC to RMOC.
9 November	Final landing sequences are uploaded to the lander.
12 November	Date of landing.

9.3 'GO OR NO GO'

The program of pre-landing operations started on 10 November 2014, with the booting up of Philae's computers. Although it had been rehearsed many times, the first attempt to start the two data processing units was unsuccessful. Fortunately, a second attempt was more successful and it seemed that operations stabilized during the following day.

Further unexpected glitches arose during preparations for the landing sequence. First, the procedure for conditioning of the primary battery ceased just one minute into the sequence. Then the valve on the tank that contained gas for the active descent system (ADS) failed to open. This meant the thruster, designed to prevent a rebound at the moment of touchdown, could not be activated.

"The cold gas thruster on top of the lander does not appear to be working, so we will have to rely fully on the harpoons at touchdown," said Stephan Ulamec, Philae lander manager at the German Aerospace Center. "We'll need some luck not to land on a boulder or a steep slope."

Hermann Böhnhardt, lead scientist for the lander, later wrote, "The lander was in a crisis, as by that time lander management – and many of us – felt it to be closer to a 'No Go' for the landing than to a 'Go' decision. Upon a suggestion by me, people from the ops and software teams at DLR Cologne reran the battery conditioning sequence at the software simulator with various software patches, while in-flight the two failing sequences were commanded to be repeated again."

After the ground and in-flight testing, the Philae team concluded that the battery conditioning problem would not prevent the separation, descent and landing from occurring.

Although the ADS problem remained, it was not considered to be a final show stopper. This was accepted by the lander management team, and, late in the night of 11-12 November, the Philae project manager gave the 'Go' for releasing the lander, followed by 'Go' for the entire landing sequence by the Rosetta project manager.

"There were various problems with the preparation activities overnight, but we've decided to go," said Paolo Ferri, ESA's head of mission operations. "Rosetta is lined up for separation."

The first 'Go/No Go' decision occurred at 19:00 UT on 11 November, with confirmation that Rosetta had left its parking orbit and was on the correct trajectory for delivering Philae. The second 'Go' was given at midnight (UT), confirming that the command sequence that would control separation and delivery was complete and ready for uploading to Rosetta. The overall status of the orbiter was also confirmed to be ready for the forthcoming descent operations. At 02:35 UT on 12 November, after verifying that Philae was ready to play its part, the third 'Go' was given.

The final maneuver by Rosetta was conducted at 07:35 UT, taking the spacecraft to a point about 22.5 km from the comet's center. Next came the final decision that confirmed the two spacecraft, the orbit, the ground stations, the ground systems, and the teams were all ready for the landing.

Despite the potential problem concerning the moment of touchdown, separation proceeded as scheduled at 08:35 UT (spacecraft time). Some 28 minutes 20 seconds later, the confirmation signal reached the Rosetta Mission Control Center at ESOC in Darmstadt, Germany.

9.4 THE DESCENT

After separation, some two hours elapsed before the lander established a communication link with Rosetta. Once this was established, the lander relayed a status report of its health, along with the first science data, which included images of the orbiter taken shortly after separation. Meanwhile, Rosetta was well placed to image Philae's free-fall to the nucleus (see Fig. 9.5).

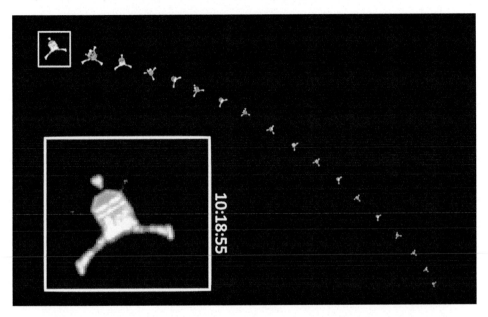

Fig. 9.5: This series of 18 OSIRIS images of Philae taken shortly after its release from the orbiter on 12 November 2014 showed that all three feet, the ROLIS descent camera boom, and the two CONSERT antennas were all deployed successfully. The time is GMT/UT. (ESA/Rosetta/MPS for OSIRIS Team UPD/LAM/IAA/SSO/INTA/UPM/ DASP/IDA)

During the descent to Agilkia, various instruments were already operational. SESAME-DIM registered a millimeter-sized, fluffy particle at an altitude of 2.4 km above the surface; it was the closest particle ever detected at a cometary nucleus.

The images acquired by ROLIS at an unprecedented resolution (down to one centimeter per pixel) did not show the expected dust deposits. Instead, they revealed that the surface of the comet in the proximity of the planned landing site was dominated by coarse debris, pebbles and rocks with dimensions ranging from several centimeters up to 5 meters (see Fig. 9.7).

The ROMAP instrument registered that 67P did not have a measurable magnetic field. This signified that at the time the comet was formed in the solar nebula, the prevailing magnetic fields were not strong enough to align the individual dust particles to permanently magnetize the material that became the comet. This was an important finding for modeling its formation.

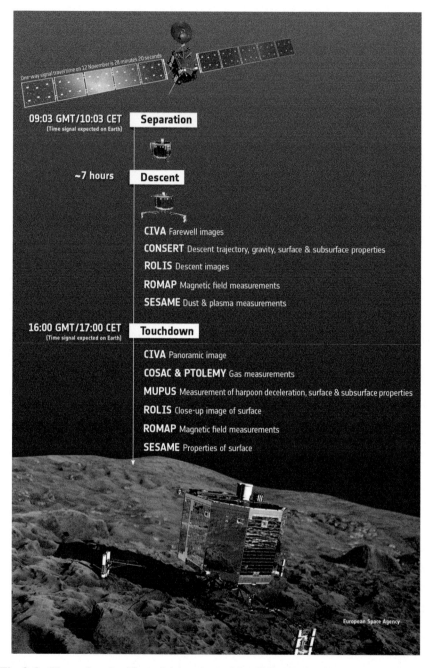

Fig. 9.6: The main scientific activities planned for Philae during its descent and touch-down. Timings are approximate. Confirmation of a successful touchdown was expected in a one-hour window centered on 16:02 GMT (CET was one hour ahead). The first image from the surface was expected about two hours later. (ESA-ATG Medialab)

Fig. 9.7: This image was taken by Philae's ROLIS imaging system from 9 meters above the Agilkia target site on the small lobe of 67P's nucleus. It was acquired at 15:33:58 UT on 12 November 2014. It measures 9.7 meters across and has a scale of 0.95 cm per pixel. There is a granular texture down to the centimeter scale, with fragments of material of diverse shapes and random orientations seen in clusters or alone. In some parts of this region the regolith is believed to be 2 meters thick but (unexpectedly) it appears to be free from very fine-grained dust deposits. Portions of the landing gear can be observed in the upper corners of the image. (ESA/Rosetta/Philae/ROLIS/DLR)

The COSAC instrument detected and analyzed cometary material during the descent and after touchdown, and the Ptolemy instrument also operated to plan (see Science From Philae below).

9.5 A TRIPLE BOUNCE

At exactly 15:34:04 UT, after a 7-hour descent, Philae touched down at the planned Agilkia landing site. The signals confirming this reached Earth, via Rosetta, some 28 minutes later.

Both Rosetta's NAVCAM and the high-resolution OSIRIS camera successfully identified the touchdown point, and Philae's downward-looking ROLIS took high-resolution images of the site from an altitude of only 9 meters.

Later analysis of a recording made by the Cometary Acoustic Surface Sounding Experiment (CASSE), whose sensors were in the feet of all three legs of the lander, indicated there was a brief, but significant 'thud' at the historic touchdown.

"The Philae lander came into contact with a soft layer several centimeters thick," said Klaus Seidensticker, lead scientist for the CASSE instrument from the German Aerospace Center's Institute of Planetary Research in Berlin. "Then, just milliseconds later, the feet encountered a hard, perhaps icy, layer."

At first, jubilation filled the Lander Control Center. Philae mission manager Stephan Ulamec announced, "Philae is talking to us! The first thing he told us was the harpoons have been fired and rewound. We are sitting on the surface."

Unfortunately, this proved not to be the case. Although it was traveling at only one meter per second, Philae's descent did not conclude with a gentle touchdown. Not only was the thruster of the active descent system inoperative, but the ice screws incorporated into the lander's feet were unable to penetrate the surface and the harpoons failed to fire.

"It seems that the problem was either with the four 'bridge wires' taking current to ignite the explosive that triggers the harpoons, or the explosive itself, which may have degraded over time," explained Ulamec later.

The first clues that the lander was not stationary, but performing an aerial ballet far above the surface, came when mission control started to receive data from Philae's instruments as they initiated their pre-programmed science measurements.

The data showed that it appeared to be rotating after the touchdown, which indicated it had lifted off the surface again. Philae had left its footprints (see Fig. 9.7) in the surface dust, but the comet's gravity (only several hundred thousandths that of Earth) was too feeble to hold it in place. Instead of settling comfortably, Philae rebounded, then embarked on an uncontrolled flight across the rugged nucleus.

The bouncing lander was later identified in OSIRIS and NAVCAM images that were taken shortly after it departed Agilkia. Somewhat later, another OSIRIS image appeared to show it above the horizon near the large depression named Hatmehit, on the comet's head.

Later analysis of the first data returned from the lander prompted the astonishing realization that it actually contacted the surface of 67P at least *three* times. This conclusion was based on data from the ROMAP magnetic field analyzer, the

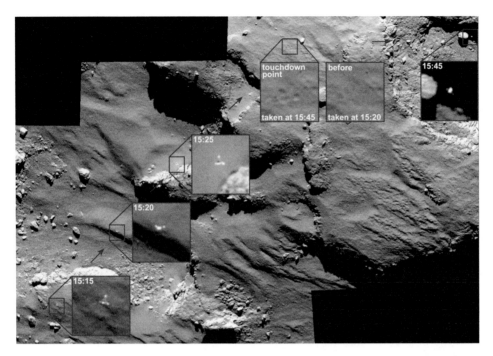

Fig. 9.8: These remarkable images show the journey of Philae as it approached (left) and then rebounded from its first touchdown on Comet 67P on 12 November. The time in UT of each image is shown. Touchdown was at 15:34 UT spacecraft time (the signal arrived on Earth at 16:03 UT). The image marked 'touchdown point' was taken afterward, and is compared with an image taken previously. The images were obtained by the OSIRIS-NAC on Rosetta at an altitude of about 15.5 km above the surface. The image resolution is 28 cm per pixel and the enlarged insets are 17 × 17 meters. The lander was moving east at a speed of about 0.5 m/s. (ESA/Rosetta/MPS for OSIRIS Team MPS/ UPD/LAM/IAA/SSO/INTA/UPM/DASP/IDA)

MUPUS thermal mapper, and the sensors in the landing gear that were pushed inward during the first impact.

Stephan Ulamec, Philae's manager at the DLR Control Center, reported that the tiny vehicle touched the surface at 15:34, 17:25 and 17:32 UT (comet time). After the first touchdown inside the predicted landing ellipse, the lander rebounded. For the next 1 hour 50 minutes it traveled about 1 km across the nucleus at a speed of 38 cm/s.

Once Philae returned to the ground, it made a smaller second hop before reaching its final resting place seven minutes later. Having slowed to only 3 cm/s, the little craft finally touched down in a region known as Abydos.

"It was in open space for hours," said Jean-Pierre Bibring, principal investigator for Philae's ÇIVA camera system. "That has never been done."

Detailed analysis released in 2015 provided a more precise account of Philae's amazing journey, based on ROLIS images and magnetic field measurements from the ROMAP instrument that provided precise timing of the various contact points.

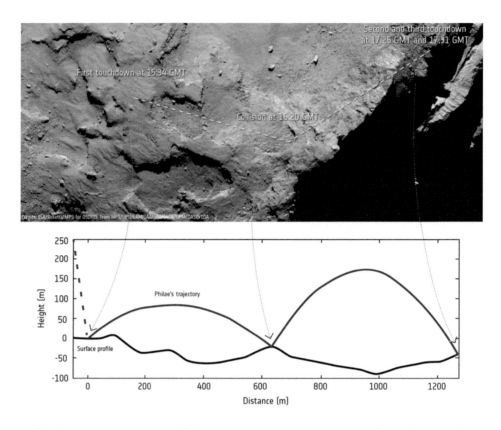

Fig. 9.9: A reconstruction of Philae's amazing journey across the nucleus of Comet 67P. (ESA; Data by Auster et al. 2015; Comet image by ESA/Rosetta/MPS for OSIRIS Team MPS/UPD/LAM/IAA/SSO/INTA/UPM/DASP/IDA)

Philae was already rotating slowly during its initial descent, but it its rate of rotation increased significantly as the momentum of its internal flywheel was transferred to the lander.

On its first giant 'hop' it grazed the rim of the Hatmehit depression at 16:20 UT, around 45 minutes after its unplanned lift-off (see Fig. 9.9). It then started to tumble, while continuing to fly over the nucleus for more than an hour longer. A third touchdown at 17:25 UT was followed by a jump of several meters that lasted about 6 minutes. Its odyssey finally ended at 17:31:17 UT in the rugged Abydos region.

Images provided by the ROLIS and ÇIVA cameras on the lander, together with telemetry and data from its instruments during almost 60 hours of surface operations, enabled scientists to assemble a picture of where it came to rest. It was tilted on its side, lodged against a cliff, and mostly in shadow in an area of rough terrain.

Showing admirable sangfroid, Marc Pircher, head of the Toulouse Space Center, operated by French space agency CNES, remarked: "This is a new site, one we would not have analyzed as a possibility."

Unfortunately, Philae's final resting place could hardly have been less favorable. As mission managers attempted to determine the precise location of the lander, it soon became apparent that the unintended landing site was cold and poorly illuminated, providing only 90 minutes of exposure to sunlight during every 12.4-hour rotation of the comet. (The site in Agilkia had offered almost 7 hours.) This restricted the lander's ability to recharge its secondary battery, with the likely outcome that it could provide only a limited amount of imagery and measurements before it ran out of power.

Fig. 9.10: The first panoramic image sent back by Philae from the surface of Comet 67P. The unprocessed image from the ÇIVA-P cameras shows a 360 degree view around the point of final touchdown. The positions of the three feet can be inferred. Superimposed is a sketch of the lander in its assumed configuration, lying on its side. (ESA/Rosetta/Philae/ÇIVA)

Mission controllers did their best to facilitate future contact with Philae by commanding it to lift itself by about 4 cm and to rotate 35 degrees, in the hope of exposing a greater area of its body-mounted solar panels.

9.6   THE SEARCH FOR PHILAE

Although Philae survived its hazardous hopping across the nucleus in good shape, the lander had disappeared from sight, and the mission team had only limited clues concerning its actual location. With Philae being forced into hibernation after about 60 hours, owing to lack of power, no additional data were available from its suite of instruments.

While Rosetta continued to study the ever-changing comet from a distance, the mission teams set to work to locate the lander using a variety of data, including imaging, magnetic field, and radio wave measurements.

The images and other data returned during Philae's first bounce were of limited help, although they gave the general direction of its travel. Fortunately, it was possible to narrow down the search using data from the CONSERT experiment, in which radio signals were sent between Philae and Rosetta after the final touchdown.

Combining the signal travel time between the two spacecraft with Rosetta's known trajectory and the shape model for the comet, the CONSERT team's simulations indicated that Philae lay inside an ellipse of roughly 16 × 160 meters, just beyond the rim of the Hatmehit depression.

In the days and weeks after landing, the OSIRIS team, led by principal investigator Holger Sierks, continued the search for Philae by carrying out a detailed study of new images as they arrived.

The first dedicated search took place in mid-December 2014, when Rosetta was about 18 km above the surface of the nucleus. At this distance, the OSIRIS-NAC had a resolution of 34 cm per pixel. However, the likelihood of success remained slim, because the body of Philae was only 1 meter across and its narrow legs projected only 1.4 meters from its center.

Furthermore, during that time the spacecraft was orbiting above the terminator, the day-night boundary. The solar cells on Philae could have been well illuminated, and yet remain hidden from Rosetta's cameras in the rugged landscape.

By taking into account the size, reflectivity, and orientation of Philae, along with the point-spread function (intrinsic resolution) of their instrument's optics, the OSIRIS team expected the search for Philae to be a significant challenge. Unsurprisingly, the inspection of the 'head' of 67P revealed many initial candidates in the form of bright spots several pixels wide.

Fig. 9.11: A July 2015 artist's impression of Philae's final resting place, based on images provided by its ÇIVA camera system. Trapped in a hollow, it is lying on its side with only two of its three feet in contact with the ground. The rough terrain limits both its supply of solar energy and its communications with the Rosetta orbiter. (CNES/A. Torres; IAS/J-P. Bibring. Illustration D. Ducros)

The CONSERT data ruled out the majority of the candidates found in OSIRIS images, but it was still not possible to rule out the various bright spots in the vicinity of the ellipse. Such an analysis was complicated by the fact that many of the bright spots were transient, depending upon changes in illumination conditions.

Nevertheless, scientists at both the Laboratoire d'Astrophysique de Marseille (LAM) and the Institut de Recherche en Astrophysique et Planétologie (IRAP) in France began to work with OSIRIS team member Philippe Lamy in a search for specific images that might solve the puzzle.

In particular, to reduce the chances of being fooled by transient glints from surface features, they examined images taken prior to and after the landing under comparable illumination. If something new appeared after landing, then it might be the lander. Scanning a broad area that encompassed the expected ellipse, they identified a promising candidate on two images taken on 12-13 December that was not on an image taken on 22 October.

"Although the pre- and post-landing images were taken at different spatial resolutions, local topographic details match well, except for one bright spot present on post-landing images, which we suggest is a good candidate for the lander," announced Philippe Lamy. "This bright spot is visible on two different images taken in December 2014 – clearly indicating it is a real feature on the surface of the comet, not a detector artefact or moving foreground dust speck."

On the other hand, the Philae candidate seemed to be just outside the ellipse indicated by the CONSERT data.

Table 9.2:   Factors Influencing Rosetta-Philae Communications

- Comet 67P had a rotation period of 12.4 hours, so Philae's location was not always visible to Rosetta. There were usually two opportunities for contact each Earth day, but their duration depended on the orientation of the transmitting antenna on Philae and the location of Rosetta along its trajectory.
- As the comet rotated, Philae was not always in sunlight, so its solar panels were not always generating enough power to receive and transmit signals. During June 2015, the contact windows varied from several tens of minutes up to three hours.
- The orbiter needed to be flying over the lander's position when the lander was awake and generating sufficient power to have its receivers and transmitter operating.
- The orientation of the lander on the surface of the comet determined how its antenna's signal pattern was projected into space, and the rugged terrain surrounding the lander could distort that antenna pattern.
- In order to establish a link, the lander's antenna pattern had to overlap with that of the orbiter. Given the constraints set by the lander antenna pattern, certain trajectories of the orbiter around the comet were more effective at receiving a strong, uninterrupted lander signal than others.
- The increasingly dusty comet environment caused Rosetta's star trackers to register false positives, placing the spacecraft in danger of losing its orientation in space. To avoid this, it was being flown at the safer altitude of about 200 km in an orbit which followed the comet's terminator, the day-night boundary.
- The strength of the signal received by the orbiter diminished with the square of the distance between it and the lander, so the chances of establishing a stable link were reduced if Rosetta was too far (over 200 km) from the comet.
- Errors or partial failures in the lander's transmission and receiver units could affect the chances of making a stable link. Philae was programmed to switch periodically between its two transmission units.
- Data were stored in two mass memories on Philae. In order to download the data in the most efficient way, a stable link-up duration of about 50 minutes was preferable. A data dump from each mass memory to Rosetta could take 20 minutes. Additional time was required to confirm that a stable link had been acquired, and for uploading new commands.

As the comet continued into the inner Solar System, its activity increased and Rosetta was no longer able to operate at low altitude, impairing the search for Philae.

On the other hand, the mission team estimated that from March 2015 onwards, the improving temperatures and illumination might enable Philae to reach the minimum of –45°C required for its systems to reboot. Accordingly, at times when the orbital geometry was predicted to be optimal, Rosetta switched on its receiver and listened for signals from the lander.

And then, to the delight of everyone, Philae finally called home!

9.7 PHILAE CALLS HOME

On the evening of 13 June 2015, after Philae had been hibernating for 211 days, signals from the lander were detected by Rosetta (see Chapter 8). A weak but solid radio link between the two craft was established for 85 seconds, enabling more than 300 'packets' – 663 kbits – of lander housekeeping telemetry to be received. However, this information had been stored by Philae days or weeks earlier, and therefore did not necessarily reflect its current status.

"We are still examining the housekeeping information at the Lander Control Center in the DLR German Aerospace Center's establishment in Cologne, but we can already tell that all lander subsystems are working nominally, with no apparent degradation after more than half a year of hiding out on the comet's frozen surface," said Stephan Ulamec, the lander project manager.

Over the ensuing days, Philae transmitted short bursts of housekeeping telemetry, including data from its thermal, power, and computer subsystems. Subsequent analysis indicated that the lander had, in fact, awakened on 26 April 2015, although it had not been able to send any signals until 13 June.

"With work done by the flight dynamics and operations team at ESOC, and based on intense planning being conducted with the mission partners today, a new orbit will be devised that ensures optimum lander communications, beginning with the next command upload tonight," said Paolo Ferri, ESA's head of mission operations.

This intensive analysis resulted in Rosetta being moved into a modified orbit for a few days, starting on 16 June, so that more, and longer, contacts with Philae could be established. This orbit included an already-planned lowering from 200 km to 180 km and 'nadir pointing' that would continuously point Rosetta's communications system at the comet.

This enabled the lander to make a further seven intermittent contacts with Rosetta on 14, 19, 20, 21, 23 and 24 June and 9 July. Housekeeping data were transferred from Philae to Rosetta on all occasions except 23 June, but the links were mostly too short and unstable to enable any scientific measurements to be commanded.

The link-up on 19 June was stable, but split into two short periods of two minutes each. On 23 June, there was a 20 second contact, but no stable link was established. In contrast, on 24 June the contact continued for 20 minutes but its quality was so patchy that only 80 packets of telemetry were received.

After an initial test command to switch on the power to the CONSERT experiment on 5 July, the lander did not respond for several days. Philae eventually communicated with the orbiter again between 17:45 and 18:07 UT on 9 July, during which it was able to transmit data from CONSERT for 12 minutes. After that the connection failed repeatedly.

The intermittent data informed the lander team that Philae's internal temperature had risen to –5°C, and warming continued as the comet moved closer to the Sun. The solar panels appeared to be receiving power for over 135 minutes during each illumination cycle, sufficient to generate electricity.

"Power levels increase during the local 'comet day' (the part of the about-12-hour rotation of when Philae is in sunlight) from 13 Watts at sunrise to above 24 Watts," said Patrick Martin, ESA's Rosetta mission manager. "It needs at least 19 Watts to switch on the transmitter."

It was quite remarkable that the lander had survived multiple contacts on 12 November and a long period of unfavorable environmental conditions that greatly exceeded the specifications of its various electronic components. The irony of this situation did not go unnoticed by the mission controllers. If Philae had landed at the Agilkia site, its mission would probably have ended in March 2015, when the temperatures rose as the comet approached the Sun. Instead, the less exposed location at Abydos had enabled it to survive much longer.

Despite the improving thermal conditions, with temperatures inside Philae soon reaching and exceeding 0°C, no further contacts were made as the comet approached perihelion in August. By then, the increasing outflows of gas and dust were impairing Rosetta's star tracker to such an extent that the spacecraft had to retreat beyond the distance at which it could detect any signals from the lander.

After perihelion, with comet activity subsiding, Rosetta was eventually able to approach the nucleus again, but despite repeated passes over the Abydos region no signals were detected, and attempts to send commands to Philae 'in the blind' failed to produce any results.

Analysis by mission engineers indicated that failures of one of Philae's two receivers and one of its two transmitters were the most likely explanation for the irregular contacts in June-July 2015 leading to a prolonged silence. Furthermore, the other transmitter was also experiencing problems.

"Sometimes it did not switch on as expected, or it switched off too early, meaning we likely missed possible contacts." said Koen Geurts, Philae's technical manager at the DLR Lander Control Center in Cologne, Germany.

Another possibility was that Philae's solar panels might be coated with dust ejected from the highly active comet around the time of perihelion, denying the lander power.

Furthermore, the attitude, and even location, of Philae might have changed since November 2014 due to cometary activity. If the direction in which its antenna was pointing had altered, this would change the expected communications window with Rosetta.

Despite the silence from Philae, the team continued to search for it in images taken as Rosetta flew closer to the comet, prior to making its own descent in September 2016.

As Matt Taylor, ESA's Rosetta project scientist said, "Determining Philae's location would also allow us to better understand the context of the incredible in-situ measurements already collected, and thus allow us to extract even more valuable science from the data. Philae is the cherry on the cake of the Rosetta mission, and we are eager to see just where the cherry really is!"

"We would be very surprised to hear from Philae again after so long," said Patrick Martin, ESA's Rosetta mission manager, "but we will keep Rosetta's listening channel on until it is no longer possible due to power constraints as we move ever farther from the Sun towards the end of the mission."

"We think we have until the end of January (2016) before the lander's internal temperature gets too cold to operate," added Koen Geurts. "It cannot work below $-51\,°C$."

However, by February 2016, the comet, with Philae, had crossed the orbit of Mars and was some 350 million km from the Sun. According to model predictions, the temperatures on the surface of the nucleus would be far below those at which Philae would be able to operate, so the mission team ceased sending commands to the lander and reluctantly accepted that it must have entered eternal hibernation.

"The chances for Philae to contact our team at our Lander Control Center are, unfortunately, getting close to zero," admitted Stephan Ulamec, "It would be very surprising if we were to receive a signal again."

Despite the general pessimism, the saga of Philae provided one more surprise.

9.8   FINDING PHILAE

On 2 September 2016, with the Rosetta mission drawing to a close, the location of Philae was finally discovered.

As the orbiter flew within 2.7 km of the surface, its OSIRIS-NAC took a remarkably detailed image that revealed Philae wedged into a poorly illuminated crevice on the nucleus. It clearly showed the body of the lander and two of its three legs (see Fig. 9.12).

Fig. 9.12: Philae's final resting place was located in OSIRIS Narrow Angle Camera images obtained on 2 September 2016 from a distance of 2.7 km. The image scale is about 5 cm per pixel. The lander's 1 meter wide body and two of its three legs can be seen. The images also provide proof of Philae's orientation. A Rosetta navigation camera image taken on 16 April 2015 is shown at top right for context, with the approximate location of Philae on the small lobe of the nucleus indicated. (Main image and lander inset: ESA/Rosetta/MPS for OSIRIS Team MPS/UPD/LAM/IAA/SSO/INTA/UPM/ DASP/IDA; context: ESA/Rosetta/NAVCAM – CC BY-SA IGO 3.0)

The ESA blog noted:

You may have noticed that instead of the lander being nicely framed in the center of the image, *which was our plan*, it is skewed all the way over on the right hand side. For this 'off pointing' we can primarily blame the comet. Due to the close distances of our fly-overs, the trajectory of the spacecraft is affected by the higher gravity drag and cometary gas existing at low altitudes along with normal maneuver uncertainties. To overcome this, and in order to point the instruments to the Philae location, the MOC Flight Dynamics team needed to predict very accurately the spacecraft position with respect to the comet for the whole commanded period. In the case of the 2 September Philae image, this trajectory prediction was performed more than 36 hours before the image was taken, achieving an accuracy of about 50 meters at the

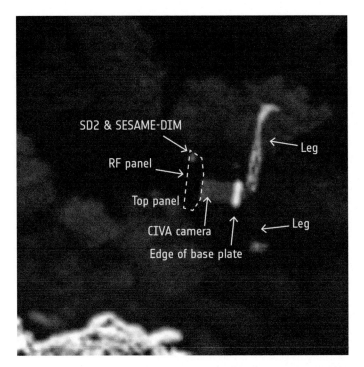

Fig. 9.13: The lander's body and two of its three legs can be seen in this OSIRIS image taken on 2 September 2016. Several of its instruments are also labeled, including one of the ÇIVA panoramic imaging cameras, the SD2 drill, and SESAME-DIM (Surface Electric Sounding and Acoustic Monitoring Experiment Dust Impact Monitor). (ESA/ Rosetta/MPS for OSIRIS Team MPS/UPD/LAM/IAA/SSO/INTA/UPM/DASP/IDA)

time of the image. So while Philae ended up not being in the center of the image, the fact that Philae was in the camera field of view at all, and not tens of degrees away, reflects the excellent performance of the Flight Dynamics navigation team.

An enlarged version of the image that revealed Philae's orientation explained why it had been so difficult to establish communication following its landing in November 2014.

The discovery entirely justified the search strategy which the mission team had conducted for almost two years. Radio ranging using the CONSERT experiment had successfully tied its position down to an area spanning several tens of meters, but a number of candidate objects spotted in lower resolution images taken from large distances had to await Rosetta's surface-skimming orbits before they could be investigated in detail.

The new image verified that the site favored by Philippe Lamy and his collaborators was the correct location (see The Search For Philae above).

"After months of work, with the focus and the evidence pointing more and more to this lander candidate, I'm very excited and thrilled that we finally have this all-important picture of Philae sitting in Abydos," said Laurence O'Rourke, who had been coordinating the search efforts over the last months at ESA along with the OSIRIS and SONC/CNES teams.

Matt Taylor, ESA's Rosetta project scientist said, "This wonderful news means that we now have the missing 'ground-truth' information needed to put Philae's three days of science into proper context – now that we know where that ground actually is!"

The discovery came less than a month before Rosetta was scheduled to drop onto the comet's surface and make its own historic touchdown (see Chapter 10).

9.9   SCIENCE FROM PHILAE

After the lander finally came to rest, the work for the team in the DLR control facility started in earnest.

In the 64 hours following its separation from Rosetta, Philae took images of the comet from above, and on, its surface, detecting organic compounds and profiling the local environment. Finally, with its power diminishing, it turned itself to optimize its orientation with respect to predicted future illumination.

Some 80% of the planned First Science Sequence was completed between Philae's release by Rosetta and its entry into hibernation after spending 57 hours on the nucleus, with the bonus of being able to collect data at several locations rather than just one (see Fig. 9.14).

After analyzing the depth profile of the lander's footprints at Agilkia, scientists concluded the feet first came in contact with a soft granular surface about 0.25 meters thick. Below this was a harder layer. This layering created a compressive strength of about 1 kilopascal. In contrast, the compression strength of the final, much harder, landing site exceeded 2 megapascals.

Close up images taken by ROLIS on the descent to Agilkia and the ÇIVA images obtained at Abydos enabled a visual comparison of the topography at these two locations to be made (see Fig. 9.15).

ROLIS images taken shortly before the first touchdown revealed a surface comprising meter-sized blocks of various shapes, coarse regolith with grain sizes of 10-50 cm, and granules less than 10 cm in size. Surprisingly, although the regolith at Agilkia was believed to be 2 meters deep in places, there seemed to be an absence of fine-grained dust deposits.

The largest boulder in the ROLIS field of view measured about 5 meters high and possessed a peculiar bumpy structure. The fracture lines running through it implied erosional forces were fragmenting the comet's boulders into smaller pieces.

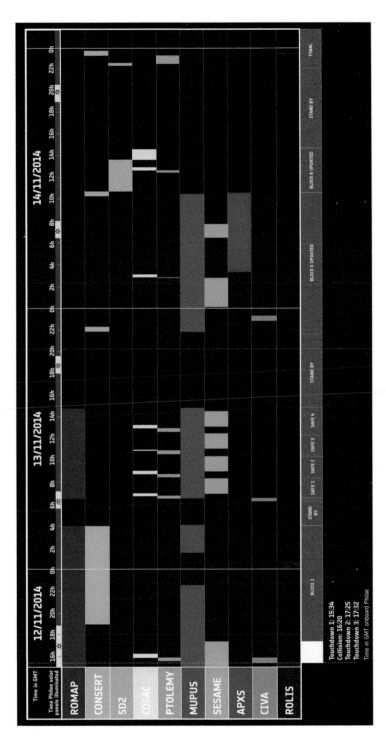

Fig. 9.14: A timeline of Philae's science operations conducted on the surface of Comet 67P between 12 and 15 November 2014. Due to its unexpected flight across the comet's surface, the planned First Science Sequence had to be adapted to suit the changed situation. It shows the approximate times (to the nearest 15 minutes) of activating each of the 10 instruments; it does not indicate the success of data acquired. (ESA)

Fig. 9.15: This 3D stereoscopic image of the Agilkia region was generated using two images acquired by ROLIS when Philae was a little less 3 km above the surface of Comet 67P. They were taken two minutes apart, about one hour prior to the initial touchdown. One of the feet can be seen at the top right. The resolution is about 3 meters per pixel. To appreciate the 3D effect, the image must be viewed with red-blue glasses. (ESA/Rosetta/Philae/ROLIS/DLR)

The boulder also had a tapered 'tail' of debris behind it, like others viewed by Rosetta from orbit. This was evidence of particles being lifted up from one part of the eroding comet and deposited elsewhere. It was reminiscent of 'wind tails' on Earth caused by eolian erosion and deposition (see Chapter 10).

The science team concluded that 67P's landscape is shaped by erosion, in part as the result of 'splashing' or saltation, which is the ejection of one or more soil particles by the impact of an incoming projectile.

Over a kilometer away at Abydos, the images revealed a much rockier, rugged terrain. The seven microcameras of the ÇIVA instrument showed details in the surrounding terrain down to the millimeter scale. They also indicated that the lander was angled against a cliff face, so that one camera was pointing at open sky (see Fig. 9.11). The images revealed fractures of all sizes in the cliff walls. Stereoscopic imagery also revealed the topography of the surroundings to a distance of about 7 meters.

The material surrounding Philae was surprisingly dominated by meter-sized blocks of dark agglomerates, perhaps comprising grains rich in organics. Brighter spots probably signified differences in mineral composition, possibly indicative of ice-rich materials. However, there was very little evidence of ice overall.

On the first touchdown at Agilkia, the 'gas sniffers' Ptolemy and COSAC analyzed samples that entered the lander. The chemical composition of the gas and dust gave information on the raw materials present in the early Solar System.

COSAC collected molecules from 10 km above the surface, then after the initial touchdown, and finally at the final resting site. These included material that entered tubes at the base of the lander after it was loosened during the first touchdown. The samples were dominated by the volatile ingredients of ice-poor grains of dust.

The results revealed 16 organic compounds comprising numerous carbon compounds and nitrogen-rich compounds, including four which had never before been detected in comets: methyl isocyanate, acetone, propionaldehyde and acetamide.

Meanwhile, Ptolemy analyzed organic compounds by sampling ambient gas entering tubes at the top of the lander. It found the main components of the coma gases – water vapor, carbon monoxide and carbon dioxide – and smaller quantities of carbon-bearing organic compounds, including formaldehyde.

Their data also indicated the presence of a radiation-induced polymer on the surface, and an absence of aromatic compounds such as benzene.

Ptolemy also showed the presence of both water and carbon dioxide, but very little carbon monoxide. This conflicted with data from the ROSINA instrument on Rosetta. ROSINA data acquired shortly prior to landing found that the concentration of carbon monoxide (although variable) was up to four times the concentration of carbon dioxide, whilst Ptolemy measured the concentration of carbon dioxide to be roughly ten times greater than the concentration of carbon monoxide.

These findings could suggest that either the composition of the gas in the coma changes by various chemical reactions as it moves away from the nucleus, or that the gas vaporized from the surface varies by location. This would mean that 67P is heterogeneous – made of different compounds in different places.

One hypothesis is that the comet grew from diverse building blocks during its formation. On the other hand, it could simply be the result of uneven heating during its travel into the inner Solar System.

Of particular interest to scientists were the compounds that play a key role in the prebiotic synthesis of amino acids, sugars and nucleobases (i.e. the ingredients for life). For example, formaldehyde is involved in the formation of ribose, which ultimately features in molecules such as DNA. The existence of such complex molecules implied that chemical processes at work as the comet was assembling, 4.5 billion years ago, could have played a key role in the formation of material that would later participate in the creation of the first simple life forms.

The MUPUS suite of instruments provided insights into the physical properties of Abydos. Because of the lander's haphazard final touchdown, the sensors were unable to penetrate the hard surface to obtain subsurface temperature readings. However, the thermal sensor on Philae's balcony showed that the temperature variation arising from the comet's 12.4 hour rotation was −180°C to −145°C. The thermal inertia calculated from the rapid rise and fall in temperature implied there was a thin layer of dust on top of a highly compact, microporous, dust-ice layer with a porosity of 30-65% (see Fig. 9.16).

The penetrating 'hammer' found that the surface and subsurface materials were substantially harder than at Agilkia, as inferred from a mechanical analysis of the first landing. The results indicated that at Abydos a layer of dust less than 3 cm thick sits on a much harder compacted mixture of dust and ice. If this harder layer is present at the Agilkia touchdown point, it must be at a greater depth.

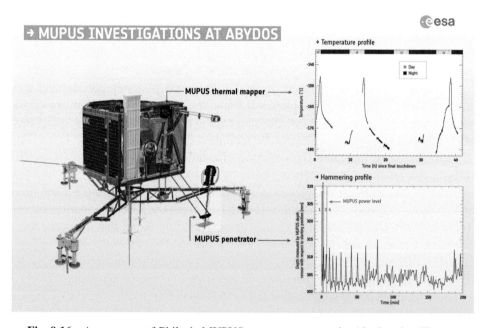

Fig. 9.16: A summary of Philae's MUPUS measurements at the Abydos site. The top graph shows the average surface temperature profile. Gaps indicate times when the instrument was not recording data. The profile shows lows of about −180°C and highs of about −145°C that correspond to the comet's 12.4 hour day. The bottom graph shows the hammering profile of the penetrator. The displacement is expressed as the position of the depth sensor with respect to its starting position above the surface. It displayed an initial displacement of about 27 mm, perhaps when passing through a thin layer of dust, followed by oscillations of 10-15 mm and smaller displacements. The data suggest the instrument was making indentations of several millimeters in a hard layer, with recoils of up to 10 mm. The red lines indicate the MUPUS power levels, which correspond to 0.49, 1.59, 2.17 and 4.23 joules, respectively. (ESA/ATG medialab; data from Spohn et al. 2015)

Beneath the surface, unique information concerning the comet's global internal structure was provided by CONSERT, which transmitted radio waves through the nucleus to Rosetta on the opposite side between 12 and 13 November 2014 (see Fig. 9.17).

The absence of a scattering pattern in the signals Rosetta received implied that the interior of the small lobe is fairly homogeneous on the scale of tens of meters throughout. Analyzing the electromagnetic measurements for the permittivity (resistance of the electrical field), implied a dust-to-ice ratio of 0.4-2.6 by volume and a very high porosity of 75-85%, consistent with a very loosely compacted mixture of dust and ice.

As mentioned earlier, CONSERT data also helped to triangulate the location of Philae's final resting place, providing a best-fit solution for an area on the small lobe that measured 21 × 34 meters.

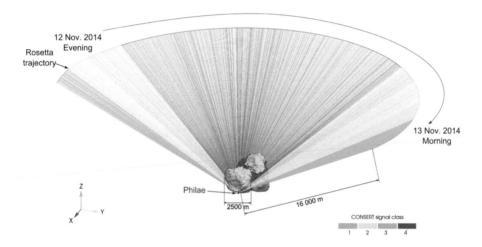

Fig. 9.17: A diagram showing the propagation of signals between Rosetta and Philae through the nucleus of comet 67P studied on 12-13 November 2014. Green is the best signal quality, decreasing to red for no signal. The signals were sent and received by the CONSERT system on both the orbiter and the lander. The time required for the signal to travel between the two, and the amplitude of the received signal yielded insights into the structure of the nucleus. In particular, the travel time depended on permittivity, which in turn was linked to the porosity, composition, temperature, and internal structure of the nucleus. The results indicated a high porosity. (ESA/Rosetta/Philae/CONSERT)

As Nicolas Altobelli, ESA's acting Rosetta project scientist explained, "These ground-truth observations at a couple of locations anchor the extensive remote measurements performed by Rosetta covering the whole comet over the last year."

Although Philae's ROMAP instrument measured a magnetic field during the descent and the subsequent bounces and touchdown, the data showed that the field strength did not depend on either its altitude or location above the surface. This signified that the field was not created by the nucleus itself.

As Hans-Ulrich Auster, co-principal investigator for ROMAP, explained, "If the surface was magnetized, we would have expected to see a clear increase in the magnetic field readings as we got closer to the surface. But this was not the case at any of the locations we visited, so we conclude that Comet 67P/Churyumov-Gerasimenko is a remarkably non-magnetic object."

Jean-Pierre Bibring, a leading lander scientist and principal investigator for ÇIVA reflected, "Taken together, the first pioneering measurements performed on the surface of a comet are profoundly changing our view of these worlds and continuing to shape our impression of the history of the Solar System."

REFERENCES

Rosetta Lander (Philae) Investigations, J.-P. Bibring et al, January 2007: https://www.research-gate.net/publication/41625911_Rosetta_Lander_Philae_Investigations

Press Kit – Landing on a Comet, 5 *November 2014*: https://sci.esa.int/documents/34878/36335/1567260126246-Rosetta_PressKit_CometLanding_November2014_v4.pdf

Philae landing press kit: http://sci.esa.int/rosetta/54816-press-kit-12-november-2014-landing-on-a-comet/#

Landing on a comet video, DLR: *https://upload.wikimedia.org/wikipedia/commons/transcoded/6/68/Landing_on_a_Comet_-_The_Rosetta_Mission.webm/Landing_on_a_Comet_-_The_Rosetta_Mission.webm.480p.vp9.webm*

Rosetta, Comet en Vue, CNES Magazine, January 2014a: http://www.cnes-multimedia.fr/rosetta/TAP_rosetta.pdf

Rosetta and Philae Landing Timeline, 7 November 2014b: http://blogs.esa.int/rosetta/2014/11/07/rosetta-and-philae-landing-timeline/

Rosetta and Philae separation confirmed, 12 November 2014: http://www.esa.int/Our_Activities/Space_Science/Rosetta/Rosetta_and_Philae_separation_confirmed

104 Hours With Philae, 19 November 2014: https://www.mps.mpg.de/3086295/Philae-Blog

Reconstructing Philae's flight across the comet (video): https://rosetta.jpl.nasa.gov/news/reconstructing-philae%E2%80%99s-flight-across-comet

Rosetta – Living With A Comet, ESA BR-321, 1 August 2015: https://sci.esa.int/web/rosetta/-/56351-esa-br-321-rosetta-living-with-a-comet

Selecting a landing site for Rosetta's lander, Philae: https://sci.esa.int/web/rosetta/-/54468-selecting-a-landing-site-for-rosettas-lander-philae

Science on the Surface of a Comet, 30 July 2015: http://blogs.esa.int/rosetta/2015/07/30/science-on-the-surface-of-a-comet/

Philae's First Days on the Comet, Science special issue (6 papers), 31 July 2015: https://science.sciencemag.org/content/349/6247

The Sound of Touchdown, 20 November 2014: http://blogs.esa.int/rosetta/2014/11/20/the-sound-of-touchdown/

Rosetta and Philae: Searching for a Good Signal, 26 June 2015: https://blogs.esa.int/rosetta/2015/06/26/rosetta-and-philae-searching-for-a-good-signal/

Understanding Philae's Wake-Up, 11 September 2015: http://blogs.esa.int/rosetta/2015/09/11/understanding-philaes-wake-up-behind-the-scenes-with-the-philae-team/

The Quest to Find Philae, 11 June 2015: http://blogs.esa.int/rosetta/2015/06/11/the-quest-to-find-philae-2/

Philae Found!, 5 September 2016: https://sci.esa.int/web/rosetta/-/58220-philae-found

The Story Behind Finding Philae, 28 September 2016: http://blogs.esa.int/rosetta/2016/09/28/the-story-behind-finding-philae/

The search campaign to identify and image the Philae Lander on the surface of comet 67P/Churyumov-Gerasimenko, L. O'Rourke et al,, Acta Astronautica, vol. 157, April 2019, p. 199-214: https://doi.org/10.1016/j.actaastro.2018.12.035

Rosetta lander – Landing and operations on comet 67P/Churyumov-Gerasimenko, Stephan Ulamec et al, Acta Astronautica, 125 (2016) p. 80-91: https://doi.org/10.1016/j.actaastro.2015.11.029

The Philae lander mission and science overview, Hermann Boehnhardt et al, Phil. Transactions of Royal Society A, published: 29 May 2017: https://royalsocietypublishing.org/doi/10.1098/rsta.2016.0248

10

A Scientific Bonanza

"Thanks to a huge international, decades-long endeavor, we have achieved our mission to take a world-class science laboratory to a comet to study its evolution over time, something that no other comet-chasing mission has attempted. The mission has spanned entire careers, and the data returned will keep generations of scientists busy for decades to come." Alvaro Giménez, ESA Director of Science.

"As well as being a scientific and technical triumph, the amazing journey of Rosetta and its lander Philae also captured the world's imagination, engaging new audiences far beyond the science community. It has been exciting to have everyone along for the ride." Mark McCaughrean, ESA's senior science adviser.

"Just as the Rosetta Stone after which this mission was named was pivotal in understanding ancient language and history, the vast treasure trove of Rosetta spacecraft data is changing our view on how comets and the Solar System formed. Inevitably, we now have new mysteries to solve. The comet hasn't given up all of its secrets yet, and there are sure to be many surprises hidden in this incredible archive. So don't go anywhere yet – we're only just beginning." Matt Taylor, ESA's Rosetta project scientist.

Although the operational side of the Rosetta mission was completed on 30 September 2016 with the historic touchdown on the surface of Comet 67P, the scientific analysis of the data from the orbiter and the Philae lander will continue for many years to come.

Hundreds of academic papers have been written about the remarkable insights gained from the mission's instruments, and many more will follow, sufficient to fill many volumes the size of this book.

© Springer Nature Switzerland AG 2020
P. Bond, *Rosetta: The Remarkable Story of Europe's Comet Explorer*,
Springer Praxis Books, https://doi.org/10.1007/978-3-030-60720-3_10

In this chapter, I have simply endeavored to summarize the principal discoveries and the most noteworthy revelations which have transformed our knowledge of planetary 'building blocks' and the birth pangs of the Solar System itself. Anyone seeking more in-depth analyses should refer to the papers listed at the end.

10.1 THE "RUBBER DUCK"

Although many thousands of comets are known, and more are discovered every year, only half a dozen of these enigmatic, icy wanderers have been examined at close quarters. Most of them are only a couple of kilometers across, and all have proved to be elongated – some more so than others. Popular descriptions of their shapes range from a peanut, to a bowling pin, to a potato.

What shape would Comet 67P assume? Data obtained by the Hubble Space Telescope back in 2003 indicated that it resembled a rugby ball with four equally spaced protruding bulges.

Imagine everyone's surprise when the nucleus hove into view in the summer of 2014 and the increasingly detailed Rosetta images revealed an object that comprised two lobes of different sizes that were in contact. Scientists and media soon likened its shape to that of a rubber duck, with the small head separated from the larger body by a narrow neck.

Small comets and asteroids with two distinct lobes appear to be commonplace, although 67P's shape was more extreme than anything previously observed. How did this come about?

Initially, scientists had two theories – they suspected either a collision of two bodies or more intense erosion in the mid-section that evolved into the neck. But analysis of high-resolution images from the Rosetta's OSIRIS instrument favored a scenario in which the two lobes were formed independently and later joined in a low-speed collision.

More than 100 terraced structures were discovered, as well as parallel layers that were clearly visible on exposed cliffs, walls and cavities. Using a 3D model of the nucleus, scientists were able to deduce the alignment and depth of the individual layers. It became clear that, although these sheet-like structures were present on both lobes, they had different characteristics. This meant that the layering developed on separate objects, not the same body. Previous spacecraft had already revealed similar onion-like structures on Comet Tempel 1 and Wild 2.

"In order to explain the measured low density and well-preserved layer structures of both the comet's lobes, the collision must have been gentle and occurred at low speed," said Ekkehard Kührt, who headed the scientific involvement of the German Aerospace Center (DLR) in the mission.

Fig. 10.1: The "rubber duck" shape of Comet 67P. (ESA/Rosetta/MPS for OSIRIS Team MPS/UPD/LAM/IAA/SSO/INTA/UPM/DASP/IDA/Jacint Roger Pérez/Emily Lakdawalla)

10.2   COMET OVERVIEW

Compared with other comets, the nucleus of 67P was very modest in size (see Appendix 1), but much more irregular in shape, having a marked bi-lobed appearance. The nucleus rotated around a north-south axis which was centered where the large lobe came into contact with the narrow neck region.

The initial rotational period determined from Rosetta's data was 12.4041±0.0001 hours, and this persisted until October 2014. It then increased, reaching a maximum of 12.4304 hours on 19 May 2015, before reducing to

12.305 hours just before perihelion on 10 August 2015. The variations were related to the changes in surface activity, as material was blasted into space by numerous jets and larger eruptions of gas/dust, initially in the northern hemisphere and then in the south.

The detailed dimensions and shape were determined by analyzing the thousands of OSIRIS and NAVCAM images. Its overall dimensions were 4.34 × 2.60 × 2.12 km, with the large lobe measuring 4.10 × 3.52 × 1.638 km and the small lobe 2.50 × 2.14 × 1.64 km.

The Radio Science Investigation (RSI) measured Doppler shifts in Rosetta's radio signal. The small increases and decreases in the spacecraft's velocity caused by the varying gravitational influence in close proximity to the comet enabled the mass of the nucleus to be calculated as roughly 10 billion tonnes.

Knowing the comet's dimensions and mass provided an estimated density of 0.53 g/cc, about half that of water. Since the nucleus is known to contain denser, rocky dust material in addition to ices, its porosity must exceed 70%. Most of the interior probably comprises weakly bonded clumps of ice and dust which are separated by small voids. An initial analysis implied four times more dust than ice by mass, and twice as much dust as ice by volume. The density was more or less consistent throughout the nucleus.

As already mentioned, the comet's gravity varied slightly across its surface, but, in general, anything moving faster than about 1 m/s (walking pace) would escape and drift off into space. This was also the orbital speed of Rosetta when it arrived at 67P in August 2014.

The popular image of a comet is a large chunk of primordial ice resembling a dirty snowball (see Chapter 1), but studies by visiting spacecraft have shown that most nuclei are very dark, rather like charcoal. The nucleus of 67P was no exception, reflecting about 4-6% of sunlight. Rosetta data indicated that the comet's icy interior was blanketed by a layer of dark dust, comprising materials such as silicates and carbon-rich compounds.

The volatiles in the near-subsurface that were exposed to any temperature increases were able to sublimate, rise to the surface and erupt into space, carrying away motes of dust.[1]

The developing coma, which grew in size as 67P approached perihelion, consisted primarily of water, carbon dioxide and carbon monoxide. However, the mass ratio between these three constituents varied significantly over the course of one 'comet day'. At times, the ROSINA mass spectrometer recorded a substantially higher proportion of water molecules, then, as the comet's rotation brought another area into sunlight, a sharp increase in carbon dioxide. Evidently the frozen materials that released the gaseous molecules were unevenly distributed across the nucleus.

---

[1] Sublimation is the change in the physical state of a substance from an ice to a gas, omitting the liquid phase.

10.3　COMET MAPPING

During two years of accompanying 67P, Rosetta mapped its entire nucleus at high resolution, resolving very small features and monitoring how the surface changed with time.

The nucleus revealed a surprising diversity of features, including cliffs and steep slopes, large depressions, small cavities or pits from which material seemed to flow, and smooth areas that showed features resembling sand dunes. The northern hemisphere was mostly covered in dust from the comet's activity (e.g. in the Ma'at and Ash regions) but the southern hemisphere was much more rugged. The surface was very dark and very dry, with little indication of water ice, but was rich in organic (carbon-based) compounds.

The mapping was complicated by the comet's duck-like shape and its seasons. The north pole was located on the larger lobe, near the neck. Furthermore, due to its double-lobed shape and the large inclination of its rotation axis (52 degrees), the northern hemisphere had a very long summer of about 5.6 years, while the southern hemisphere endured a long, dark winter.[2]

Consequently, the first maps of 67P showed only the northern hemisphere. Nineteen distinct topographic regions were identified. Given the ancient Egyptian theme of the mission, these regions were named for Egyptian deities, and grouped according to the dominant terrain type. Regions on the small lobe were given female names, while those on the large lobe and on the neck received male names.

Five basic categories of terrain type were recognized:

- Dust-covered (Ma'at, Ash and Babi).
- Brittle materials with pits and circular structures (Seth).
- Large-scale depressions (Hatmehit, Nut and Aten).
- Smooth terrains (Hapi, Imhotep and Anubis).
- Exposed, more consolidated "rock-like" surfaces (Maftet, Bastet, Serqet, Hathor, Anuket, Khepry, Aker, Atum and Apis).

In May 2015, several months before the comet reached perihelion, the southern hemisphere became increasingly illuminated (see Chapter 8). Its brief, but relatively warm summer lasted until darkness returned in March 2016.

The change in illumination temporarily revealed the topography of the southern hemisphere. Unfortunately, this coincided with the time of maximum warming around perihelion and the greatest comet activity, and it was unsafe for Rosetta to make close approaches. The detailed geological investigation of the southern hemisphere therefore had to wait until the spacecraft resumed close approaches in early December 2015.

---

[2] This situation is comparable to the weeks of complete winter darkness experienced in Earth's polar regions.

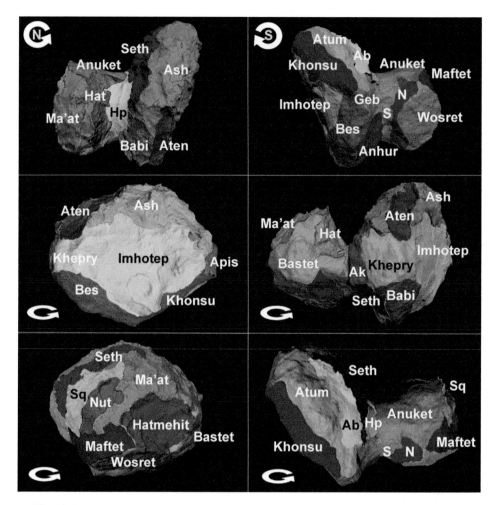

Fig. 10.2: Six different viewing angles of the nucleus, showing its named regions: Hapi (Hp), Hathor (Hat), Sobek (S), Neith (N), Aker (Ak), and Serqet (Sq). Circular arrows indicate the direction of rotation. The two upper panels are northern and southern polar views. The other four panels are equatorial views at approximately 90 degree intervals. (ESA/Rosetta/MPS for OSIRIS Team MPS/UPD/LAM/IAA/SSO/INTA/UPM/DASP/ IDA; El-Maarry et al., Astron. & Astrophys., 2017)

Seven regions were identified in the southern hemisphere – named Anhur, Bes, Geb, Khonsu, Sobek, Neith and Wosret – bringing the total number of defined regions to 26 (see Fig. 10.2).

Bes was on the edge of the large lobe in contact with the Imhotep region, and spanned almost all of the south polar region of the comet. Anhur and Geb were cliff regions on the large lobe. Khonsu was near the equator on the underside of the large lobe. Wosret covered the southern face of the smaller lobe. Sobek was located on the neck. Neith represented the main cliff on the small lobe.

The two hemispheres appeared very different, primarily because the southern regions lacked wide-scale smooth terrains, dusty deposits and large depressions (see Fig. 10.3). This seemed to be a consequence of the uneven distribution of the comet's seasons. That is, during its short summer the south experienced enhanced erosion with powerful jets blasting out gas and dust, with much of that dust ending up on the northern hemisphere.

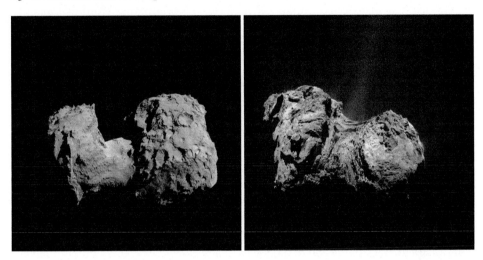

Fig. 10.3: The northern (left) and southern (right) hemispheres of 67P displayed very different surface characteristics. (ESA/Rosetta/MPS for OSIRIS Team MPS/UPD/LAM/IAA/SSO/INTA/UPM/DASP/IDA/Emily Lakdawalla)

10.4   CHANGES UPON CHANGES

The surface of 67P was very dynamic, particularly near perihelion. The spectacular changes included major outbursts of gas and particles, newly created and growing fractures, collapsing cliffs, and enormous rolling boulders. Material in motion buried some features on the surface and exhumed others.

The division of the surface into rough and smooth terrains was illustrated by Philae's multiple landings. It provided in-situ measurements of the smooth terrain at its first touchdown point in Agilkia and the hard surface at its final resting point in Abydos.

The rough terrain was predominantly exposed bedrock, while the smooth terrain represented disaggregated, transported remnants of compacted material that had fragmented.

The principal factor responsible for the breakdown of the solid surface material was the heat from the Sun, which varied both as the comet orbited the Sun and as it spun on its axis. These variations produced thermal stresses sufficient to fracture the bedrock at all observable scales. Heat penetrating deep into fractures further weakened the nucleus.

10.5 COMPACT AND FLUFFY DUST

Scientists believe that the dusty grains detected in the months after Rosetta's arrival had been left on the nucleus's surface from its previous perihelion passage. As it receded from the Sun, the outflow of gas had reduced to the level that it was no longer able to lift dust grains off the surface. While the dust lay undisturbed on the surface, gas production continued very slowly, originating from ever deeper beneath the surface during the years the comet was farthest from the Sun. In essence, the nucleus was 'drying out' on the surface and just below it. Only when the activity began to increase again did the removal of this accumulated surface layer resume, with dust being carried aloft to feed the inner atmosphere, or coma, of the comet. Once all of the old dust had been removed, fresher material of a more icy composition was exposed at the surface.

One of the key areas of research for Rosetta was an investigation by three experiments of this dusty coating and its expulsion into space.

On arrival, the GIADA (Grain Impact Analyzer and Dust Accumulator) instrument detected numerous showers of dust, each consisting of hundreds of small particles. It was particles of dust floating nearby Rosetta that caused the 'false stars' that impaired navigation when it was in close proximity to the nucleus.

GIADA found two main classes of dust grains within the coma: 'compact' particles that had sizes in the range 0.03-1 mm, and fluffy aggregates ranging in size from 0.2 to 2.5 mm. The individual compact particles had bulk densities in the range 0.8-3.0 g/cc, which was consistent with a variety of non-volatile (rocky) minerals or mixtures of minerals. On the other hand, the effective densities of the larger aggregates were less than 1 g/cc because they comprised many sub-micrometer sized grains separated by voids that produced fluffy, highly porous structures composed mostly of empty space.

Although more fluffy particles were detected than compact ones, these were a minor fraction of the total mass of dust being lost by the comet. Many of them arrived in short-lived showers which lasted between 0.1 and 30 seconds. Several of these showers were detected by GIADA each week during the early months of Rosetta's near-nucleus operations.

Another experiment, the COmetary Secondary Ion Mass Analyzer (COSIMA), was designed to collect, image, and measure the composition of dust particles. It enabled scientists to study the way that large dust grains, at least 0.05 mm across, fragmented or shattered on striking the instrument's target plate at speeds as low as 1-10 m/s.

The fact that they broke apart so readily meant that the individual grains were not well bound together. Moreover, if they had contained ice, they would not have shattered – instead, the icy component would have evaporated from the grain shortly after contacting the collecting plate, leaving voids in what remained. If a pure grain of water ice had struck the detector, then only a dark patch would have been seen.

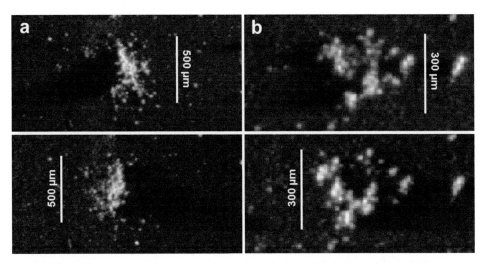

Fig. 10.4: Two examples of dust grains collected by the COSIMA instrument in the period 25-31 October 2014. Both grains were collected at a distance of 10-20 km from the nucleus. (a) A dust particle (nicknamed Eloi) that crumbled into a rubble pile when collected. (b) A dust particle (nicknamed Arvid) that shattered. Both grains are shown twice, under different grazing illumination, with the top image being illuminated from the right and the bottom one from the left. The shadows indicate Eloi rose about 0.1 mm above the target plate and Arvid only 0.06 mm. These fluffy grains are believed to come from the dusty layer which built up on the surface while the comet was far from the Sun. (ESA/Rosetta/MPS for COSIMA Team MPS/CSNSM/UNIBW/TUORLA/IWF/IAS/ESA/BUW/MPE/LPC2E/LCM/FMI/UTU/LISA/UOFC/vH&S)

The fluffy dust particles collected between August and October 2014 were found to be rich in sodium, sharing the characteristics of 'interplanetary dust particles' in meteor streams which originate from comets, such as the annual Perseids associated with Comet 109P/Swift-Tuttle. This material was likely not subject to processing prior to being accreted into a comet and, as such, may pre-date the birth of the Sun.

Compositional analysis of dust particles collected in May 2015 showed evidence of calcium-aluminum-rich inclusions (previously found in Stardust samples) and organic matter in grains of dust collected in May and October 2015. These had not been observed in particles from the comet earlier in the mission.

A third instrument, MIDAS, collected dust grains and then physically scanned them using an Atomic Force Microscope which gave extremely fine resolution – even greater than GIADA and COSIMA. It employed a very fine tip, rather like an old-fashioned record player needle, that was scanned across a particle. The deflection of the needle, and therefore the height of the sample, was measured to make a 3D picture. This enabled scientists to determine the structure of the particle and hypothesize about how it might have formed.

Even at scales of a few tens of micrometers down to a few hundred nanometers, the grains of dust analyzed by MIDAS appeared to be aggregates of numerous smaller grains. It detected both small, tightly packed 'compact' grains and larger, more porous, loosely arranged fluffy ones. The grains also appeared to be several times longer in one direction than in the others. The results bolstered the (likely) link between comet dust and interplanetary dust particles, the raw material from which the rocky planets were formed.

The slow speeds of the particles (much less than the escape velocity of 67P) implied a strong deceleration of the dust on approaching the spacecraft, possibly due to both the spacecraft and the dust particles being negatively charged, producing electrostatic repulsion. The amount of deceleration experienced by any particle was related to its charge and mass.

The fluffy aggregates were able to collect about 20 times more electric charge than compact particles of equivalent radius, so these were more readily decelerated. Indeed, if their charge-to-mass ratio was sufficient, they could be halted or repelled away from the spacecraft. The more charge a fluffy dust particle picked up, the larger the internal disruptive forces and the greater the chances of it losing integrity on approaching the spacecraft. The denser compact particles were not greatly influenced, but the fluffy aggregates were slowed and disrupted, creating the showers and sub-showers that GIADA detected.

10.6 JETS AND ERUPTIONS

By their very nature, periodic comets are transient objects that shed a great deal of material whenever they venture close to the Sun. As they vaporize in the warmth, they create cocoons of gas and dust, called comas, along with tails of debris which are aligned away from the Sun (see Chapter 1). 67P was no exception. Comparisons of differences in the Doppler shift of Rosetta's signals early and late in the mission gave an estimated mass loss of 10.5 million tonnes, or 0.1% of the total mass, which was equivalent to removing a layer of material 70 cm thick from the entire surface of the nucleus during each perihelion passage.

As expected, the nucleus showed little activity far from the Sun, but increasing amounts of gas and dust were ejected as the comet moved toward the inner Solar System and experienced higher temperatures. The jets switched on and off like clockwork as the comet rotated in just over 12 hours, synchronizing with the rise and fall of the insolation.

Jet-like features were observed erupting from a number of regions, including Hapi, Hathor, Anuket and Aten. In the case of the jets from the Hapi region, activity persisted across many rotations but the morphology of the jets evolved, sometimes during a single rotation.

In addition to the regular jets and flows of material streaming from the nucleus, there were dozens of violent eruptions. Rosetta's cameras captured 34 outbursts

during the three months centered around perihelion on 13 August 2015. These outbursts were much brighter than the normal jets, taking the form of sudden, brief,
high-speed releases of dust. They were typically detected only in a single image,
indicating a lifetime shorter than the interval between images of 5-30 minutes. In
those few minutes, a typical outburst appeared to release 60-260 tonnes of material.

Fig. 10.5: This short-lived outburst was captured by the OSIRIS-NAC at 13:24 UT on 29
July 2015 (middle image). The image at left, taken at 13:06 UT, does not include any visible signs of the jet. Residual traces of activity are very faintly visible in the final image at
13:42 UT (right). The jet is estimated to have come from a location on the comet's neck,
in the rugged Anuket region, and to have had a minimum speed of 10 m/s. (ESA/Rosetta/
MPS for OSIRIS Team MPS/UPD/LAM/IAA/SSO/INTA/UPM/DASP/IDA)

The outbursts around the time of perihelion occurred once every 30 hours on
average, about 2.4 comet rotations. Based on the appearance of the dust flow, scientists classified them into three categories. One type was associated with a long,
narrow jet that extended far from the nucleus; the second had a broad base that
expanded more laterally; the third was a complex hybrid of the other two.

The lack of frequent images meant it was difficult to determine whether these
three shapes of plume corresponded to different mechanisms, or were actually
evolutionary stages of a single process. However, if there was only one process
operating, then the logical sequence began with a long, narrow jet of dust ejected
at high speed, most likely from a confined space. As the local surface around the
vent was modified, a larger fraction of fresh material would be exposed, broaden-
ing the base of the plume. Finally, when the source region had been changed so
much that it was no longer able to support the narrow jet, this left a broad plume.

Just over half of the events occurred in regions where the early morning Sun
started to warm the surface after many hours of darkness. The rapid change in local
temperature is thought to trigger thermal stresses in the surface. This can lead to a
sudden fracturing and exposure of volatile material which rapidly heats up and
vaporizes explosively. The other events occurred after local noon – following several hours of illumination. These outbursts were attributed to a different cause,
when the cumulative heat build-up reached pockets of ices buried beneath the
surface, again causing sudden heating, vaporization, and gaseous eruption.

Occasionally 67P would spring a surprise, as happened in mid-March 2015 when OSIRIS saw a jet of material blasting from the dark, night side of the nucleus. By analyzing the brightness fluctuations along the unexpected dust jet, scientists were able to estimate that the particles were speeding away from the comet at a velocity of at least 8 m/s; similar to speeds calculated for eruptions on the day hemisphere.

It was also noted that most of the outbursts seemed to originate from places where there were changes in texture or topography in the local terrain, such as steep cliffs, pits or alcoves. The presence of boulders or other debris around the sources of the outbursts confirmed that these areas were particularly susceptible to erosion.

Whilst slowly eroding cliff faces are believed to be responsible for some of the regular, long-lived jet features, a weakened cliff edge might suddenly collapse at any time, night or day, to expose substantial amounts of fresh material, and this could lead to an outburst even when not in sunlight. The above-mentioned night-time outburst may have been related to the collapse of a cliff.

Fig. 10.6: This OSIRIS-WAC image taken on 3 July 2016 captured a plume of dust which originated from a depression in the Imhotep region. (ESA/Rosetta/MPS for OSIRIS Team MPS/UPD/LAM/IAA/SSO/INTA/UPM/DASP/IDA)

On 3 July 2016, Rosetta serendipitously flew through a plume and five of its instruments were able to study it. The jet originated in a 10 meter wide circular area, within a depression, where frozen water was exposed at the surface. This suggested the outburst was the result of sudden sublimation of frozen gases on the surface, and that as the gas streamed into space it entrained the dust, which formed the visible jet. However, this process alone could not fully explain the event. Dust production in the outburst was measured at approximately 18 kg per second. This was a lot 'dustier' than predicted by conventional models. It indicated that energy must have been released from beneath the surface to support the plume, perhaps from cavities filled with compressed gas. According to this theory, after sunrise the radiation warmed the surface and opened cracks that allowed the gas to escape.

According to another theory, the outburst was caused by a deposit of 'amorphous ice' beneath the surface. In this type of frozen water, the molecules are arranged in a less orderly manner than in 'crystalline ice'. As heat stimulates a phase transition from amorphous to crystalline, it releases a lot of latent energy. By this theory, the result was sufficient to crack the surface and produce a jet.

10.7   PITS AND GOOSE BUMPS

High-resolution OSIRIS images from distances of only 10-30 km enabled at least some of the dust jets to be traced back to rounded, active pits. It was the first time this had ever been seen.

Eighteen quasi-circular pits were evident in the northern hemisphere, but no similar features were seen in the southern hemisphere – which received significantly more solar heating when at perihelion. The pits were several tens to several hundreds of meters in diameter, as much as 210 meters deep, and often had a smooth, dusty floor.

Some of the pits were sources of continuing activity, with jets rising from the fractured areas of their inner walls. The active pits were more steep-sided, and probably younger, whereas pits without any observed activity were shallower and likely to be dormant or extinct. Middle-aged pits exhibited boulders on their floors that had fallen from the sides, while the oldest pits had degraded rims and were filled with dust.

As the finale to its mission the Rosetta spacecraft landed in the pitted Ma'at region. The target was adjacent to a 130 meter wide, well-defined pit officially listed as Ma'at 2 and informally named Deir el-Medina by the mission team (see Fig. 8.12 and Fig. 10.7).

Scientists think such pits were formed when a cavity slowly grows in the porous subsurface. When the roof becomes too thin to support its own weight (even under the weak gravity of the comet) it collapses. Exposing the fractured interior allows new material to sublimate, causing further erosion of the pit. Three theories have been suggested for how the precursor void may have originated.

Perhaps the least likely scenario is that they have existed since the comet formed, billions of years ago, as a result of very low-speed collisions between

Ma'at_01   Ma'at_02   Ma'at_03

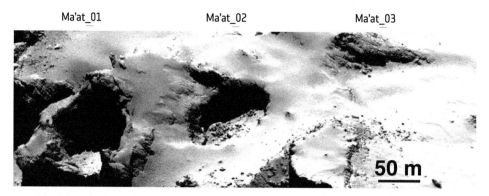

50 m

Fig. 10.7: Pits Ma'at 1, 2 and 3 show differences in appearance that may reflect their history of activity. Pits 1 and 2 are active, but no activity was observed from pit 3. The young, active pits are more steep-sided, and pits lacking any observed activity are shallower and seemingly filled with dust. Middle-aged pits tend to exhibit boulders on their floors from mass-wasting of the sides. From left to right, the pits measured 125, 130 and 140 meters wide, respectively, and 65, 60 and 50 meters deep. The OSIRIS-NAC image was taken from a distance of 28 km. (ESA/Rosetta/MPS for OSIRIS Team MPS/UPD/LAM/IAA/SSO/INTA/UPM/DASP/IDA)

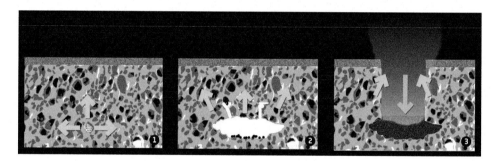

Fig. 10.8: Pits on Comet 67P could form through sinkhole collapse. The grey surface layer is a mixture of dust and ices. Heat causes subsurface ices to sublimate (1, blue arrows), forming a cavity (2). When the roof is too weak to support its own weight, it collapses and produces a deep, circular pit (3, red arrow). Sublimation of the newly exposed material in the pit walls accounts for the observed activity (3, blue arrows). (ESA/Rosetta/J-B Vincent et al. 2015)

primordial building blocks tens to hundreds of meters in size. The collapse of the roof above such voids could then be triggered by weakening of the surface, perhaps as a result of sublimation, seismic shaking or the impact of boulders ejected from elsewhere on the comet.

Rather more likely is direct sublimation of pockets of volatile ices, such as carbon dioxide and carbon monoxide, that are heated by the warmth of sunlight penetrating an insulating surficial layer of dust.

Alternatively, sublimation could be driven by the energy liberated by water ice changing from amorphous to crystalline structures, with the released latent heat sublimating the more volatile surrounding carbon dioxide and carbon monoxide ices.

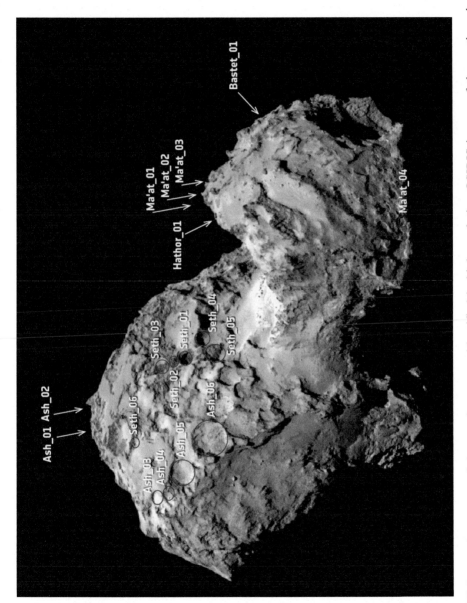

Fig. 10.9: Eighteen pits – some of them active – were identified in high-resolution OSIRIS images of the northern hemisphere (arrows and red circles). The pits were named after the region where they resided. The context image was taken on 3 August 2014 from a range of 285 km. The image resolution is 5.3 meters per pixel. (ESA/Rosetta/MPS for OSIRIS Team MPS/UPD/LAM/IAA/SSO/INTA/UPM/DASP/IDA)

If either of the latter two processes is the driving force, then the patchy distribution of the pits may indicate an uneven distribution of ices within the first few hundred meters of the comet's surface.

The features of the pit walls varied significantly, displaying fractured material and terraces, horizontal layers and vertical striations, and/or globular structures nicknamed 'goosebumps'.

40 m

Fig. 10.10: Close-ups by the OSIRIS-NAC of the surface texture nicknamed 'goosebumps'. The characteristic scale of all the bumps is approximately 3 meters, extending over regions greater than 100 meters. They are observed on very steep slopes and on exposed cliff faces, but the way in which they were formed is uncertain. (ESA/Rosetta/MPS for OSIRIS Team MPS/UPD/LAM/IAA/SSO/INTA/UPM/DASP/IDA)

These intriguing, lumpy structures, typically 1-3 meters across, were also seen on very steep slopes and on exposed cliff faces. Rosetta scientists believe these could represent small lumps of primitive material that came together to form the comet more than 4 billion years ago (see A Planetary Building Block below).

"Our thinking is that accreting gas and dust would have formed little 'pebbles' at first that grew and grew until they got to the size of these goosebumps, about 3 meters in size, and for whatever reason, they could not grow any further," said

Holger Sierks, principal investigator for the OSIRIS cameras. "Eventually, they'd have found a region of instability and clumped together to form the nucleus."

10.8   COLLAPSING CLIFFS

Much of the surface of 67P was remarkably craggy and rugged. Some of the cliffs were very steep, with slopes occasionally exceeding 45 degrees. The most spectacular example was the cliff of Hathor, bordering the neck. This 900 meter tall cliff was almost vertical. It revealed vertical linear fractures and roughly perpendicular linear features whose alignment with small terraces was suggestive of inner layering (see Fig. 10.15). Scientists suspect this spectacular cliff is the result of erosion and that it is exposing the internal structure of the comet's head.

The process of 'scarp retreat' appears to be common on comets. There had been examples on Comet Tempel 1, but Rosetta was able to monitor such changes on an ongoing basis, and at a higher resolution.

As Comet 67P approached perihelion, in-situ weathering was occurring all across the nucleus, where consolidated materials were weakened by processes such as heating and cooling cycles on daily or seasonal timescales, resulting in fragmentation. In combination with the heating of subsurface ices that caused outflows of gas, this sometimes resulted in the sudden collapse of cliff walls. Imagery showed scarps in several smooth plains retreating at rates of up to several meters per day around perihelion.

One example was a cliff collapse in the Seth region of the large lobe. The first close images of 67P taken in September 2014 revealed a fracture, 1 meter wide and some 70 meters long, on a prominent cliff edge that was subsequently named Aswan.

As perihelion approached, sporadic and short-lived, but high-speed releases of dust and gas increased in frequency. One outburst that could be traced back to the Seth region was seen on 10 July 2015 by Rosetta's navigation camera.

The next time the Aswan cliff was observed, five days later, a bright, sharp slope was spotted where the previously identified fracture had been, along with many new meter-sized boulders at the foot of the 134 meter high cliff.

"The last time we saw the fracture intact was on 4 July. In the absence of any other outburst events recorded in the following ten day period, this is the most compelling evidence that we have that the observed outburst was directly linked to the collapse of the cliff," said Maurizio Pajola, who led a study of the Aswan cliff.

The event on 10 July also provided a unique opportunity to study how pristine water ice that had been buried tens of meters beneath the surface evolved as the exposed material turned to gas in the ensuing months.

At first, the exposed cliff face was calculated to be at least six times brighter than the overall average surface of the nucleus. By 26 December 2015 it had faded by half, suggesting much of the ice had vaporized by that time. Images taken on 6 August 2016 showed most of the new cliff face had faded to the average brightness, with only one large, reflective block remaining.

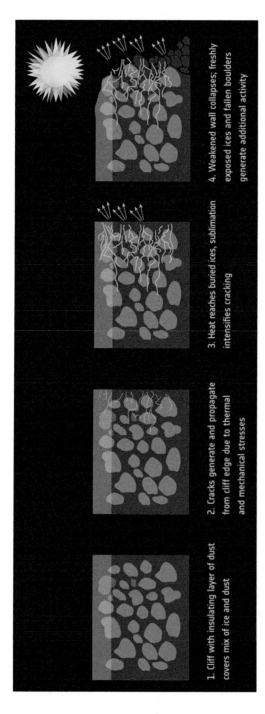

1. Cliff with insulating layer of dust covers mix of ice and dust

2. Cracks generate and propagate from cliff edge due to thermal and mechanical stresses

3. Heat reaches buried ices, sublimation intensifies cracking

4. Weakened wall collapses; freshly exposed ices and fallen boulders generate additional activity

Fig. 10.11: Much of the comet's regular activity may be linked to the steady erosion of cliff walls that are initially fractured by thermal or mechanical erosion. These fractures propagate into the underlying mixture of ice and dust. As the ices sublimate, gases escape through the fractures, which act like nozzles to focus the gas flows. They pick up dust along the way and create distinct collimated jets. Continued cracking, heating and sublimation eventually leads to sudden collapse of the cliff wall – the likely source of more-transient outburst events. The debris that falls to the foot of the cliff exposes previously hidden material that contributes to the observed outflow. (ESA, based on J.-B. Vincent et al. 2015)

In addition, the team had a clear 'before and after' look at how the crumbling material settled at the foot of the cliff. By counting the number of new boulders present after its collapse, the team estimated 99% of the fallen debris had collected at the foot of the cliff, with only 1% of it being lost to space.

This corresponded to around 10,000 tonnes of removed cliff material, compared with at least 100 tonnes that did not reach the ground – consistent with estimates made for the volume of dust in the eruptive plume. What is more, the size range of the new debris of 3-10 meters, was consistent with the distributions at the foot of several other cliffs on the nucleus.

"We see a similar trend at the foot of other cliffs that we have not been so fortunate to have before-and-after images, so this is an important validation of cliff collapse as a producer of these debris fields," said Maurizio Pajola.

But what prompted the cliff to collapse at that moment? An earlier analysis argued that rapid daily changes in heating and longer-term seasonal changes can both produce thermal stresses sufficient to cause fracturing. And exposing volatile materials can trigger a rapid outburst that will cause a weakened cliff to collapse.

Even though the Aswan cliff region had been experiencing large temperature changes in the months prior to its collapse, the collapse occurred at local night – ruling out a sudden extreme temperature change as the immediate trigger. Instead, the daily and seasonal thermal stresses may have propagated fractures deep into the cliff, predisposing it to collapse.

"If the fractures permeated volatile-rich layers, heat could have penetrated the deeper layers, causing a loss of deeper ice," explained Maurizio Pajola. "The gas released by the vaporizing material could further widen the fractures, leading to a cumulative effect that eventually led to the cliff collapse."

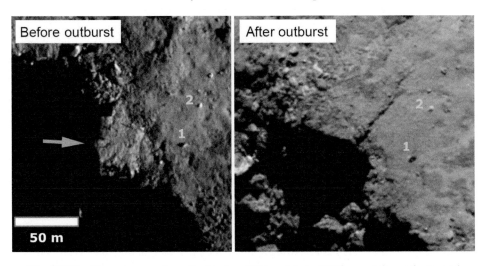

Fig. 10.12: These images from September 2014 and June 2016 were taken prior to and after an outburst event caused by the collapse of a cliff at the boundary between the

northern and southern hemispheres of the small lobe. The blue arrow points to the scarp which collapsed. Two boulders (labeled 1 and 2) are marked for orientation. (ESA/Rosetta/MPS for OSIRIS Team MPS/UPD/LAM/IAA/SSO/INTA/UPM/DASP/IDA (CC BY-SA 4.0)

A completely different process may have been responsible for a 500 meter long fracture, first noted in August 2014, that ran through the comet's neck in the Anuket region. By December 2014 it had extended by about 30 meters. This was probably a consequence of the increasing spin rate as the comet neared perihelion.

Furthermore, in images taken in June 2016, a new 150-300 meter long fracture was identified parallel to the original fracture. Nearby, a 4 meter boulder was found to have moved about 15 meters, as determined by a comparison of images in March 2015 and June 2016. It is not known whether the extension of the fracture and the movement of the boulder were related or were due to different processes.

During perihelion passage, the southern hemisphere experienced a brief summer, producing increased levels of activity and more intensive erosion than elsewhere on the nucleus. This caused an even larger collapse event, linked to a bright outburst observed on 12 September 2015. The scarp in question was located along the divide between the northern and southern hemispheres. Inspection of images obtained before and after the event indicated that the cliff was intact in May 2015. At some time after that (presumably about 12 September) an area of about 2,000 square meters collapsed.

Other debris nearby suggested that large erosion events had happened there in the past. The loose material included blocks of sizes ranging up to tens of meters; substantially larger than the boulder population created by the collapse of the Aswan cliff, which was mainly several meters in diameter. This variability in the size distribution of the fallen debris could be due to different cliff-collapse processes or differences in the strength of the layered materials.

10.9   BOUNCING BOULDERS

Large boulders were present all across the nucleus, but the fact that the southern hemisphere had three times as many as in the north indicated it had more intense thermal fracturing and associated activity.

Most were found in taluses – loose rocks piled at the foot of a slope and probably created by rock falls from nearby cliffs. A smaller number were isolated boulders, sometimes in smooth terrains.

Some rocks had moved quite large distances. One set of images showed a boulder, about 10 meters across, which had left indentations in soft, smooth surface material as it made a series of jumps and then rolled across the surface, bound by the comet's low gravity. This 230 tonne boulder appeared be the largest fragment in a landslide from a nearby 50 meter high cliff.

An even larger boulder, some 30 meters in size and with a mass of 12,800 tonnes, was found to have moved an impressive 140 meters in the Khonsu region, on the larger lobe of the nucleus (see Fig. 10.13). This seems to have moved during the perihelion period, perhaps propelled by several outburst events that were detected near its original position.

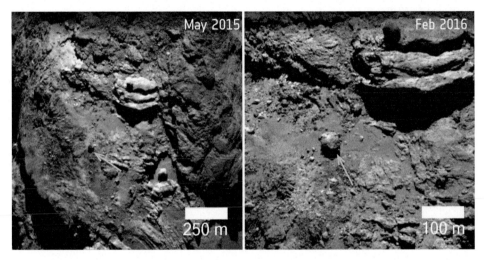

Fig. 10.13: In the lead up to perihelion in August 2015 a boulder 30 meters wide with a mass estimated as 12,800 tonnes was found to have moved 140 meters in the Khonsu region of the nucleus. The arrows point to the boulder. In the right-hand image, a dotted circle outlines its original location. The OSIRIS images were taken on 2 May 2015 (left) and 7 February 2016 (right), with spatial resolutions of 2.3 meters per pixel and 0.8 meters per pixel, respectively. (ESA/Rosetta/MPS for OSIRIS Team MPS/UPD/LAM/ IAA/SSO/INTA/UPM/DASP/IDA)

The massive rock's motion could have been triggered in one of two ways: either the material on which it was sitting eroded away, allowing it to roll down slope, or a sufficiently forceful gaseous outburst propelled it to its new location. Several outburst events were noted near the original position of the boulder during perihelion.

Even more remarkable were images obtained by the OSIRIS-NAC on 30 July 2015, showing a boulder in the process of being ejected into space,

Fig. 10.14: An OSIRIS image of part of a large fracture that runs through the Hapi region in the neck of 67P. The arrow shows where it appears as a chain of pits. (ESA/ Rosetta/MPS for OSIRIS Team MPS/UPD/LAM/IAA/SSO/INTA/UPM/DASP/IDA/ Emily Lakdawalla)

never to return. Its exact size could not be determined because its distance from the camera was not known, but it was probably between 1 and 50 meters across.

Studying the history of such boulders highlighted the characteristics of the surface material of the nucleus and the strength of the blocks, which were typically 100 times weaker than freshly packed snow, a characteristic of little consequence due to the comet's minuscule gravitational field.

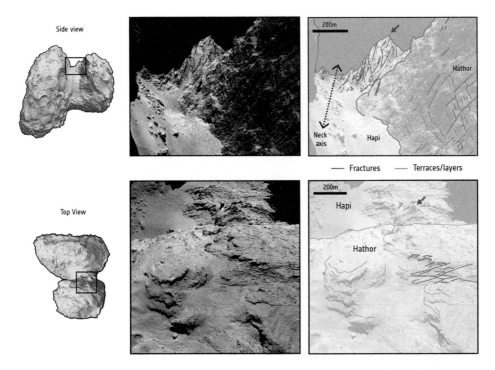

Fig. 10.15: These images show how the neck region of 67P has been affected by a process called mechanical shear stress. The shape of the nucleus is shown on the left, from top and side perspectives. The locations of the four frames on the right are marked by black boxes (the neck). The red arrow points to the same spot in both images, viewed from a different perspective. The two central frames are OSIRIS images. The frames on the right highlight different geological features. Red lines trace fracture and fault patterns that result from the irregular shape and rapid rotation of the nucleus. Green lines indicate layers of terracing. (ESA/Rosetta/MPS for OSIRIS Team MPS/UPD/LAM/ IAA/SSO/INTA/UPM/DASP/IDA; C. Matonti et al. 2019)

10.10   A CRACKING COMET

One of the most significant findings was the large fractures in the nucleus. In addition to cliff collapses, Rosetta saw randomly oriented fractures in many of the exposed cliff walls.

It appears likely that the dominant process influencing the landscapes of 67P is the positive feedback loop of fracturing induced by the thermal cycles arising from the 12.4 hour day and 6.5 year elliptical path around the Sun, with the fractures promoting failures that expose fresh volatiles to continue the cycle. Stresses and flexing generated by the increased spin rate in the lead-up to perihelion may also have contributed. Another possible factor is flexing induced by solar gravitational tides.

One major example was a 500 meter long crack, first noticed in August 2014, that ran across the comet's neck – the site of most eruptions of gas and dust. It was in the smooth, boulder-strewn Hapi region, close to the Hathor cliff face, and extended into the Anuket region (see Fig. 10.14). In places, it was a meter wide and parts of it were filled with dust. This fracture was found to have increased in width by about 30 meters by December 2014. In images taken in June 2016, a new 150-300 meter long fracture was identified parallel to the original one.

Images of the neck showed that this region, in particular, was affected by a geological process known as mechanical shear stress – a force which is more familiar on Earth in earthquakes or glaciers. It occurs when two bodies, or blocks, push and move past one another in different directions. In the case of 67P, the evidence of the fracture and fault patterns implies this stress was caused by the rapid rotation and irregular shape of the nucleus.

The neck region appeared to experience the greatest mechanical stresses, or internal torques, from solar heating and tidal forces as the comet rotated and approached perihelion. It has been suggested that as the neck is further weakened by frequent perihelion passes, the two lobes of the nucleus may eventually separate.

Detailed analysis by Marco Franceschi et al in April 2020 revealed the presence of fractures of many different sizes all over the nucleus. Such fracturing is usually characteristic of brittle rocky material, but that would be difficult to reconcile with the apparent low strength of the material. The existence of a hardened layer at the surface has therefore been proposed.

Nevertheless, many fractures were several hundred meters in length and extended deep into the layered surface structure. Such fractures could not be interpreted as cracks resulting from thermal contraction.

Furthermore, the occurrence of long fractures was not limited to the neck region, where they could be attributed to the torques imparted by the spin of the nucleus. The fact that they were widespread hinted at an origin related to an event that globally influenced the comet, possibly even the collision that produced the two-lobed nucleus.

As Christophe Matonti of Aix-Marseille University, France, explained, "We found networks of faults and fractures penetrating 500 meters underground and stretching out for hundreds of meters. These geological features were created by shear stress … This is hugely exciting – it reveals much about the comet's shape, internal structure, and how it has changed and evolved over time." (See also A Planetary Building Block below)

Fig. 10.16: An OSIRIS-NAC view of a smooth terrain on 67P. Taken from a distance of only 10 km with a resolution of about 20 cm per pixel, it shows the granular material of the plains. (ESA/Rosetta/MPS for OSIRIS Team MPS/UPD/LAM/IAA/SSO/INTA/ UPM/DASP/IDA)

On a much smaller scale, the nucleus displayed over 6,300 polygons in consolidated terrains and grouped in localized networks. They were present all across the nucleus, including on pit walls and lineaments. About 1.5% of the observed surface was covered by polygons, 90% of them in the 1-5 meter size range. Their variety of morphologies depended on the width and depth of the troughs. The networks consisted of nodes at which three or four cracks intersected. The cracks were consistent with diurnal or seasonal cycles which induced thermal contraction in a hard, consolidated sintered layer of water ice at a depth of a few centimeters. There were more evolved polygons with deeper and larger troughs where the temperature and the diurnal and seasonal temperature range were at their highest.

Analysis indicated that the polygons were fairly young features that probably formed after the comet was first deflected into the inner Solar System, perhaps 100,000 years ago. If they were a result of the reduction of its perihelion distance from 2.7 to 1.3 AU in 1959, then they would be just several decades old.

10.11   SMOOTH TERRAIN

The northern hemisphere of 67P was dominated by smooth, dust covered regions. In contrast, the southern hemisphere possessed such plains only in small, isolated areas.

The smooth plains were small pools of granular materials embedded in other regions, in local gravitational lows where the terrain was generally flat or gently sloping.

Any large object that impacted such material would leave an imprint. OSIRIS saw the tracks of bouncing boulders (see Bouncing Boulders), while the Philae lander imaged the footprints of its first touchdown in the Agilkia region (see Chapter 9).

One circular hollow that may have represented the sole impact crater on 67P, showed signs of dust beginning to fill the small depression.

Philae also recorded a brief 'thud' when its legs touched the dusty plain. This was interpreted as the lander coming into contact with a soft layer of regolith comprising unconsolidated solid material several centimeters thick, with a harder (perhaps icy) layer underneath.

The comet's smooth plains were generally thought to be covered by a layer of loose carbon-rich material, perhaps as much as 20 cm thick, which represented the transported remnants of formerly consolidated bedrock. In a jet event, the non-volatile material that does not achieve escape velocity will fall back and drape the rocky surface, eventually smoothing it out. This was particularly apparent on the body of the nucleus, and within gravitational lows where the underlying consolidated material appeared to outcrop from a blanket of smooth material.

Some of the smooth material gave rise to the most dynamic features observed on the comet, particularly around the time of perihelion.

The seasons were the main influence on the transfer of dust across the surface. Because the southern summer was shorter and more intense than in the north, its surface erosion was three times greater and the more intense activity ejected dust at higher speeds. Although some of it escaped to space, during the months around perihelion Rosetta saw material ranging from dust grains several millimeters in size right up to meter-sized objects falling on the northern part of the nucleus to make a fresh layer as much as several meters deep.

"If we were standing on the comet, I could imagine this dust movement to be like a very calm and gentle pumice fall, albeit with dust up to giant beach ball size, on a night with a sustained breeze, but too tenuous to be felt, coming from the south," said Rosetta scientist Marco Fulle.

Whilst some of the dust was redistributed around the comet in this way, other particles were ejected at speeds sufficient to escape the gravity of the comet. The GIADA instrument noted 'showers' of dust which lasted 0.1-30 seconds, and some of these were imaged by the OSIRIS cameras.

But this dust cannot explain the variable distribution of smooth areas on the nucleus. One of the largest smooth terrains was in the Imhotep region (see Fig. 8.8). Smooth terrains covered about one third of the region, an area of almost a square kilometer. Located near the equator, Imhotep never experienced protracted darkness as the comet orbited the Sun.

Close up images showed variations in the size distribution of surface material in the smooth areas, but most of it was relatively small, with the largest being clumps up to several tens of centimeters across.

Most of the smooth terrains were local gravitational low points. All of them were in flat areas where the gravitational slope did not exceed 15 degrees. The largest, at the center of Imhotep, was remarkably flat, with gravitational slopes of less than 5 degrees, and very near the lowest gravitational point of the region.

The thickness of the smooth terrains appeared to vary across the region, but the actual depths could not be determined. The underlying rocky terrains were revealed on the margins, where the thickness of a blanket shallowed. Some of the associated terraces were gentle, but others were very steep.

Smooth terrains are probably relatively undisturbed areas that evolve slowly, so material has time to settle and accumulate. The absence of fractures in the smooth terrains suggests either that they are more recent than the fractures, or that the loose material from which they are formed is unable to retain fractures.

In a November 2015 paper, A.-T. Auger et al described a possible scenario for the formation and evolution of smooth terrains on Imhotep:

The fine material comes from the cliffs on the border of the basins where mass wasting occurs. It is then transported by gravity downslope to a flat surface where it stays. The wideness of the smooth area can be explained by the progressive retreat of the cliffs over a long time, probably some tens to hundreds of perihelion passages. The more distant the fine material from the cliff, the older the deposit is. We suggest that the linear features visible in smooth terrains reveal the topography of the rocky terrain underneath. Some might also be scars of the previous location of the cliffs. This scenario is more suitable with an erosion of cliff runs by events limited in time, such as the passage at perihelion that strongly increases the activity of the comet. Air-fall deposits are also not excluded, but probably only account for a small fraction of the smooth terrains.

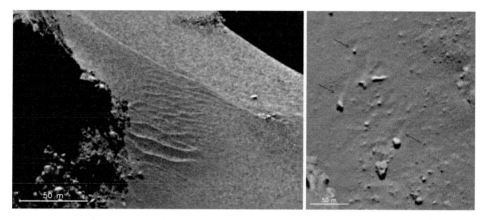

Fig. 10.17: Features in the Hapi region showed evidence of local, gas-driven transport that produced dune-like ripples (left) and boulders with 'wind tails' (right) produced where the boulder served as a natural obstacle to the direction of the gas flow and created a streak of dust 'downwind' of it. The images were obtained by OSIRIS-NAC on 18 September 2014. (ESA/Rosetta/MPS for OSIRIS Team MPS/UPD/LAM/IAA/SSO/INTA/UPM/DASP/IDA)

The smooth terrains also exhibited more Earth-like features, including small dunes and wind tails – the accumulation of loose material on the leeward side of boulders (see Fig. 10.17). If the 'atmosphere' on 67P is almost non-existent, how can winds blow across the nucleus? The consensus opinion is that the eruptions of gas-laden jets create strong 'gusts' which can drive dust particles into dunes.

As Nicolas Thomas from Switzerland's University of Bern explained, "The trick, we think, is that there are very strong winds there – 300 m/s – and that these winds can, even though the density of the gas is very low, push particles around to make the dunes."

The smooth areas were not always modified by the deposition and transport of loose material alone. Some OSIRIS images of the Imhotep region showed that, over time, erosion removed some of the smooth material to exhume previously buried features. In one location, a depth of about three meters had been removed, most likely through the sublimation of underlying ices.

10.12   THE CHAMELEON COMET

One of the interesting side effects of the changes in eruptive activity was an alteration in the color of the nucleus and the coma during Rosetta's two years of exploration. The subtle color changes were associated with the amount of water ice that was exposed on the surface and the amount of material injected into the coma.

In August 2014, the spacecraft rendezvoused with the comet whilst still a long way from the Sun and fairly inactive. At such distances, the nucleus was covered in layers of dust and little ice was visible. As a result, the surface appeared red when it was analyzed using the VIRTIS instrument. Although there was little dust surrounding the comet at that time, any particles that were present contained water ice and thus appeared more blue.

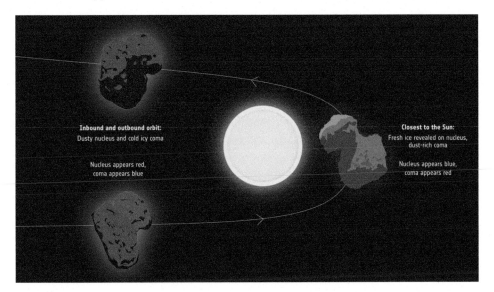

Fig. 10.18: Two years of data from Rosetta's VIRTIS instrument showed subtle changes in color as 67P traveled closer to the Sun, passed perihelion, and then receded again. When far from the Sun the nucleus was redder than the surrounding particles in the coma, which were dominated by water ice grains measuring about 100 micrometers across. As the comet drew closer to the Sun, the nucleus became bluer because fresh ice was revealed. In contrast, the coma grew redder, as more sub-micrometer dust grains made of organic matter and carbon were ejected. When the comet moved away from the Sun, the jet activity decreased and the nucleus once again became redder than the coma. (ESA)

As the comet moved closer to the Sun, it crossed the Solar System's frost line, which is the heliocentric distance at which ices become gases. Within about 3 AU, any exposed ices were heated sufficiently to sublimate directly into gases.

As the heating increased, the hidden water ice began to sublimate and it started to eject more dust grains. This exposed layers of pristine ice, which made the nucleus turn bluer in color, as seen by VIRTIS.

The situation was reversed in the coma. As the comet crossed the frost line, the ice in the dust grains that surrounded the nucleus sublimated quickly, leaving mostly dehydrated dust which reddened the coma.

Once 67P headed back into the outer Solar System, VIRTIS showed that the colors reversed, with the nucleus becoming redder and the coma bluer.

10.13   THE WATER CYCLE

Most of 67P's surface was dry and dark, with a widespread coating of organic material, but patches of water ice were detected from time to time and water vapor appeared in the jets and coma.

VIRTIS revealed that local gas and dust jets were accompanied by recurring patterns of water ice on the neck of the comet that followed a day-night cycle. At certain times of the cometary day, water vapor merged from the interior to the surface, froze in the shadowed regions, then sublimated when exposed to sunlight and eventually escaped to space (see Fig. 10.19).

The water cycle involved the sublimation of water ice on, or close to, the surface during the local day, causing water vapor to escape to space. As the nucleus rotated and day turned into night, the surface temperature dropped sharply, but the subsurface water ice remained warm and continued to sublimate, making its way through the porous interior to the surface, where it froze. At the start of the next day, the ice began to sublimate again – starting with the fresh ice that accumulated on the surface overnight.

During its two year sojourn at the comet, Rosetta's instruments were in a unique position to monitor how much water vapor was released into space, and how the water production rate varied with heliocentric distance.

The overall pattern of change was more or less as expected from numerous studies of other comets. The data showed an overall increase of the production of water, from several tens of thousands of kilos per day when Rosetta first reached the comet in August 2014, to almost 100 million kg per day around perihelion in August 2015. In addition, data from ROSINA established that the peak in water production was followed by a rather steep decrease in the months following perihelion (see Fig. 10.20).

However, a more complex pattern was discovered when ROSINA scientists took into account measurements made locally, at specific points around the comet. They found that owing to its complex shape and seasonal cycle, the outgassing from the nucleus was far from uniform.

The data revealed that, during the first several months of observations, when the comet was between 3.5 and 1.7 AU from the Sun, water was predominantly produced in the northern hemisphere. In May 2015, when about 1.7 AU from the Sun, the equinox marked the end of the 5.5 year northern summer and the start of its short but intense southern summer. Scientists had expected the peak of water production to drift slowly from the northern hemisphere to the southern hemisphere, but the data revealed an abrupt change, perhaps because the shape of the nucleus produces highly variable illumination conditions and even self-shadowing effects.

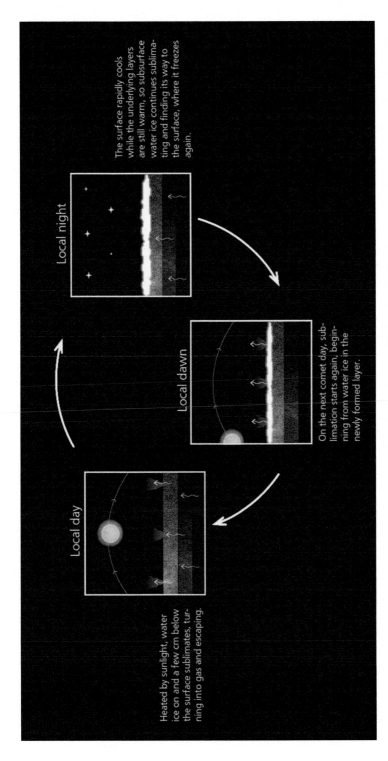

Fig. 10.19: The day-night cycle of sublimation and freezing of water ice on 67P. During the local day, water ice on and just below the surface sublimates and the gas escapes. During the local night the surface rapidly cools while the underlying layers are still warm, so subsurface water ice continues to sublimate and find its way to the surface, where it freezes. On the next comet day, sublimation restarts with the fresh ice that accumulated on the surface overnight. (ESA/Rosetta/VIRTIS team)

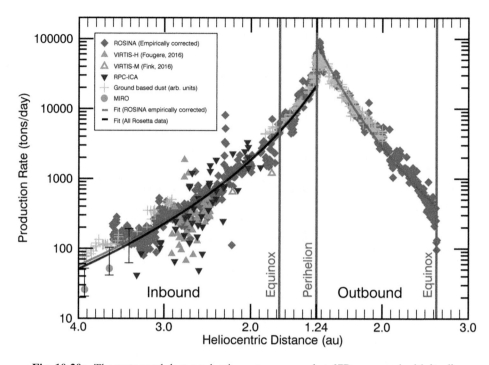

Fig. 10.20: The water and dust production rate measured at 67P compared with its distance from the Sun (in astronomical units, AU). Note the peak around the time of perihelion, the time of maximum heating. The data are from-DFMS (blue diamonds), MIRO (yellow circles), VIRTIS-H (solid green triangles), VIRTIS-M (unfilled green triangles), and RPC-ICA (red triangles). The dust production rate estimated from ground-based observations is indicated by tan crosses. The data covers the period June 2014 to May 2016. (ESA, image adapted from Hansen et al. 2016)

As expected, the production of water peaked between the end of August and early September 2015, about three weeks after the perihelion passage at 1.24 AU on 13 August.

Some contradictions were found when measurements by VIRTIS and MIRO were compared. MIRO data indicated a rising trend in the water production rate from June to September 2014, whereas the first months of ROSINA data, starting in August, showed an almost constant rate during that period.

"This could be explained if a sudden surge in the water production happened around the time of the first MIRO measurement, a few weeks before Rosetta's rendezvous with 67P, and the beginning of ROSINA observations," said Kenneth C. Hansen of the University of Michigan, the lead author of a 2016 paper in Monthly Notices of the Royal Astronomical Society.

The scientists also compared the comet's production rate of water to that of dust, as measured by robotic telescopes across the globe, including Chile, Hawaii and the Canary Islands.

As Hansen noted, "The correlation between the production rate of water and dust, both before and after perihelion, is impressive, suggesting that the gas-to-dust ratio remained constant over this long period."

Based on the water production rate, the team estimated that the comet lost 6.4 million tonnes of water to space during the period monitored by Rosetta, with the peak mass loss about three weeks after perihelion.

Allowing for other gas molecules and the dust, the total mass loss could have been roughly 10 times greater than that. It was the equivalent of losing a layer 2-4 meters thick from the entire nucleus.

10.14   THE MYSTERY OF EARTH'S OCEANS

One of the leading hypotheses for Earth's formation says that it was so hot when it formed 4.6 billion years ago that any original water ought to have boiled off. However, today, two thirds of the surface is covered by oceans. Where did this water come from?

One popular theory favors the delivery of water after the planet had cooled down, most likely in a bombardment of comets and asteroids. The relative contribution of each type of object is, however, still open to debate. Indeed, the proposition that comets were the primary source of the abundant water on Earth has been cast into doubt by measurements of the compositions of numerous comets.

The key to determining where the water originated lies in its 'flavor', in terms of the relative proportion of deuterium (a form of 'heavy' hydrogen that has a neutron) to normal hydrogen. By comparing the deuterium-to-hydrogen (D/H) ratio in Earth's oceans with that for different comets or asteroids, it is possible to deduce which class of objects might be the source of our water.

Previous measurements of the D/H ratio in comets have shown a wide range of values. Of the 11 comets for which measurements have been made, only the Jupiter-family Comet Hartley 2 matched the composition of Earth's water, based upon observations by ESA's Herschel Space Observatory in 2011.

Comet 67P is a Jupiter-family comet which is thought to have formed in the Kuiper Belt, out beyond Neptune (see Chapter 1). However, Rosetta muddied the debate still further by finding the D/H ratio (measured by the ROSINA instrument) to be more than three times greater than for Comet Hartley 2. Indeed, it was even higher than measured for any comet from the even more remote Oort Cloud as well.

"This surprising finding could indicate a diverse origin for the Jupiter-family comets," said Kathrin Altwegg, principal investigator for ROSINA. "Perhaps they formed over a wider range of distances in the young Solar System than we previously thought. Our finding also rules out the idea that Jupiter-family comets contain solely Earth ocean-like water, and adds weight to models that place more emphasis on asteroids as the main delivery mechanism for Earth's oceans."

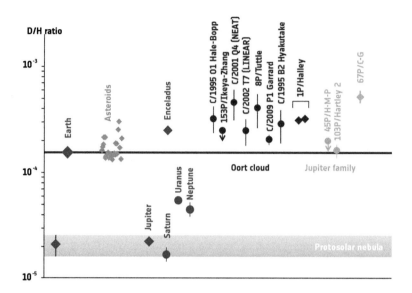

Fig. 10.21: The deuterium-to-hydrogen (D/H) ratio in water for various bodies in the Solar System. The data points indicate planets and moons (blue), chondritic meteorites from the asteroid belt (grey), comets that emerged from the Oort Cloud (purple), and Jupiter-family comets (pink). 67P, Rosetta's Jupiter-family comet, is yellow. Diamonds represent in-situ data and circles are data that were obtained by astronomical methods. The lower portion of the graph is the D/H for molecular hydrogen in the atmospheres of the giant planets (Jupiter, Saturn, Uranus, Neptune) and an estimate of the typical value in molecular hydrogen for the protosolar nebula, from which our Solar System formed. The horizontal blue line shows the D/H ratio in Earth's oceans, which is 1.56×10^{-4}. Rosetta's ROSINA measured the ratio for 67P's water vapor and found it to be 5.3×10^{-4}, which is more than three times greater than for Earth's oceans. (ESA. Data from Altwegg et al. 2014 and references therein)

10.15   HIDDEN ICES

As has already been mentioned, when Rosetta first arrived at 67P the surface of the nucleus was remarkably dark and dry, with very little evidence of ices. However, this began to change as the comet approached perihelion and its activity increased.

One type of ice that Rosetta's instruments were seeking was carbon dioxide (known as 'dry ice' on Earth). This is known to be one of the most abundant species in cometary nuclei, but owing to its high volatility the ice is generally hidden at depth.

In late March 2015, scientists were surprised (and excited) when VIRTIS detected spectral absorption lines that indicated the presence of a very large patch of carbon dioxide ice – the first ever detection of this type of ice on a comet's nucleus. It was in Anhur, a region in the southern hemisphere that had recently exited from local winter. The 80×60 meter patch was observed on two consecutive days, but when the region was next observed, three weeks later, it had sublimated and disappeared.

Analysis revealed that the layer consisted of a few percent of carbon dioxide ice mixed with a darker blend of dust and organic material. The data suggested that the grains of ice and darker material were separated, not mixed. Assuming that all of the ice had turned into gas, scientists estimated that the patch had contained 57 kg of carbon dioxide; the equivalent of a layer 9 cm thick.

In the comet's environment, carbon dioxide froze at −193°C, far below the temperature where water turns into ice. Above this temperature, it turns directly from a solid to a gas. This limits the opportunities for finding carbon dioxide as ice on the surface.

Scientists believe the icy patch formed several years earlier, when the comet was still far from the Sun and its southern hemisphere was experiencing a long winter. At that time, some of the carbon dioxide still outgassing from the interior would have condensed on the surface, where it remained frozen. It vaporized only as the local temperature finally rose again in April 2015.

If so, this indicates a seasonal cycle of carbon dioxide ice that unfolds over the 6.5 year orbit, as opposed to the daily cycle of water ice that VIRTIS also saw shortly after Rosetta's arrival (see Fig. 10.18).

By contrast, the distribution of water ice beneath the dusty surface of the comet seemed to be widely spread, although not uniformly so, with small patches appearing and disappearing with changes in the level of activity. Occasionally, larger and thicker portions of water ice that dated back to a previous approach to the Sun were uncovered.

Rosetta detected numerous small, temporary, patches of water ice on several regions of the nucleus. In September 2014 alone, imaging found 120 bright areas that were up to 10 times brighter than the average surface brightness and were probably exposures of water ice. These were located on various types of terrains: smooth, rough, and below overhangs or near cliffs.

Many were found in clusters, but others seemed to be isolated boulders that displayed bright patches on their surfaces. The clusters, comprising several tens of meter-sized boulders spread across tens of meters, were typically in debris fields at the base of cliffs, and were most likely due to recent erosion or the collapse of a cliff that revealed fresher material from beneath the dusty surface. As perihelion approached, some exposures were seen that were considerably larger.

In April 2015, OSIRIS images revealed two large patches of water ice in the Anhur region, each larger than an Olympic swimming pool and much larger than any exposures of water ice previously seen. These newly detected patches comprised 20-30% of water ice mixed with a darker material in a layer of solid ice up to 30 cm thick. One of them was likely beneath the sheet of carbon dioxide ice that was observed by VIRTIS about one month before.

"We had already seen many meter-sized patches of exposed water ice in various regions of the comet, but the new detections are much larger, spanning some 30 × 40 meters each, and they persisted for about 10 days before they completely disappeared," said Sonia Fornasier of LESIA-Observatoire de Paris and Université Paris Diderot, France, who was lead scientist for a study of seasonal and daily surface color variations.

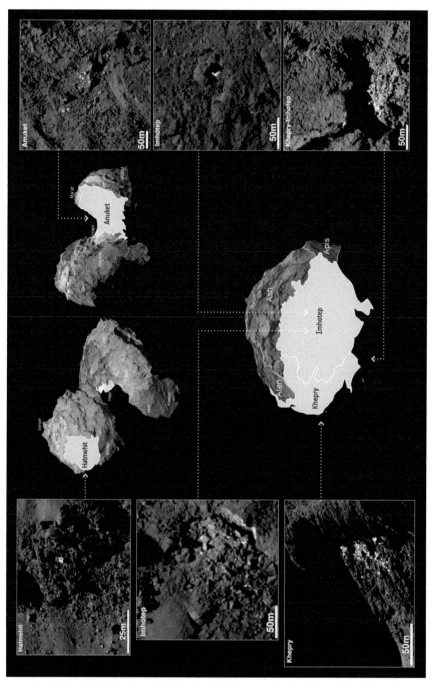

Fig. 10.22: Six bright, icy patches on 67P imaged by the OSIRIS-NAC in September 2014. The insets point to the broad regions in which they were discovered (not specific locations). Top left: a boulder showing icy patches in Hatmehit. Middle left: a cluster of icy features in Imhotep. Bottom left: a cluster in Khepry. Top right: a cluster in Anuket. Middle right: a bright feature in Imhotep. Bottom right: a cluster close to the Khepry-Imhotep boundary. (ESA/Rosetta/MPS for OSIRIS Team MPS/UPD/LAM/IAA/SSO/INTA/UPM/DASP/IDA)

Another example that was associated with the Aswan cliff collapse in the Seth region of the large lobe has already been mentioned (see Collapsing Cliffs). It exposed a bright, sharp slope and deposited numerous meter-sized boulders at the foot of the 134 meter high cliff.

Ice-rich areas such as these appeared as very bright portions of the surface and reflected light that was bluer in color than the redder surroundings. Experiments with mixtures of dust and water ice found that as the concentration of ice increases, the reflected light becomes gradually more blue, until it reaches a point where equal amounts of light are reflected in all colors.

Not all water ice exposures were short-lived. Numerous water-ice-rich deposits were seen to survive more than a few months. High spatial resolution multispectral imaging allowed large, ice-rich features to be monitored for up to 2 years. How could the ice survive for so long on the surface of the comet? One idea was that it was mixed with quite low amounts of dust. In at least one case though, it was located in a hollow that was protected from strong illumination by a nearby boulder.

10.16   THE CHEMICAL ZOO

Rosetta's instruments detected an abundance of gases flowing from the nucleus. Water vapor, carbon monoxide and carbon dioxide were the most abundant gases in the coma, contributing 95% of the volatiles. They also detected numerous other carbon compounds, nitrogen compounds and sulfur compounds, as well as the noble gases argon, xenon and krypton that do not easily react with other molecules.

One of the most surprising discoveries was the presence of molecular oxygen, this being the first detection of this gas in a comet.

Oxygen is the third most abundant element in the Universe, but it is surprisingly difficult to detect, even in star-forming clouds, because it is highly reactive and breaks apart to bind with other atoms and molecules.

ROSINA data from September 2014 to March 2015 indicated an abundance of 1-10% relative to the amount of water, with an average of 3.80±0.85%. The amount of molecular oxygen detected showed a strong correlation with the amount of water at any given time, suggesting that their origin on the nucleus and release mechanism were linked.

By contrast, it was poorly correlated with carbon monoxide and molecular nitrogen, even though they all have a similar volatility. Over the six-month investigation, the O_2/H_2O ratio remained constant, despite the decreasing heliocentric distance. Neither did it vary with Rosetta's longitude or latitude above the comet. At no time was O_3 (ozone) detected.

ROSINA also made the first measurement of molecular nitrogen at a comet. Nitrogen had only been seen previously in compounds such as hydrogen cyanide and ammonia.

For some time, scientists were perplexed by a mysterious infrared absorption feature observed by VIRTIS, but after ground-based experiments involving 'artificial' cometary surfaces they realized the chemical signature was predominantly due to salts of ammonium (NH_4^+) mixed with dust. The widespread presence of these salts implied the presence of considerably more nitrogen on 67P than first thought. The findings agreed with the ROSINA detection of gases produced by the sublimation of ammonium salts on dust grains ejected from the comet.

Although comets such as 67P may not have been the principal source of the water on Earth, Rosetta and its Philae lander detected a wide variety of chemical species, from simple atoms to increasingly complex molecules, including some ingredients that were crucial for the origin of life on Earth.

In all, ROSINA discovered over 60 molecules, 34 of which had never before been found on a comet. They included a number of organic compounds that are regarded as ingredients in the origin of terrestrial life. The prime example was the amino acid glycine, which is commonly found in proteins. Other molecules that were detected, such as ammonia, methylamine and ethylamine, are precursors to the formation of glycine.

There were also many carbon-chain molecules that are frequently used as fuel on Earth. These ranged from methane and ethane to propane, butane, pentane and heptane. In addition, there were many alcoholic compounds and aromatic ring compounds ranging from benzene to naphthalene.

Also present were acetylene, hydrogen cyanide, cyanogen, formaldehyde, hydrogen sulfide and other smelly molecules that are rich in sulfur, plus phosphorus, which plays a key role in DNA and cell membranes. And there were nonvolatile species 'sputtered' from the nucleus by energetic solar wind particles, including sodium, potassium, silicon and magnesium. There were also silicon isotope ratios that had never before been measured.

Early during the near-comet operations, Kathrin Altwegg, principal investigator for ROSINA, described the chemical zoo that her instrument had already detected.

"The perfume of 67P is quite strong, with the odor of rotten eggs (hydrogen sulfide), horse stable (ammonia), and the pungent, suffocating odor of formaldehyde. This is mixed with the faint, bitter, almond-like aroma of hydrogen cyanide. Add some whiff of alcohol (methanol) paired with the vinegar-like aroma of sulfur dioxide and a hint of the sweet aromatic scent of carbon disulfide, and you arrive at the 'perfume' of our comet."[3]

---

[3] This description was rather tongue in cheek, since these molecules were only present in small amounts.

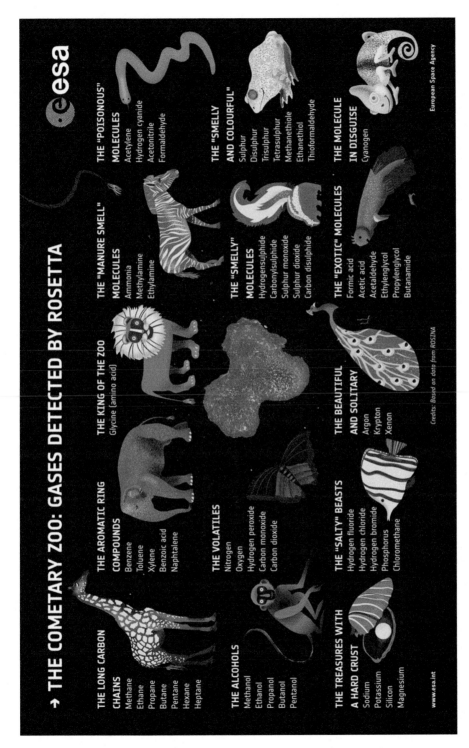

Fig. 10.23: A tongue-in-cheek artistic representation of the chemical zoo found by ROSINA in the gas around 67P. (ESA)

Inspired by Philae's detection of hydrogen sulfide, ammonia and hydrogen cyanide (which smell of rotten eggs, cat urine and bitter almonds respectively) Colin Snodgrass of the Open University in Milton Keynes, UK, and colleagues, commissioned The Aroma Company to develop a "perfume" named EAU de Comète.

"If you could smell a comet, this is what you would get," said Snodgrass, as he handed out samples at the 2016 Royal Society summer exhibition in London.

10.17   INTERIOR

The Philae lander made in-situ measurements of the nucleus, finding the surface to be a rigid crust 10-50 cm thick that was a mix of ice and dust, barely held together by ice grains sticking to each other. Below that was much fluffier material. In plain terms, 67P was crunchy on the outside and soft on the inside.

As already noted, Doppler measurements of Rosetta's radio signals showed that 67P's density is much lower than ice. The high porosity suggested there were voids. But neither CONSERT nor RSI found any evidence to support this. Instead, the investigations found that the porosity is due to the nucleus being a more or less uniform mixture of ice and fluffy dust grains, rather than a honeycomb structure of large voids.

The CONSERT experiment was able to 'sound' the small lobe of the nucleus by beaming a radio signal through the comet between the orbiter and Philae. The way in which the signals were slowed down by the ice and dust gave information about the nature of its interior. When the results were compared against theoretical models, scientists found that, rather than being a compacted solid, the dust in the interior was fluffy. This was consistent with the similar fluffy grains detected in the comet's dust jets by the COSIMA and GIADA instruments.

The CONSERT data indicated the presence of denser dust, or a higher dust-to-ice ratio, close to the surface – on average within a depth of 15 meters. At greater depths, the average density decreased, probably due to a higher ratio of ice to dust. These results were supported by those of the RSI experiment, which measured the distribution of mass inside the comet by analyzing the manner in which its gravity pulled the spacecraft, producing variations in the frequency of the signals received at Earth. Large voids would have been detectable as a tell-tale drop in the spacecraft's acceleration, but no such changes were recorded.

Although there was nothing to suggest internal differentiation of the icy material, there was evidence of stratification, both on the surface in the form of layers in cliff faces and terraces, and under the surface. Two concentric layering patterns were observed, on both of the lobes, resembling layers in an onion. A recent analysis has suggested that the widespread fracturing and layering dates to the gentle collision that gravitationally bound two objects together as the twin-lobed nucleus amalgamated to create 67P, billions of years ago (see A Planetary Building Block below).

10.18   PARTICLE ENVIRONMENT AND MAGNETIC FIELD

In order to measure the plasma environment of 67P, Rosetta had a set of instruments known as the Rosetta Plasma Consortium (RPC), while the Philae lander was equipped with the Rosetta Lander Magnetometer and Plasma Monitor (ROMAP). These were to investigate the nature of the surface and the surrounding plasma and electromagnetic environment.[4]

Although previous missions to comets had indicated the presence of a magnetic field around the nucleus, fly-by missions yielded only snapshots. Rosetta's unique ability to fly alongside a comet for two years permitted scientists to observe, for the first time, the 'birth' and evolution of the magnetosphere – seeing it form, expand around the spacecraft, then diminish as the comet retreated from the Sun. Furthermore, the magnetosphere changed in size and shape.

When a comet is far from the Sun and inactive, it travels through interplanetary space that is pervaded by the solar wind, a flow of electrically charged particles that carry a magnetic field.

As the comet starts to warm up, volatile substances vaporize and surround the nucleus with an atmosphere (coma). Ultraviolet sunlight can knock electrons from the molecules in the coma. So can collisions with solar wind particles. The electrons become part of the solar wind. The newly formed ions are influenced by the solar wind's electric and magnetic fields, and can be greatly accelerated.

When the comet approaches near enough to the Sun, its coma becomes so dense and ionized that it is electrically conductive. When this occurs, the atmosphere starts to resist the solar wind, and the magnetosphere is born as a shielded region around the nucleus. This slows the solar wind, diverting its flow, and preventing it from directly interacting with the nucleus.

The first cometary ions were unambiguously identified in Rosetta data returned on 7 August 2014 – one day after its official arrival at the comet. But for the first 9 months of near-comet operations, the RPC continued to detect the stream of charged particles from the solar wind. It was not until May-June 2015, shortly prior to perihelion, that the spacecraft moved inside the magnetosphere, within which it remained until December.

The early data from the RPC experiment indicated that, because of the comet's low activity, the ions were not fully 'picked up' by the solar wind. They traveled perpendicularly to the magnetic field, as a cross-field electric current that was unstable and caused oscillations in the magnetic data. Scientists were able to represent this data in audio form. As time went by, and larger amounts of gas poured into the coma and were ionized, the interaction with solar wind particles increased and the "singing comet" changed its tune.

---

[4] Plasma is a gas comprising charged particles (positive ions and free electrons).

Between April and June 2015, separate plasma regions with different velocity, temperature and density started to form around the nucleus. The comet's activity was not yet sufficient for the boundaries to be stable, so this chaotic phase produced oscillations in the magnetic field of 30-40 nanoteslas on time scales of seconds to hours.

The outermost boundary between the solar wind and the comet's magnetosphere is known as a 'bow shock' because it is a shock wave produced by the sudden slowing of supersonic solar wind particles. Rosetta did not detect a fully-fledged bow shock – because the spacecraft was too near the comet for most of its mission – but it was able to observe the formation of a bow shock.

A few months before perihelion, the spacecraft crossed the newly forming shock wave several times. It detected signs of an infant bow shock around 50 times closer to the nucleus than anticipated. This shifting boundary was observed to be asymmetric, wider than the fully developed bow shocks observed at other comets, and moving in unexpected ways. It was the first time a bow shock in such an early stage of formation had been detected anywhere in the Solar System. Rosetta also monitored the fading of the bow shock in the months that followed perihelion.

The comet interacted with the solar wind and plasma environment in unexpected ways. As the level of activity grew, the RPC-ICA instrument noticed a surprisingly large deflection of the comet's water ions when they interacted with the solar wind. When the water ions were added to the solar wind, they were accelerated in the direction of the solar wind's prevailing electric field and, owing to conservation of momentum, the solar wind particles were deflected in the opposite direction (rather like colliding billiard balls).

"When the deflection reached 90 degrees we thought that would be the maximum," explained Hans Nilsson of the Swedish Institute of Space Physics. "But the solar wind ions kept being deflected, reaching 180 degrees and moving back towards the Sun."

One of the most remarkable measurements was the record-breaking magnetic field strength of 280 nanoteslas on 3 July 2015.

"As far as we know, this is the highest magnetic field directly measured by any mission in interplanetary space," said RPC team member Christopher Carr. "It is especially remarkable given the background magnetic field carried by the solar wind would be only 1-2 nanoteslas and that the comet itself is unmagnetized!"[5]

Another surprise in June 2015 was the detection of a giant magnetic cavity, a region that had zero magnetic field. The RPC-MAG team observed several hundred such magnetic-field-free regions during the mission. These were formed when the solar wind (and its magnetic field) was unable to penetrate the dense coma around the comet.

---

[5] Earth's field ranges between approximately 25,000 and 65,000 nanotesla.

A similar feature had been observed at Comet Halley, but 67P's cavity occurred much closer to the nucleus than expected. However, since 67P was much less active than Halley, scientists predicted that such a diamagnetic cavity could form only in the months around perihelion and would extend only 50-100 km from the nucleus.

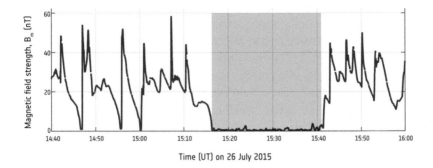

Fig. 10.24: The decrease in magnetic field strength measured by the RPC-MAG instrument at 67P on 26 July 2015 at a distance of about 170 km from the comet. The measurement reveals a 'diamagnetic cavity' – a region devoid of magnetic field – caused by the interaction of the comet with the solar wind. The cavity was much bigger than expected. The reason for that is probably a dynamic perturbation, or instability, that propagates and gets amplified along the boundary between the solar wind and the magnetic-field-free cavity, prompting the cavity to expand and allowing Rosetta to detect it. (ESA/Rosetta/RPC/IGEP/IC)

Rosetta made almost 700 detections of regions which had no magnetic field from June 2015 onwards, as perihelion approached. To the surprise of the scientists, its instruments were able to detect a magnetic-field-free bubble at a distance of 170 km from the nucleus.

Although one of the cavity detections, on 29 July 2015, occurred in conjunction with a strong outburst of gas and dust, this was an isolated example. Almost all of the other observations of magnetic-field-free regions were not accompanied by any appreciable increase of outgassing.

The most likely explanation seems to lie in the dynamic nature of the cavity boundary. Small oscillations can arise in the pile-up region of the solar wind, where it encounters the magnetic-field-free region on the Sun-facing side of the comet. If these oscillations propagate, and are amplified along the boundary in the direction opposite the Sun, they could readily cause the cavity to expand.

Such a traveling instability would also explain why the measurements of magnetic-field-free regions were sporadic and mainly spanned several minutes;

one on 26 July lasted 25 minutes and the longest one, in November, was about 40 minutes. The short duration of the detections was because the magnetic cavities repeatedly passed by the spacecraft.

After perihelion, as the comet moved away from the Sun and its outgassing and rate of dust production declined, Rosetta moved closer to the nucleus and the magnetometer continued to find magnetic-field-free regions for several months (the last detection was in February 2016). Meanwhile, as expected, the magnetic oscillations died down and became more erratic.

Scientists were very interested to discover whether 67P's nucleus was generating its own magnetic field. The ideal opportunity to find out came when Philae descended for a landing in November 2014.

The lander's unexpected bounces off the surface prior to coming to rest enabled the ROMAP instrument to obtain several measurements of the comet's magnetic field (both on its surface and just above) instead of at a single location. Meanwhile, the RPC-MAG on board the orbiter was measuring the magnetic field at an altitude of about 17 km.

The complementary magnetic field data ruled against local magnetic anomalies. If large chunks of material on the surface of 67P had been magnetized, ROMAP would have recorded additional variations in its signal as Philae passed over. By carefully reconstructing Philae's trajectory, scientists could combine the magnetic field data with the comet's shape. After careful analysis, Hans-Ulrich Auster and Philip Heinisch, with colleagues at the Institute for Geophysics and Extraterrestrial Physics at the Universität Braunschweig in Germany, found that the comet did indeed have a very weak magnetic field.

The upper limit of 0.9 nanotesla for the magnetic field at the surface of the nucleus implied an upper limit of 4 microtesla for the magnetic field existing in the solar nebula from which the comet was formed.

10.19   A PLANETARY BUILDING BLOCK

Understanding how comets change and evolve over time provides important insights into the types and abundance of ices in comets, and how long comets can spend within the inner Solar System before losing all their ice and becoming dry, dusty husks. Detailed studies of comets such as 67P also help us to understand the conditions that prevailed in the Solar System some 4.5 billion years ago.

Understanding the formation and evolution of our Solar System and its planets was one of the primary objectives of the Rosetta mission. Scientists were not disappointed. The flood of data from the orbiter and lander yielded insights into the origin of comets, the formation of planets and the origin of life.

Like other members of the family of comets whose orbits have been modified by Jupiter, 67P is believed to have spent billions of years far beyond the orbit of Neptune, in the Kuiper Belt. As examples of the primitive material that formed in the cold outer regions, these objects have remained largely unchanged, in particular by solar heating and catastrophic collisions. Only when its orbit was perturbed by a gravitational interaction with a distant star or other massive object, was 67P deflected inward toward the Sun, where it fell into the gravitational clutches of Jupiter.

Studies of the composition of 67P's coma established that the elemental abundances of the comet's dust matched the elemental abundances of the Solar System as a whole, suggesting that the comet is a very primitive body which dates back to the time the Solar System formed. Furthermore, nearly half of the mass of 67P's dust comprised complex carbon-rich molecules similar to those found in primitive meteorites known as carbonaceous chondrites.

Rosetta's instruments also detected molecular oxygen, molecular nitrogen, and isotopes of the noble gases neon, krypton etc. These data, together with the value of the deuterium-to-hydrogen (D/H) ratio in the cometary water, implied that the millions of objects that lurk unseen in the Kuiper Belt were formed far from the nascent Sun, in a very cold region of the protoplanetary nebula.

Theoretical simulations indicate that the D/H ratio should change with distance from the Sun, and also with time during the first few million years of Solar System history. Since the ratio does not change during a comet's life, it is an indicator of the conditions prevailing when it was being formed, which, in turn, is a clue to where it was most likely born relative to the Sun.

The D/H ratio has has been measured for 11 comets, mainly via remote observations, with Rosetta's comet having the highest value of them all (see Fig. 10.21). This suggests that comets formed across a much broader range of heliocentric distances in the young Solar System than had previously been thought.

Nitrogen and oxygen in particular, require low temperatures in order to become trapped in the ices of a comet, therefore the fact that they are observed outgassing today together with water vapor suggests they were incorporated into the comet during its formation.

The detection of molecular nitrogen was particularly important, because it is believed to have been the most common form of this gas available when the Solar System was forming. In the colder outer regions, molecular nitrogen was likely the main source of the nitrogen which was incorporated into the gaseous planets and into the icy building blocks that became the nuclei of comets.

Rosetta's observations of the physical components of 67P's nucleus provided insights into the processes by which micrometer-sized dust and ice particles grew into kilometer-sized objects such as comets.

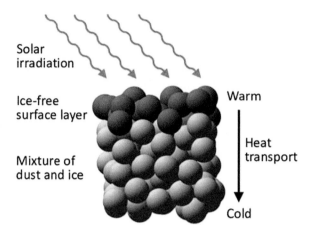

Fig. 10.25: A schematic representation of the porous surface structure of 67P. The nucleus is composed of millimeter-sized 'dust pebbles'. The interior is a mixture of dust and ice (light blue spheres). Only the uppermost layers, exposed to direct sunlight, do not contain any ice (dark grey spheres). (Maya Krause, TU Braunschweig)

As mentioned above (see Pits and Goosebumps), the results indicated that the nucleus consists of 'dust pebbles' that range in size from millimeters to centimeters. These were formed in the solar nebula when dust and ice particles collided and stuck together. According to researchers led by Jürgen Blum of the Technical University of Braunschweig, in Germany, this favors a particular model for the growth of larger bodies in the young Solar System.

In their model, the pebbles were concentrated so strongly by an instability in the solar nebula that a bound clump of these tiny dust aggregates collapsed inward by mutual gravitational attraction. Since this was a fairly gentle process the dust agglomerates were not destroyed, but combined into a larger body, with an even greater gravitational attraction – the foundation of the comet.

Blum's team regard this process as the missing link between the well-established formation of 'dust pebbles' ('planetary building blocks' formed in the solar nebula by collisions that cause dust and ice particles to stick together) and the gravitational accretion of much larger planetesimals into planets.

This pebble-collapse formation model explains many observations, such as the high porosity of the comet's nucleus and the amount of gas escaping from the interior.

Meanwhile, other Rosetta data indicated that the two lobes of 67P formed separately from the same material and subsequently became gravitationally bound in a gentle collision that took place at less than 1 m/s.

After spending billions of years in the cold depths of the Kuiper Belt, it was deflected inward to the zone occupied by the giant planets, and started to be shaped by mechanical shear stress. This defined both its surface and its interior, particularly the neck where the two lobes joined, subjecting it to fracturing and thinning (see Fig. 10.15 and Fig. 10.26).

Some tens of thousands of years ago, the comet began to travel closer to the Sun, so that the sublimation of its ices became the main process shaping the nucleus. Jets of gas loaded with dust were blasted into space, eroding the surface and redepositing fine material on the surface during each visit to the inner Solar System. In addition to reducing the size and mass of the nucleus, the various stresses further weakened the neck and may ultimately cause the nucleus to break apart.

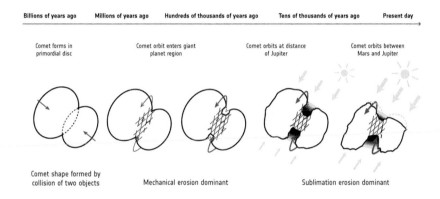

Fig. 10.26: The likely evolution of Rosetta's dual-lobed comet. 67P is thought to have been formed 4.5 billion years ago in the primordial disk of the Solar System, as two small objects slowly collided and stuck together. After remaining in the remote Kuiper Belt for billions of years, 67P entered the giant planet region hundreds of thousands to millions of years ago. By this time, a form of geological erosion named mechanical shear stress had begun to shape its surface and interior. Rosetta data show that this stress peaked in the neck which connects the two lobes. The neck fractured and thinned over time, as shown by the cross-hatched lines. In the period from tens of thousands of years ago to the present day, erosion by sublimation was dominant in shaping the comet's surface and interior. This kind of erosion takes place as the Sun warms ices within the comet, causing the ice to turn to gas and escape to space, carrying dust motes with it. This activity further weakened the comet's neck. All these stresses were increased as the perihelion distance shifted inwards from the orbit of Jupiter towards that of Mars. It is important to recognize that the red arrows don't imply cometary rotation; instead, they represent shear deformation, and therefore illustrate the torque generated at the neck. (C. Matonti et al, 2019)

REFERENCES

Rosetta publications archive, ESA: http://sci.esa.int/rosetta/31062-publications/?farchive_objecttypeid=15&farchive_objectid=30995&fareaid_2=13

ESA Rosetta website: http://rosetta.esa.int/

ESA Rosetta blogs: https://blogs.esa.int/rosetta/

View Rosetta's Comet: http://sci.esa.int/comet-viewer/

Rosetta Mission (Facebook): https://www.facebook.com/RosettaMission/

Catching a comet, Science special issue, Vol. 347, Issue 6220, 23 January 2015a: http://sci.esa.int/rosetta/56403-science-special-issue-catching-a-comet/

Philae's first days on the comet, Science special issue, Vol. 349, Issue 6247, 31 July 2015: http://sci.esa.int/rosetta/56291-science-special-issue-philaes-first-look/

Rosetta mission results pre-perihelion, Astronomy & Astrophysics special issue, Vol. 583, November 2015: http://sci.esa.int/rosetta/56496-rosetta-papers-in-astronomy-and-astrophysics/

From Giotto to Rosetta – a collection from the ESLAB 50 Symposium, Monthly Notices of the Royal Astronomical Society, Vol. 462, Issue Suppl_1, November 2016: https://academic.oup.com/mnras/issue/462/Suppl_1

Comets: A new vision after Rosetta and Philae, Monthly Notices of the Royal Astronomical Society, Vol. 469, Issue Suppl_2, July 2017: https://academic.oup.com/mnras/issue/469/Suppl_2

Rosetta-Philae: The Adventure Continues, CNES Magazine #71, January 2017: https://cnes.fr/sites/default/files/drupal/201702/default/cnesmag_71_gb_web-simple.pdf

ESA Planetary Science Archive for Rosetta: https://archives.esac.esa.int/psa/#!Table%20View/Rosetta=mission

Impressions of Rosetta's legacy, ESA, 22 December 2016: http://sci.esa.int/rosetta/58680-impressions-of-rosetta-s-legacy/#

Science news – Rosetta: https://www.sciencemag.org/tags/rosetta

Science special issue, *Catching a Comet.* (First results from the Rosetta orbiter instruments.) Science, Vol. 347, Issue 6220, 23 January 2015b: https://sci.esa.int/web/rosetta/-/56403-science-special-issue-catching-a-comet

Science special issue, *Philae's First Look.* (6 papers), Vol. 349, Issue 6247, 31 July 2015: https://sci.esa.int/web/rosetta/-/56291-science-special-issue-philaes-first-look

Astronomy & Astrophysics special issue, *Rosetta Mission Results Pre-perihelion* (48 papers). Astron. & Astrophys., Vol. 583, November 2015: https://sci.esa.int/web/rosetta/-/56496-rosetta-papers-in-astronomy-and-astrophysics

Monthly Notices of the Royal Astronomical Society special issue, *Results from the "ESLAB 50 Symposium – Spacecraft at Comets from 1P/Halley to 67P/Churyumov-Gerasimenko"*, MNRAS, Vol. 462, Issue Suppl_1, November 2016: https://academic.oup.com/mnras/issue/462/Suppl_1

Rosetta end of mission, grand finale – press kit, 25 September 2016: http://sci.esa.int/rosetta/58334-rosetta-end-of-mission-grand-finale-press-kit-september-2016/#

Max-Planck-Institute for Solar System Research: https://www.mps.mpg.de/en/Rosetta

OSIRIS Image of the Day Archive: https://planetgate.mps.mpg.de/Image_of_the_Day/public/IofD_archive.html

Living With A Comet – Summary, 29 September 2016: https://blogs.esa.int/rosetta/2016/09/29/livingwithacomet-blog-post-series-summary/

VIRTIS Maps Comet 'Hot Spots', 8 September 2014: http://blogs.esa.int/rosetta/2014/09/08/virtis-maps-comet-hot-spots/

Rosetta Measures Production of Water at Comet Over Two Years, 27 September 2016a: https://blogs.esa.int/rosetta/2016/09/27/rosetta-measures-production-of-water-at-comet-over-two-years/

How Rosetta's Comet Got Its Shape, 28 September 2015: https://sci.esa.int/web/rosetta/-/56543-how-rosetta-s-comet-got-its-shape

The Nucleus of comet 67P/Churyumov-Gerasimenko – Part I: The global view – nucleus mass, mass-loss, porosity, and implication, Pätzold, Martin, et al, MNRAS, Vol. 483, Issue 2, February 2019, p. 2337-2346: https://doi.org/10.1093/mnras/sty3171. https://academic.oup.com/mnras/article-abstract/483/2/2337/5210098

Geomorphology of Comet 67P/Churyumov-Gerasimenko, Samuel P. D. Birch, et al., MNRAS, Vol. 469, Issue Suppl_2, July 2017, p. S50-S67: https://doi.org/10.1093/mnras/stx1096 Published: 6 May 2017. https://academic.oup.com/mnras/article/469/Suppl_2/S50/3800696

Regional surface morphology of comet 67P/Churyumov-Gerasimenko from Rosetta/OSIRIS images: The southern hemisphere, M. R. El-Maarry et al, Astron. & Astrophys., Vol. 593, September 2016: https://doi.org/10.1051/0004-6361/201628634. Published online 30 September 2016. (Corrigendum published online 26 January 2017, https://doi.org/10.1051/0004-6361/201628634e)

The Rosetta mission orbiter science overview: the comet phase, M. G. G. T. Taylor, et al, Philos Trans A Math Phys Eng Sci. 2017 July 13; 375(2097). Published online 29 May 2017. https://www.ncbi.nlm.nih.gov/pmc/articles/PMC5454230/

Summer Fireworks on Rosetta's Comet, 23 September 2016: https://sci.esa.int/web/rosetta/-/58317-summer-fireworks-on-rosetta-s-comet

Comet sinkholes generate jets, 1 July 2015: https://sci.esa.int/web/rosetta/-/56116-comet-sinkholes-generate-jets

Getting to know Rosetta's comet, 22 January 2015: https://sci.esa.int/web/rosetta/-/55295-getting-to-know-rosetta-s-comet

Rosetta Watches Comet Shed Its Dusty Coat, 26 January 2015: https://sci.esa.int/web/rosetta/-/55341-rosetta-watches-comet-shed-its-dusty-coat

Before and after: Unique Changes Spotted on Rosetta's Comet, 21 March 2017: http://www.esa.int/Science_Exploration/Space_Science/Rosetta/Before_and_after_unique_changes_spotted_on_Rosetta_s_comet

Dust-to-Gas and Refractory-to-Ice Mass Ratios of Comet 67P/Churyumov-Gerasimenko from Rosetta Observations, Mathieu Choukroun, et al, Space Science Reviews, vol. 216, Article number: 44 (2020). https://link.springer.com/article/10.1007/s11214-020-00662-1

Collapsing Cliff Reveals Comet's Interior, 21 March 2017: https://www.esa.int/Science_Exploration/Space_Science/Rosetta/Collapsing_cliff_reveals_comet_s_interior

Comet's Collapsing Cliffs and Bouncing Boulders, 18 September 2019: https://www.esa.int/Science_Exploration/Space_Science/Rosetta/Comet_s_collapsing_cliffs_and_bouncing_boulders

Geomorphology of the Imhotep region on comet 67P/Churyumov-Gerasimenko from OSIRIS observations, A.-T. Auger et al, Astron. & Astrophys., vol. 583, November 2015: https://www.aanda.org/articles/aa/full_html/2015/11/aa25947-15/aa25947-15.html . Published online 30 October 2015 https://doi.org/10.1051/0004-6361/201525947

Global-scale brittle plastic rheology at the cometesimals merging of comet 67P/Churyumov-Gerasimenko, Marco Franceschi et al, PNAS: https://www.pnas.org/content/117/19/10181 . First published 27 April 2020, https://doi.org/10.1073/pnas.1914552117

Meter-scale thermal contraction crack polygons on the nucleus of comet 67P/Churyumov-Gerasimenko, A.-T. Auger, et al, Icarus, Vol. 301, February 2018, p. 173-188: https://www.sciencedirect.com/science/article/abs/pii/S0019103516300410 https://doi.org/10.1016/j.icarus.2017.09.037

Rosetta Reveals Comet's Water-Ice Cycle, 23 September 2015a: https://sci.esa.int/web/rosetta/-/56513-rosetta-reveals-comet-s-water-ice-cycle

Rosetta Measures Production Of Water At Comet Over Two Years, 27 September 2016b: https://blogs.esa.int/rosetta/2016/09/27/rosetta-measures-production-of-water-at-comet-over-two-years/

Icy Surprises at Rosetta's Comet, 17 November 2016: http://www.esa.int/Science_Exploration/Space_Science/Rosetta/Icy_surprises_at_Rosetta_s_comet

The Colour-Changing Comet, 7 April 2016: https://blogs.esa.int/rosetta/2016/04/07/the-colour-changing-comet/

Rosetta and the Chameleon Comet, 5 February 2020: https://sci.esa.int/web/rosetta/-/rosetta-and-the-chameleon-comet

Exposed Water Ice Detected on Comet's Surface, 24 June 2015: https://www.esa.int/Science_Exploration/Space_Science/Rosetta/Exposed_water_ice_detected_on_comet_s_surface

An Orbital Water-Ice Cycle On Comet 67P From Colour Changes, Gianrico Filacchione et al, Nature, Vol. 578, p. 49-52, 5 February 2020: https://www.nature.com/articles/s41586-020-1960-2

Long-term Survival of Surface Water Ice on Comet 67P, N. Oklay, et al, MNRAS, Vol. 469, Issue Suppl_2, July 2017, p. S582-S597: https://doi.org/10.1093/mnras/stx2298 . Published: 11 September 2017. https://academic.oup.com/mnras/article/469/Suppl_2/S582/4111159

Rosetta Fuels Debate On Origin Of Earth's Oceans, 10 December 2014: https://www.esa.int/Science_Exploration/Space_Science/Rosetta/Rosetta_fuels_debate_on_origin_of_Earth_s_oceans

The Great Pit of Deir el-Medina, 9 September 2016: https://blogs.esa.int/rosetta/2016/09/09/the-great-pit-of-deir-el-medina/

COSIMA Watches Comet Shed Its Dusty Coat, 26 January 2015: https://blogs.esa.int/rosetta/2015/01/26/cosima-watches-comet-shed-its-dusty-coat/

GIADA Investigates Comet's "Fluffy" Dust Grains, 9 April 2015: http://blogs.esa.int/rosetta/2015/04/09/giada-investigates-comets-fluffy-dust-grains/

Imaging Tiny Comet Dust in 3D, 31 August 2016: http://blogs.esa.int/rosetta/2016/08/31/imaging-tiny-comet-dust-in-3d/

Synthesis of the morphological description of cometary dust at comet 67P/Churyumov-Gerasimenko, C. Güttler et al, Astron. & Astrophys., Vol. 630, October 2019: https://doi.org/10.1051/0004-6361/201834751 . Published online 20 September 2019: https://www.aanda.org/articles/aa/full_html/2019/10/aa34751-18/aa34751-18.html

Rosetta Finds Comet Plume Powered From Below, 26 October 2017: http://www.esa.int/Science_Exploration/Space_Science/Rosetta/Rosetta_finds_comet_plume_powered_from_below

Rosetta Makes First Detection of Molecular Nitrogen at a Comet, 19 March 2015b: http://www.esa.int/Science_Exploration/Space_Science/Rosetta/Rosetta_makes_first_detection_of_molecular_nitrogen_at_a_comet

First Detection of Molecular Oxygen at a Comet, 28 October 2015: https://www.esa.int/Science_Exploration/Space_Science/Rosetta/First_detection_of_molecular_oxygen_at_a_comet

Ammonium Salts Found on Rosetta's Comet, 13 March 2020: http://www.esa.int/ESA_Multimedia/Images/2020/03/Ammonium_salts_found_on_Rosetta_s_comet#.Xnya7AaxfFI.link

The Cometary Zoo, 29 September 2016: https://blogs.esa.int/rosetta/2016/09/29/the-cometary-zoo/

Watching the Birth of a Comet Magnetosphere, 22 January 2015: http://blogs.esa.int/rosetta/2015/01/22/watching-the-birth-of-a-comet-magnetosphere/

Rosetta Finds Magnetic Field-Free Bubble At Comet, 11 March 2016c: https://blogs.esa.int/rosetta/2016/03/11/rosetta-finds-magnetic-field-free-bubble-at-comet/

Can the Magnetic Field of Rosetta's Comet Tell Us How the Outer Solar System Formed?, 17 June 2019: https://astrobiology.nasa.gov/news/can-the-magnetic-field-of-rosettas-comet-tell-us-how-the-outer-solar-system-formed/

A Comet's Life – a New Sonification of RPC Data, 29 September 2016: https://blogs.esa.int/rosetta/2016/09/29/a-comets-life-a-new-sonification-from-rosettas-rpc-data/

Rosetta's Ancient Comet, Rubin, Martin, and Tubiana, Cecilia, Planetary Report, June 2019: https://www.planetary.org/blogs/guest-blogs/2019/rosettas-ancient-comet.html

Philae science results: Comet 67P is Crunchy on the Outside, Soft on the Inside, Emily Lakdawalla, Planetary Society, 9 May 2018: https://www.planetary.org/blogs/emily-lakdawalla/2018/0509-philae-science-results-comet.html

Comet Mission Reveals 'Missing Link' In Our Understanding Of Planet Formation, 25 October 2017: https://www.eurekalert.org/pub_releases/2017-10/ras-cmr102517.php

Rosetta's Comet Sculpted by Stress, 18 February 2019: https://sci.esa.int/web/rosetta/-/61136-rosetta-s-comet-sculpted-by-stress

Evidence For The Formation Of Comet 67P/Churyumov-Gerasimenko Through Gravitational Collapse Of A Bound Clump Of Pebbles, Jürgen Blum, et al, MNRAS, Vol. 469, Issue Suppl_2, July 2017, p. S755–S773: https://doi.org/10.1093/mnras/stx2741. https://academic.oup.com/mnras/article/469/Suppl_2/S755/4564447

Appendix 1
Comet Missions Past and Future

International Cometary Explorer (ICE) (NASA)

Launched 12 August 1978. ICE achieved the first ever comet encounter. It was originally known as ISEE-3 (International Sun-Earth Explorer 3). Having completed its mission, it was reactivated and diverted to pass through the tail of Comet Giacobini-Zinner on 11 September 1985. It also observed Halley's Comet from a distance of 28 million km in March 1986.

https://solarsystem.nasa.gov/missions/isee-3-ice/in-depth/

Vega 1 and Vega 2 (USSR)

Launched 15 and 20 December 1984. Two identical Russian probes. Each released a lander as it flew past Venus, prior to rendezvousing with Halley's Comet. Both imaged the nucleus while other experiments studied its nearby dust, plasma, gas, energetic particles and magnetic field. Vega 1 passed by the nucleus at a range of 8,890 km on 6 March 1986, and three days later its partner did so at 8,030 km.

http://www.iki.rssi.ru/ssp/VEGA/

Sakigake and Suisei (JAXA/ISAS)

Launched 8 January 1985 and 19 August 1985, respectively. Japan's inaugural deep space missions. Suisei flew to within 151,000 km of Halley's Comet on 8 March 1986 to study its interactions with the solar wind. Sakigake flew by at a distance of 7 million km on 11 March 1986 to study radio and plasma waves.

http://www.isas.jaxa.jp/en/missions/spacecraft/past/suisei.html
http://www.isas.jaxa.jp/en/missions/spacecraft/past/sakigake.html

© Springer Nature Switzerland AG 2020
P. Bond, *Rosetta: The Remarkable Story of Europe's Comet Explorer*,
Springer Praxis Books, https://doi.org/10.1007/978-3-030-60720-3

Giotto (ESA)

Launched 2 July 1985. Europe's first deep space mission. Giotto obtained the most detailed pictures yet taken of a comet when it flew past the nucleus of Comet Halley at a distance of less than 600 km on 13 March 1986. The images showed a black, potato-shaped object with bright, active regions that were firing jets of gas and dust into space. Giotto was damaged by an impact with a large dust particle. Although some instruments were no longer operational, including its camera, it was redirected to another target. It became the first spacecraft to visit two comets when it passed within 200 km of Comet Grigg-Skjellerup on 10 July 1992.

https://sci.esa.int/web/giotto
https://www.esa.int/esapub/bulletin/bulletin125/bulletin125.pdf

Deep Space 1 (NASA)

Launched 25 October 1998. The first spacecraft in NASA's New Millennium program. The primary mission was to test 12 new advanced technologies, particularly ion propulsion. Most of these technologies were validated during the first few months of flight. It then approached to within 26 km of asteroid 9969 Braille on 28 July 1999. The few pictures returned showed its longest side to be about 2.2 km and its shortest side to be about half of that across.

Despite the loss of its star tracker navigation system, the mission was extended and DS1 completed a successful fly-by of Comet Borrelly on 22 September 2001, passing within 2,200 km of the nucleus. Analysis of the 30 or so black-and-white images of the bowling-pin-shaped nucleus established that it was about 8 km long. Other instruments examined the gas and dust in the surrounding coma and studied the interaction of the comet with the solar wind. It found the nucleus was not in the center of the coma, as had been expected. The coma's 'lopsided' shape was due to a huge jet of material shooting into space from one side of the nucleus.

https://solarsystem.nasa.gov/missions/deep-space-1/in-depth/

Stardust (NASA)

Launched 7 February 1999. A NASA Discovery-class mission. It was the first spacecraft to bring samples of comet material back to Earth. The primary goal was to fly by Comet Wild 2, collecting samples of dust from its coma. On its way to the comet, the spacecraft captured interstellar particles that originated outside the Solar System. The mission included a 45.7 kg sample return capsule shaped like a blunt-nosed cone. The samples were to be collected using a low density, microporous, silica-based substance known as 'aerogel', which was attached to deployable panels on the spacecraft designed to preserve the fragile materials.

On 2 January 2004, the spacecraft entered the coma, passing about 250 km from the nucleus. The imaging system took 72 pictures of the nucleus. On 15 January 2006, the Stardust sample return capsule was released and parachuted to a designated landing zone in the Utah desert. It delivered more than 10,000 particles larger than 1 micrometer from the comet.

The pictures of Wild 2 showed pinnacles 100 meters tall and craters more than 150 meters deep. Some craters had a rounded central pit surrounded by ejected material, others had a flat floor and straight sides. One crater, called Left Foot, had a diameter of about 1 km, one-fifth the width of the nucleus. More than two dozen jets were erupting simultaneously. The largest of these bombarded the vehicle with about a million particles per second. One key result was the detection of glycine, a fundamental building block of life. This was the first time an amino acid had been found in a comet.

The Stardust mother craft flew past Earth and was retargeted to Comet Tempel 1, which had been stuck by the Deep Impact mission. As Stardust/NExT (New Exploration of Tempel 1) it flew past this target on 15 February 2011 at a range of 181 km, sending back 72 images of the nucleus that identified changes made six years earlier. It identified the crater produced by the Deep Impact probe. This was the first time a comet had been revisited.

https://solarsystem.nasa.gov/missions/stardust/in-depth/
https://www.nasa.gov/mission_pages/stardust/main/index.html
https://stardust.jpl.nasa.gov/home/index.html

Comet Nucleus Tour (Contour) (NASA)

Launched 1 July 2002. A NASA Discovery mission to make a detailed study of the interiors of at least two comet nuclei, compiling topographical and compositional maps, and collecting data on the structure and composition of their comas. Encounters were planned with Comet Encke (12 November 2003) and Comet Schwassmann-Wachmann 3 (19 June 2006), with one additional encounter with Comet d'Arrest as a possibility. Unfortunately, the spacecraft was destroyed when its engine exploded while attempting to leave Earth orbit.

https://solarsystem.nasa.gov/missions/contour/in-depth/

Deep Impact (NASA)

Launched 12 January 2005. A NASA Discovery mission to Comet Tempel 1. It consisted of two craft, which separated as the comet was approached. One of these was a 350 kg copper 'impactor'. It crashed into the sunlit side of the comet on 3 July 2005. The impact took place at 10 km per second and produced a crater estimated at 150 meters in diameter and 25 meters deep. This was the first time any human-made object had collided with a comet. At the same time, the main space-craft flew past the comet at a range of about 500 km, taking pictures and infrared spectral mapping data of the impact site (obscured by a dust cloud), the ejecta plume, and the entire nucleus.

For its extended mission, known as EPOXI, Deep Impact was redirected to Comet Hartley 2, which it flew past at a range of 694 km on 4 November 2010. Some images enabled scientists to identify jets of dust with particular features on the nucleus. This encounter determined the peanut-shaped nucleus to be spinning around one axis while simultaneously tumbling around another. It had two lobes, which were different in composition, with water ice and methanol, carbon dioxide, and possibly ethane. The release of these gases occurred at different locations on the comet. Carbon dioxide driven jets erupted from the ends of the comet, particularly the smaller end. Water vapor was released from the 'waist' with very little, or no carbon dioxide or water ice. The smooth waist also seemed to be the site of material deposited on the surface. The surface was scattered with brighter blocks about 50 meters high and 80 meters wide that were several times more reflective than the surface average.

The spacecraft was also used for the remote study of distant objects such as Comet Garradd in early 2012 and Comet ISON in early 2013.

https://solarsystem.nasa.gov/missions/deep-impact-epoxi/in-depth/
https://www.nasa.gov/mission_pages/deepimpact/main/index.html#.Xn46TmDgqUk

Gallery: Comparing Comets

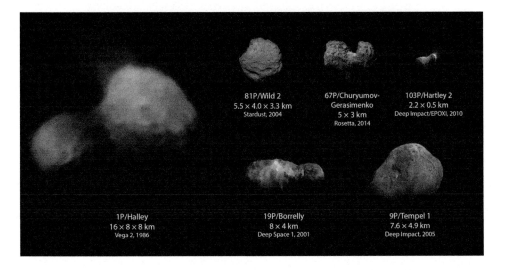

Fig. Ap1.1 Comet gallery. A scale comparison of comets that have been visited by spacecraft. (Halley: Russian Academy of Sciences/Ted Stryk. Borrelly: NASA/JPL/Ted Stryk. Tempel 1 and Hartley 2: NASA/JPL/UMD. Churyumov-Gerasimenko: ESA/Rosetta/NAVCAM/Emily Lakdawalla. Wild 2: NASA/JPL. The montage was prepared by Emily Lakdawalla/Planetary Society)

Borrelly: The 8 km long bowling-pin-shaped nucleus of Comet Borrelly possesses a variety of terrains and surface textures, mountains and fault structures.

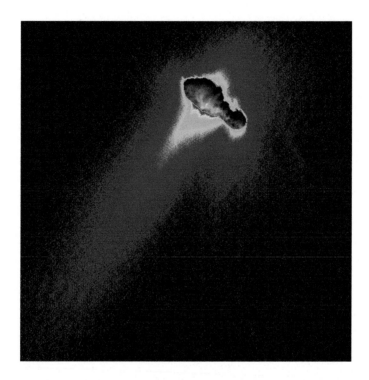

Fig. Ap.1.2 Comet Borrelly. (NASA/JPL)

Wild 2: Stardust imaged several large depressions on the nucleus while jets of gas and dust erupted from a very active surface. The egg-shaped nucleus measured about 3 × 4 × 5 km. Of all the comet nuclei so far observed at close quarters, this most closely resembles Comet 67P, with its topographically variable surface and numerous smooth-floored circular hollows.

Fig. Ap.1.3 Comet Wild 2. (NASA/JPL)

Tempel 1: The nucleus was about 5 × 7 km. Deep Impact's impactor imaged smooth regions and numerous circular features that were probably impact craters.

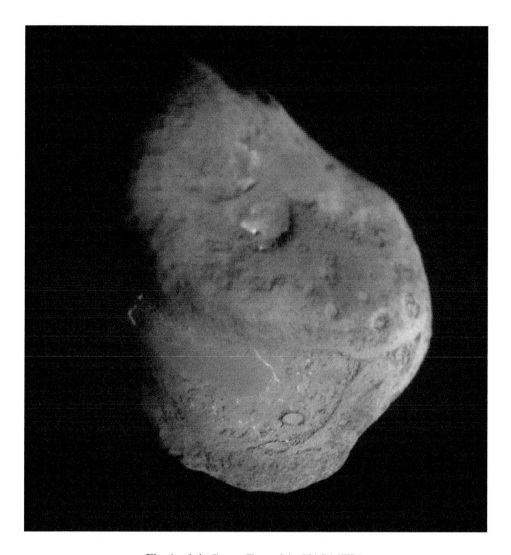

Fig. Ap.1.4 Comet Tempel 1. (NASA/JPL)

Hartley 2: The double-lobed nucleus was about 2 km long and 0.5 km across its waist. The smooth waist contrasted with more rugged terrain on the two lobes, as well as blocks ranging up to 50 meters high and 80 meters across.

Fig. Ap.1.5 Comet Hartley 2. (NASA/JPL/UMD)

Halley: The Soviet Vega spacecraft and ESA's Giotto imaged a potato-shaped nucleus about $16 \times 8 \times 8$ km, shrouded in dust and gas ejected by numerous jets. It had a dark surface with extremely varied topography, including hills, mountains, ridges, depressions, and at least one crater.

The Future: ESA's Comet Interceptor

On 19 June 2019, ESA's Science Program Committee announced its intention to launch a probe that will visit a comet heading sunward from the outer Solar System.

The €150 million spacecraft, known as Comet Interceptor, will be the space agency's first so-called "fast" or F-class mission in its Cosmic Vision Program. Taking under a decade from selection to launch, the spacecraft is expected to launch in 2028. The mission is expected to be completed within five years of launch.

Comprising three spacecraft, it will be the first ever mission to visit a truly pristine comet or other interstellar object visiting the inner Solar System. It is envisaged as a revolutionary new type of mission that will be launched before its primary target has been detected or identified.

One exciting possibility is that the mission will be able to intercept a comet or asteroid that originates from well beyond our Solar System. All previous missions, including ESA's pioneering Giotto and Rosetta, encountered comets with orbital periods of less than 200 years that have approached the Sun many times and, consequentially, have undergone significant changes.

However, in recent years, two interstellar objects have been discovered that were not gravitationally bound to the Sun. The first, named 1I/'Oumuamua, was spotted racing through the Solar System in September 2017. Unfortunately, its discovery occurred too late to arrange a fly-by mission, and the object left the Solar System with many unanswered questions. The surprises included an absence of cometary activity and what appeared to be an extremely elongated shape. A second, comet-like interstellar object, known as 2I/Borisov, was discovered in August 2019.

The appearance of two such intruders in rapid succession has led researchers to suggest that interstellar visitors could be fairly frequent. Yale astronomers Gregory Laughlin and Malena Rice have predicted that a few large objects and hundreds of smaller ones could be entering the Solar System each year.

Günther Hasinger, ESA's Director of Science, commented, "Pristine or dynamically new comets are entirely uncharted and make compelling targets for close-range spacecraft exploration to better understand the diversity and evolution of comets."

Comet Interceptor will be launched with ESA's ARIEL satellite, which will scrutinize the atmospheres of extrasolar planets. It will travel out to the L2 Lagrange point – 1.5 million km 'behind' Earth as viewed from the Sun – where it will lie in wait, for several years if need be, until a suitable comet is spotted on an inward trajectory.

Recent advances in ground-based surveys mean that the sky can be scanned for much fainter objects. Astronomers will search for a candidate target using sky surveys such as Pan-STARRS, ATLAS and the Large Synoptic Survey Telescope (LSST), which is now under construction in Chile. By routinely detecting inbound comets well outside the orbit of Saturn, LSST will provide much longer advance warning of their approach.

"With LSST observing more than three magnitudes deeper than current surveys, we will routinely pick up new comets at a greater distance than ever before," said mission team member Michele Bannister from Queen's University Belfast.

"So for the mission target, we'll have years to study it with telescopes on Earth before Comet Interceptor flies across to it. It's possible the mission target could be discovered even before Comet Interceptor launches."

Once an object has been selected, Comet Interceptor will head toward it, completing a fly-by while the inbound object is still approaching 1 AU. The three modules of Comet Interceptor will separate several weeks prior to interception, each with a complementary science payload to provide different perspectives of the comet's nucleus and its gas, dust, and plasma environment. This will enable astronomers to build a detailed picture of its composition and shape, creating a 3D profile of a 'dynamically new' object that contains unprocessed material from the dawn of the Solar System.

The proposed instruments for the three spacecraft are:

Spacecraft A (ESA)

- CoCa: Comet Camera – to obtain high resolution images of the comet's nucleus at several wavelengths.
- MANIaC: Mass Analyzer for Neutrals and Ions at Comets – a mass spectrometer to sample the gases released from the comet.
- MIRMIS: Multispectral InfraRed Molecular and Ices Sensor – to measure the heat radiation emitted by the comet's nucleus and to study the molecular composition of the gas coma.
- DFP: Dust, Field, and Plasma – to understand the charged gases, energetic neutral atoms, magnetic fields, and dust surrounding the comet.

Spacecraft B1 (JAXA)

- HI: Hydrogen Imager – an ultraviolet camera to study the cloud of hydrogen gas surrounding the target.
- PS: Plasma Suite – to study the charged gases and magnetic field around the target.
- WAC: Wide Angle Camera – to take images of the nucleus around closest approach from a unique viewpoint.

Spacecraft B2 (ESA)

- OPIC: Optical Imager for Comets – mapping of the nucleus and its dust jets at different visible and infrared wavelengths.
- EnVisS: Entire Visible Sky coma mapper – to map the entire sky within the head and near-tail of the comet to identify changing structures within the dust, neutral gas, and ionized gases.
- DFP: Dust, Field, and Plasma – a subset of the DFP sensors on Spacecraft A.

The instrument suite will draw on heritage from earlier missions, including a camera derived from one on the Trace Gas Orbiter of the ExoMars mission, along with a mass spectrometer, and dust, fields and plasma instruments similar to those that flew on Rosetta.

On 6 February 2020, following an internal assessment of the results of the Phase 0 studies, ESA agreed to start the Definition Phase (Phase A) for Comet Interceptor, with a Request for Information being released to industry in preparation for issuing the spacecraft Invitation To Tender later in the year. The Definition Phase is expected to be completed in 2022.

The international Comet Interceptor team is headed by Geraint Jones (UCL Mullard Space Science Laboratory, UK) and Colin Snodgrass (University of Edinburgh, UK).

REFERENCES

ESA's New Mission to Intercept a Comet, 19 June 2019: https://www.esa.int/Science_Exploration/Space_Science/ESA_s_new_mission_to_intercept_a_comet
http://sci.esa.int/cosmic-vision/61416-esa-s-new-mission-to-intercept-a-comet/
Comet Interceptor proposal team: http://www.cometinterceptor.space/
The European Space Agency's Comet Interceptor lies in wait, Snodgrass, C., Jones, G. H., Nature Communications, 10, 5418 (2019). https://doi.org/10.1038/s41467-019-13470-1
Comet mission given green light by European Space Agency, Physics World, 21 June 2019: https://physicsworld.com/a/comet-mission-given-green-light-by-european-space-agency/

Appendix 2
The Rosetta Team

From the preliminary studies of a comet mission in the 1970s and 1980s to Rosetta's historic touchdown on the nucleus of a comet and the subsequent analysis of the resulting data, many thousands of engineers, scientists and others participated in this unique endeavor.

Page limitations prevent me from naming all of these people, and fully acknowledging their contributions, but here are biographical summaries of some of the project managers, project scientists, principal investigators, and others who made important contributions.

Project Managers

The project managers during mission development and implementation of Rosetta were:

- John Credland (July 1994 to July 1997).
- Michel Verdant (August 1997 to 1998).
- Bruno Gardini (1997 to 2000).
- John Ellwood (2000 to 2004).

Mission Managers (overall mission responsibility, taking over from the project manager for the Operations Phase):

- Gerhard Schwehm (2004 to May 2013).
- Fred Jansen (June 2013 to January 2015).
- Patrick Martin (February 2015 to 2020).

Project Scientists (scientific mission responsibility throughout the mission life cycle):

- Gerhard Schwehm (1989 to 2006); before that he was Study Scientist (1985 to 1989).
- Rita Schultz (2007 to 2013).
- Matthew Taylor (2013 to 2020).

© Springer Nature Switzerland AG 2020
P. Bond, *Rosetta: The Remarkable Story of Europe's Comet Explorer*,
Springer Praxis Books, https://doi.org/10.1007/978-3-030-60720-3

A more exhaustive list of Rosetta team members at the end of the mission is available at: https://sci.esa.int/web/rosetta/-/43058-mission-team#PIs_Rosetta_Orbiter.

Selected Biographies

Klim Churyumov: Co-Discoverer of Comet 67P

Klim Ivanovich Churyumov was born on 19 February 1937 in Nikolaev, Ukraine, and passed away on 15 October 2016. He is best known as the co-discoverer of Comet 67P/Churyumov-Gerasimenko, which became the target of the Rosetta mission.

Klim was an outstanding student, eventually entering Kiev University to study physics. In his third year, he was disappointed to be assigned to the faculty of optics, rather than theoretical physics, but he continued to attend lectures on theoretical physics even though the authorities disapproved, and was eventually moved to the faculty of astronomy, where there were vacant places.

Over the following years he studied the physics of comets under the tutelage of well-known comet researcher Prof. Sergej K. Vsekhsvyatskij. At first, this involved using the telescopes of the Astronomical Observatory of Kiev University. Further research was undertaken from observatories in the southern Soviet Union: Azerbaijan, Armenia, Uzbekistan, Turkmenistan, Crimea and in particular Alma-Ata (Kazakhstan).

In 1969, Klim co-discovered the object which was later to become the target for the Rosetta mission. He was leading an expedition to the observatory of the Institute of Astrophysics in Alma-Ata, along with Svetlana Gerasimenko and Ludmila Chirkova, the latter a photographic laboratory assistant. While they were comparing photographic plates taken at different times, they discovered a new comet, later designated 67P.

He went on to co-discover another comet in 1986, in collaboration with astronomer Valentin Solodovnikov. Named Churyumov-Solodovnikov, it was making its first passage through the inner Solar System.

Klim was a leading scientist on the Soviet Vega missions to Halley's Comet. Asteroid (2627) Churyumov was named in his honor. He won numerous awards and medals for his scientific work, and in 2006 became a Corresponding Fellow of the National Academy of Sciences of Ukraine. He appeared on Kiev TV speaking about astronomy, and wrote some 1,000 popular scientific articles as well as seven popular science books.

He was present at Kourou during the launch of Rosetta in 2004 and also attended an event at ESOC in November 2014 to coincide with Philae's historic landing on 67P.

Svetlana Gerasimenko: Co-Discoverer of Comet 67P

Svetlana Gerasimenko was born in the Ukrainian village of Baryshevka (near Kiev) in 1945. She is best known as co-discoverer of Comet 67P/Churyumov-Gerasimenko, which became the target of the Rosetta mission.

While at school she developed an interest in astronomy. This coincided with the period when space exploration was starting. She dreamed about becoming a cosmonaut or studying objects in space. In 1963 she enrolled as a student at Kiev University and studied astronomy in the physics faculty. Five years later, she entered the postgraduate training program led by Prof. Sergej K. Vsekhsvyatskij and began to study comets.

In September 1969, Svetlana joined an expedition to the Alma-Ata Institute of Astrophysics in Kazakhstan, under the leadership of Klim Churyumov, who had been working for several years at the department of astronomy in Kiev University.

Utilizing a 50-cm-diameter Maksutov telescope owned by the Institute of Astrophysics of the Academy of Sciences of Kazakhstan, Svetlana made a fortuitous discovery. While exposing and developing photographic plates of the night sky, she found a new comet on what seemed to be a flawed plate. In October, back in Kiev, Svetlana and Klim Churyumov confirmed the discovery of Comet 67P, which was later named after them.

In 1973, Svetlana moved to Dushanbe and started working at the Institute of Astrophysics in the Academy of Sciences of the Republic of Tajikistan, continuing to study comets and other small Solar System objects. In 1975, she gained a medal for "discovery of new astronomical objects" and in 1995 asteroid 3945 was named in her honor.

She was present at Kourou during the launch of Rosetta in 2004, and also attended an event at DLR Lander Control Center in November 2014 to coincide with Philae's historic landing on 67P.

John Ellwood: Rosetta Project Manager

John Ellwood grew up in the coastal town of Dartmouth in S.W. England. After leaving school in 1965, he gained a scholarship from the UK Atomic Energy Authority (UKAEA) to go to Cambridge University. On completing his degree in mechanical engineering, he spent another five years with UKAEA, working on military space systems and obtaining a PhD at London University in the structural dynamics of spacecraft.

In 1974, he moved to the European Space Research and Technology Center (ESTEC) in The Netherlands, where he worked as a structural engineer on an experimental communications satellite called, rather mundanely, the Orbital Test Satellite, and on the first payloads for the new Ariane rocket.

In the early 1980s, he became principal mechanical engineer for the European contributions to the NASA-ESA Hubble Space Telescope. His primary responsibilities were the ESA Faint Object Camera and the two giant solar panels that supplied the spacecraft's electrical power. After serving as Phase A (Initial Study) manager for two major science missions (the SOHO solar satellite and the XMM-Newton X-ray satellite) in 1989 he joined the Cluster project – another ESA Cornerstone science mission for the Horizon 2000 program. After the loss of the inaugural Ariane 5 launch vehicle in 1996 carrying the four Cluster spacecraft, Ellwood was appointed Cluster 2 project manager when the project was reinstated in 1997. He oversaw the launches of pairs of satellites on two Russian rockets in 2000.

Around the same time, he became the Rosetta project manager, just as the first hardware was beginning to be delivered and tested. He successfully reorganized the project and ensured that Rosetta was ready for the first launch opportunity in January 2003. After a postponement, he helped to prepare the mission for launch in early 2004 to a different comet. Then he became the project manager for the ATV (Automated Transfer Vehicle) that was to ferry supplies the International Space Station. The first mission, named 'Jules Verne', launched in 2008 with a cargo mass of 7 tonnes.

After 4 years in Paris, he returned to ESTEC in The Netherlands and in 2009 became Head of the Projects Department in the Directorate of Science and Robotic Exploration, where he remained until retirement in 2012. Since then, he has continued to have some consultancies with ESA, in particular on ExoMars, and the European Southern Observatory on its European Extremely Large Telescope (E-ELT).

He lives in Dartmouth, spending a lot of time sailing and being involved with various local charities and organizations. He has participated in several of the grueling Fastnet races, where ocean-going yachts race from Plymouth to Ireland and back.

Gerhard Schwehm: Project Scientist

Gerhard Schwehm was born in Ludwigshafen am Rhein, Germany, in 1949. While attending Ruhr University, Bochum, he was asked to work on a space science research project which involved the dynamics of interplanetary dust. Schwehm became the experiment manager and co-investigator for a photopolarimeter to fly on NASA's Ulysses spacecraft, but that mission was canceled and the instrument never flew.

After obtaining a doctorate in applied physics, he started his own consultancy company in Darmstadt, working on environmental models and attempting to improve astronomical data. During this time, he scanned photographic plates taken of Halley's Comet in 1910, in order to improve knowledge of its orbital parameters. This background stood him in good stead when the European Space Operations Center (ESOC) in Darmstadt sought to recruit a scientist with a background in cometary physics. In 1985, he joined the ESA Space Science Department as the Agency's first planetary scientist, and was soon assigned to support the Giotto project. He also joined a NASA-ESA study group that investigated missions to primitive bodies.

Appointed as deputy project scientist for Giotto in April 1985, he attended the first meeting of a NASA-ESA group commissioned to develop plans for a Comet Nucleus Sample Return mission.

Launched in July 1985, Giotto – ESA's first deep space mission – made history with its high speed fly-by of Halley's Comet. On surviving this ordeal, Giotto was eventually directed to a second target, Comet Grigg-Skjellerup. Schwehm became the project scientist for this Giotto Extended Mission, and also the project study scientist for ESA's next comet mission, which would become Rosetta. In 1993, the Comet Nucleus Sample Return was accepted as an ESA Cornerstone with Schwehm as its project scientist. Over the next decade, he was responsible for bringing together the most complex scientific payload ever flown on space mission, with 21 experiments onboard the Rosetta orbiter and lander. In December 2002, when an Ariane 5 failure resulted in postponement of Rosetta's launch towards Comet Wirtanen, Schwehm was involved in selecting Churyumov-Gerasimenko as the new target.

In 2001-2006, in addition to his responsibilities as Rosetta project scientist, he was appointed Head of ESA's Planetary Mission Division. He also became the mission manager for ESA's Smart 1 and Proba 2 missions. He was Head of Solar System Science Operations Division at ESAC in Madrid for 2007-2011.

Since retiring from ESA in March 2014, he has worked as a freelance consultant, providing advice on space instrument development, project, mission and science management.

Matt (Matthew) Taylor: Rosetta Project Scientist

Matt Taylor was born in Manor Park, N.E. London, in 1973. He gained his undergraduate physics degree at the University of Liverpool, and a PhD from Imperial College London. His early career focused on space plasma measurements, working in Europe and the U.S. on data from ESA's Cluster mission. In 2005, this led to a post at ESA, when he became the deputy project scientist for Cluster and the project scientist for the ESA-China Double Star mission.

Although he was not a specialist in cometary science, he was appointed the project scientist on the Rosetta mission in summer 2013. He successfully united the science team behind the goals of the mission, in order to solve the problems associated with insertion into orbit around a comet, and then the pursuit of the mission's objectives during its time in close proximity to Comet 67P.

He also devoted considerable energy to promoting Rosetta to international audiences through social media, public lectures, and TV and radio interviews. He became instantly recognizable by his tattoos. After the spacecraft successfully awoke from hibernation in 2014, new tattoos of Rosetta and its Philae lander were placed on his thigh.

Taylor was awarded the 2018 Service Award for Geophysics by the Royal Astronomical Society for his outstanding contribution to the Rosetta mission. The citation stated, "Rosetta was a historic achievement that would not have been anywhere near as successful, or visible, without Dr. Taylor's dedication, hard work and talent."

He is currently a study scientist for ESA's Comet Interceptor mission and the Lunar Orbital Platform-Gateway.

Stephan Ulamec: Philae Lander Project Manager

Born in Salzburg, Austria, in 1966, Stephan Ulamec studied geophysics at the University of Graz in Austria, where he received his doctorate in 1991. He was then a Research Fellow at the European Space Agency's Technology and Research Center (ESTEC) in The Netherlands until 1993 working on the preparation of planetary science missions, most notably Rosetta.

Since then he has been based at the German Aerospace Center (DLR) in Cologne, where he has played a major role in the development and operations of space systems and instruments. He also worked at the Institute for Space Simulation, the Space Operations and Astronaut Training Center, and the Institute for Planetary Science. He was appointed project manager, as well as systems engineer and technical project manager, for the Philae lander. Leading the Philae team from its Lander Control Center, he oversaw the first ever landing on a comet in 2014.

He is also the payload manager of MASCOT, a small lander for Japan's Hayabusa 2 mission to asteroid Ryugu. The surface package landed early October 2018. He is lead scientist on a rover for the planned MMX Mission to the Martian moon Phobos, which is to be launched in 2024. He is also involved in the planning and design of future NASA and JAXA missions to comets and asteroids.

He was awarded the Yuri Gagarin Medal of the Russian Cosmonautics Federation. Asteroid (11818) Ulamec is named in his honor.

Sylvain Lodiot: Rosetta Spacecraft Operations Manager

Born in Dar-es-Salaam, Tanzania, in 1975, Sylvain Lodiot spent most of his childhood in Africa. He followed what he calls the "classical French path" in education, completing his Bac C (baccalauréat) in 1993, followed by Higher School Preparatory Classes until 1996, and then attended an engineering school in Grenoble until 1999, with one year in the UK.

He first joined the Rosetta flight control team in January 2001. He followed the preparations for launch until June 2003 but missed the launch in 2004, having decided on a career change. Realizing that he missed spacecraft operations, he rejoined the Rosetta flight control team in ESOC in January 2007, just before the Mars swing-by. He then followed the mission until its end in September 2016, apart from spending one year on another project while Rosetta was in hibernation. After it awakened, he became the spacecraft operations manager in April 2014.

From a spacecraft operations point of view, Rosetta was very unique. Operations were never routine, because there was always a change, an issue to address (sometimes major, such as a failing reaction wheel or a propellant leak), a new phase to prepare (e.g. hibernation, asteroid fly-bys, planet fly-bys), and new operations to execute that had never been attempted before (e.g. comet approach in 2014).

Lodiot led the preparation for the Steins asteroid fly-by – a first for Europe – as well as the preparation and execution of the deep space hibernation phase. There were several stressful moments too, such as an unknown software bug triggering a safe mode on the spacecraft just before an Earth fly-by. Sylvain says he learned a lot from that event, particularly how to deal with, and manage stress.

During the landing of Philae on the comet, he was in the ESOC main control room at night, seeing live science from Philae appear on the consoles, something he describes as a magical experience. "It took many months to accept that Rosetta was no longer there," he wrote later. "After so many days and nights with Rosetta, so many calls to come in and fix issues, so many moments of joy too, it was like losing a child. It was an immense honor to be part of Rosetta, and in particular to lead the flight control team in the last years of the mission."

Since 2017, he has been the spacecraft operations manager for Solar Orbiter. This spacecraft was launched in February 2020 and he is looking forward to many new, exciting challenges.

Claudia Alexander: U.S. Rosetta Project Scientist

Claudia Alexander was born in Canada in 1959, but grew up in Santa Clara, California. She was inspired by Carl Sagan's TV series 'Cosmos', which she watched perhaps a dozen times, and by President Kennedy's declaration, "…we will go into space, not because it is easy, but because it is hard…"

She really wanted to go to the University of California at Berkeley, but her parents said they would pay the fees only if she majored in something "useful" like engineering – a subject she "hated". However, while on an engineering internship at NASA Ames Research Center, she often "sneaked over" to the space building. When her supervisor realized she was enjoying her time there, he placed her with Dr. Ray T. Reynolds in the Space Science Division. For her master's degree from the University of California, Los Angeles, she completed a dissertation on the impact of solar cycle variations on the ionosphere of Venus, using the Pioneer Venus Orbiter data.

She earned her PhD from the University of Michigan (U-M) in 1993, and was among the first 20 African Americans to graduate with a PhD in an astronomy/physics-related field. She was also named U-M Woman of the Year in Human Relations that same year. In 2007 she set up the Claudia Alexander Scholarship for undergraduate students at U-M.

Claudia was very well known for her role in NASA's Galileo and Cassini projects. In fact, she was the last project manager of the Galileo mission to Jupiter. She was deeply involved with the Rosetta mission as U.S. Rosetta project scientist. She was passionate about outreach, including engaging amateur astronomers through the ground-based observing campaign of Rosetta's target comet, 67P/Churyumov-Gerasimenko. She also wrote science fiction stories.

In 2003 she was awarded the Emerald Honor for Women of Color in Research & Engineering by Career Communications Group, Inc.

Claudia died on 11 July 2015 following a 10 year battle with breast cancer. Alexander Gate, a feature on Comet 67P, is named in her honor.

According to James A. Slavin, Chairman the Climate and Space Sciences and Engineering Department at U-M, "She was a great scientist and leader, an inspiration to us and to all of the students who followed, a benefactor to our student fellowship program, and one of those singular, truly beautiful people who energize everyone with whom they come in contact."

Hans Balsiger: ROSINA Principal Investigator

Hans Balsiger was born in Bern, Switzerland, in 1937. He was educated at the University of Bern, gaining his master's degree in physics in 1964 and his doctorate in 1967.

He held various junior positions at the Physikalisches Institut in the University of Bern from 1963-1968, then in 1968-1970 he held an ESRO-NASA postdoctoral fellowship in the Space Science Department of Rice University in Houston, Texas. On his return to the University of Bern, he became a lecturer in physics and then assistant professor in 1979. That same year, he was appointed the scientific assistant to the Chairman of the European Science Foundation's Space Science Committee.

Five years later, he became associate professor of physics, followed by a full professorship in 1990. From 1993-2002, he was the Director of the Physikalisches Institut. One of his first projects was to help in the design of an upper atmosphere mass spectrometer at the University of Bern. He retired in 2003.

He participated in a number of space science missions. In addition to being a co-investigator and project manager for an experiment on the GEOS satellite, he was also co-investigator for experiments on the first International Sun-Earth Explorer (ISEE 1); the Ulysses mission to study the solar poles; the Dynamics Explorer and AMPTE satellites; and the Wind, Polar, SOHO and Cluster missions to explore the Sun-Earth interaction.

His interest in comets began with ESA's Giotto mission, when he was principal investigator for the Ion Mass Spectrometer that investigated the composition of comets Halley and Grigg-Skjellerup. He later became principal investigator for Rosetta's ROSINA mass spectrometer.

In 1986, he served as Chairman of ESA's Solar System Working Group, then Chairman of ESA's Space Science Advisory Committee 1987-1990. He was Chairman of the ESA Science Program Committee 1996-1999.

He was awarded the Prix Cortaillod from the University of Neuchâtel in 1981.

Luigi Colangeli: GIADA Principal Investigator

Luigi Colangeli was born in Lecce, Italy, in 1958. He graduated in physics at the University of Lecce in 1982. He attained his PhD in astrophysics at the University of Bari in 1986, with a thesis on the simulation in a laboratory of materials representative of cosmic dust.

He was a Research Fellow in the Space Science Department at ESA-ESTEC from 1987-1988, where he was involved in the scientific and organizational definition of the Rosetta mission. He was a researcher in physics at the Engineering Faculty, University of Cassino, Italy, from 1988-1993, then an associate astronomer at the Astronomical Observatory of Capodimonte, Naples from 1993-2001.

Since 2001 he has been a full professor of astronomy at INAF-Capodimonte. He was Director of the Observatory 2005-2010. During 2002-2004 he was a member of the ESA Solar System Working Group. He was Head of ESA's Solar System Missions Division, based at ESTEC in the Netherlands, 2010-2013. From October 2013 to October 2016 he was the Head of ESA's Scientific Program Coordination Office. Since then he has served as the Head of the Science Coordination Office in ESA's Directorate of Science.

His research has focused on the realization of laboratory experiments for the study of the properties of materials present in interstellar and circumstellar environments and in the Solar System. He has designed various instruments for space missions dedicated particularly to the study of cosmic dust.

He was principal investigator of the GIADA experiment on Rosetta 2001-2010, after which he remained as co-investigator. He was also involved as a co-investigator for the VIRTIS and COSIMA experiments on Rosetta.

As co-principal investigator 2004-2010, he led the development of the high resolution imager on the SIMBIO-SYS experiment on ESA's BepiColombo mission to Mercury, becoming a co-investigator after joining ESA staff. He is also a co-investigator for the MERTIS (Mercury Radiometer and Thermal Imaging Spectrometer) instrument on BepiColombo.

Angioletta Coradini: VIRTIS Principal Investigator

Angioletta Coradini was born in Rovereto, Italy, in 1946. She was one of the most important figures in the field of space sciences for almost forty years, before she succumbed to cancer on 5 September 2011.

After spending her youth in Rovereto, Naples and Turin, she settled in Rome and in 1970 she graduated with honors in physics from the 'La Sapienza' University. After an initial period of collaboration at the university, she took up service at the National Research Council (CNR) at the Institute of Space Astrophysics (IAS) in the Department of Planetology, which was led by Marcello Fulchignoni, where she obtained a permanent position in 1975.

Her first geological investigations, carried out in the Gulf of Cagliari, earned her considerable international credit, and led to her becoming one of the first non-American scientists to whom NASA entrusted examination of the lunar samples obtained during the Apollo missions. Her broad scientific interests prompted involvement in almost all fields of research on the Solar System, and she published models for the formation of the Solar System and the giant planets Jupiter and Saturn.

From 1987-1995, she was Director of the CNR National Group of Astronomy. In 1997 she became a member of ESA's Solar System Working Group. She was director of the Institute of Physics of Interplanetary Space (IFSI) 2001-2011, managing the transition of the Institute itself from the CNR, to which it had belonged, to the new National Institute of Astrophysics (INAF). She was also involved in development of instrumentation for space probes. Starting in the 1990s, she collaborated with ESA and NASA to provide imaging spectrometers for the determination of the chemical composition of planetary surfaces.

The first of these instruments was VIMS (Visual and Infrared Mapping Spectrometer), which was launched with the Cassini-Huygens mission in 1997. Subsequently, she led the design of VIRTIS for Rosetta, with similar instruments on Mars Express and Venus Express. She was the lead for instruments on NASA's Dawn mission to dwarf planet Ceres and asteroid Vesta, the European ExoMars mission, and NASA's JUNO mission to Jupiter.

In 2007 she was awarded the David Bates Medal by the European Geosciences Union (EGU). From 1993 she was a member of Russian Academy of Sciences. Asteroid 4598 is named in her honor. The Angioletta Coradini Mid-Career Award is an annual award given by NASA's Solar System Exploration Research Virtual Institute (SSERVI) to a researcher for significant, lasting accomplishments related to exploration science.

Eberhard Grün, Rosetta Interdisciplinary Scientist

Eberhard Grün was born in Bad Nauheim, Germany, in 1942. He graduated in physics at the University of Heidelberg in 1968 and gained his PhD at the Max-Planck-Institute for Nuclear Physics (MPIK) in Heidelberg in 1970.

From 1970 until 1974, he was a research assistant in the MPIK, studying interplanetary dust and developing dust detectors for space missions. During this period, he spent six months as a visiting scientist at NASA Goddard Space Flight Center focusing on data analysis of cosmic dust experiments, and another six months at NASA Ames Research Center doing research on hypervelocity impact phenomena. In 1974, he became a senior research scientist at the MPIK.

In 1981, he returned to the U.S. for six months as a senior research associate of the National Research Council (NRC), while working at the Jet Propulsion Laboratory. This was followed in 1982 by six months at the Lunar and Planetary science Institute (LPI) in Houston. Back at the University of Heidelberg, he became a lecturer in 1983. He became a professor there in 1989. He was a researcher at the Hawaii Institute of Geophysics and Planetology from 2000-2007, while remaining a senior research scientist at the MPIK.

Grün was an interdisciplinary scientist for the Rosetta mission, and a co-investigator for the Rosetta lander, as well as for the COSIMA, CONSERT and RSI experiments on the orbiter. It was Grün who first proposed the name 'Rosetta' for the comet mission (see Chapter 3).

In 2007, he became a research associate at the Laboratory for Atmospheric and Space Physics (LASP) of the University of Colorado in Boulder, USA. He also remains an active emeritus professor at the MPIK in Heidelberg.

Grün received the Gerard P. Kuiper Prize of the Division for Planetary Sciences (DPS) of the American Astronomical Society in 2002, in recognition of "outstanding contributions to the field of planetary science". In 2003, he was awarded the David Robert Bates Medal from the European Geosciences Union (EGU). In 2006, he received the Space Science award of the COSPAR for outstanding contributions to space science. He received the Gold Medal of the Royal Astronomical Society in 2011 for his leading role in the science of space dust for over 30 years. Asteroid 4240 was named in his honor.

Samuel Gulkis: MIRO Principal Investigator

Samuel Gulkis was born in West Palm Beach, Florida, in 1937. At the University of Florida, he graduated as a bachelor of aeronautical engineering with high honors (1960), a master's in physics (1963) and a doctorate in physics (1965).

His first post was a research associate in Radio Astronomy with Cornell University, when he was based at the 1,000 foot radio telescope at Arecibo, Puerto Rico (1965-1967). He was later appointed assistant professor of astronomy at Cornell University, Ithaca, New York. From 1968 onwards, he held various senior positions at the Jet Propulsion Laboratory in Pasadena, California, becoming a senior research scientist in the Earth and Space Sciences Division in 1981. He also served as the program scientist for Solar System Exploration in the Space and Earth Science Program Directorate. Apart from his work at JPL, he served as a distinguished visiting scientist at the Division of Radiophysics of CSIRO (the Commonwealth Science and Industrial Research Organization) in Epping, New South Wales, Australia, during 1981, and as a visiting scientist at the Paris Observatory, Section de Meudon, from 1987-1988. Over more than fifty years, he carried out research in radio and submillimeter astronomy, specializing in the physics of the magnetosphere of Jupiter, the atmospheres of the major planets, comets and experimental cosmology.

He served as a science co-investigator on three NASA space missions (Voyager, COBE, and JUNO) and principal investigator for the MIRO instrument on ESA's Rosetta mission from 1989-2015, when he announced his retirement. He continued to work full time as a research scientist in the Astrophysics and Space Science section at JPL. He served on various NASA Advisory Committees on Planetary and Space Astronomy, and served on the Board of Editors for the journal Icarus. He received two NASA Exceptional Scientific Achievement Awards, and, as an original member of the COBE science team, the Gruber award for Cosmology in 2006.

Horst Uwe Keller: OSIRIS Principal Investigator

Horst Uwe Keller was born in Prague in 1941. After attending high school in Hamburg from 1953-1960, he studied physics in Hamburg and Munich. He received his diploma in physics from the University of Munich and worked as a PhD student at the Max-Planck-Institute for Physics and Astrophysics.

After completing his doctorate in 1971, he worked as a scientist at the Max-Planck-Institute, and from 1972-1975 he worked at the Laboratory for Atmospheric and Space Physics of the University of Colorado in Boulder, USA. In 1976 he moved to the Max-Planck-Institute for Aeronomie in Katlenburg-Lindau, Germany, where, in 1987, he was promoted to the position of senior scientist. His fields of research covered the physics of comets, stellar evolution and atmospheres, interplanetary matter, radiative transfer, planetary atmospheres and atmosphere-surface interaction.

Keller was awarded the 2008 Christiaan Huygens Medal for his significant achievements in spacecraft investigations of the Solar System, through his innovation and development of cameras and related instrumentation. He participated in a number of rocket experiments, deep space missions, and observations by satellites. He was principal investigator for the Halley Multicolor Camera on ESA's Giotto, and for the Camera System on ESA's Rosetta. He was also in charge of the infrared spectrometer on SMART 1, ESA's first mission to the Moon. In more recent times, he led the German team that provided the cameras for the NASA Dawn mission to Vesta and Ceres. His team also provided hardware for several Mars missions and for the Cassini-Huygens mission to Saturn and its primary satellite, Titan.

Within ESA, he served as a member of the Comet Consultant Group and the Space Telescope Working Group, and participated in a number of other committees.

Jochen Kissel: COSIMA Principal Investigator

Jochen Kissel was born in Heidelberg, Germany. He studied at the University of Karlsruhe from 1962-1967 to obtain a master's degree. He went on to gain his PhD at the University of Heidelberg in 1971 with a thesis about ions in the Earth's upper atmosphere, then took a post as a research associate at the Max-Planck-Institut für Kernphysik in Heidelberg.

In 1976, he became Head of the Dust Accelerator Laboratory at the Institute, a post which he held to 1988. During that time, he spent a year (1978-1979) as a guest researcher with TRW-Systems in Redondo Beach, California, and Martin Marietta Aerospace in Denver, Colorado. While in the USA, he also served on NASA's Comet Science Working Group and its Galileo Meteoroid Protection Working Group. In March 1998, he moved to the Max-Planck-Institut für Extraterrestrische Physik at Garching, Germany.

Over the years, he has worked on numerous space science missions. Starting in the 1970s, he was a co-investigator on the German AEROS aeronomy satellite, the German-NASA Helios solar probe, the ESA Ulysses mission to investigate the solar poles, and the NASA Galileo mission to Jupiter.

In the 1980s, he was principal investigator for the PIA Dust Mass Spectrometer on the ESA Giotto comet probe to Halley's Comet, and instruments carried by the Soviet Vega missions. His links with the Soviet space program continued when he was a principal investigator for the LIMA-D experiment on the Phobos missions to Mars. From 1986-1992 he was PI for the CoMA experiment on NASA's canceled CRAF comet mission. He was then PI for the CIDA experiment carried by NASA's Stardust mission. Two years later, he became PI for a similar experiment on NASA's ill-fated CONTOUR mission (see Appendix 1). He joined a working group for the definition of an ESA-only comet rendezvous mission in the early 1990s, and in 1996 was made principal investigator for the COSIMA experiment on Rosetta.

He became a member of the German Physical Society in 1968 and has been a corresponding member of the International Academy of Astronautics (IAA) and a corresponding member of the New York Academy of Sciences. He was awarded honors by ESA, NASA, and the Space Research Federation of the USSR.

Wlodek Kofman: CONSERT Principal Investigator

Wlodek Kofman was born in Leningrad, USSR, in January 1945, but returned to his parents' home country of Poland later that year.

He received a master's degree in electronics engineering in 1968. Whilst studying at Warsaw Polytechnic, he was imprisoned without trial for five months in the wake of student strikes and demonstrations in March 1968. Unable to continue his studies in Poland, he applied for refugee status in France, which he obtained in 1969. He moved to the University of Grenoble, where he got his doctorate in 1972, and the These d'Etat degree in physics from the National Polytechnic Institute of Grenoble and Grenoble University in 1979.

Kofman began as a researcher at CNRS in 1976, became Research Director at CNRS in 1985, and Research Director Emeritus at CNRS in 2011. His researches included experimental and theoretical ionospheric investigations and using radio and radar techniques to study planetary surfaces and subsurfaces. He became involved in the EISCAT incoherent scatter radar project in 1979, studying the auroral ionosphere. He also worked on the development and application of radar techniques to study glaciers and subsurface features. He served as President of the EISCAT Council 1997-1999.

He was principal investigator of the CONSERT experiment on Rosetta, a co-investigator on the MARSIS experiment on Mars Express, on the LRS radar on the JAXA Kaguya mission, the RIME radar on ESA's forthcoming JUICE mission, and the REASON radar of NASA's Europa Clipper mission. He has collaborated with many foreign scientists in various institutes in Europe, the United States, Canada, and Japan, and undertaken long stays in these institutes, including as a visiting professor at the University of Western Ontario in Canada (1995) and Nagoya University in Japan (2006). He is today a distinguished visiting scientist at the NASA Jet Propulsion Laboratory in California.

From 1990-1999 he served as Deputy Director of Grenoble Observatory. He then founded the Grenoble Planetology Laboratory. It merged with the Grenoble Laboratory of Astrophysics in 2009 to become the Institute of Planetology and Astrophysics in Grenoble (IPAG).

He was science adviser at Direction de la recherche de l'Ecole Polytechnique in Paris (2008-2013) and a member of ESA's Space Science Advisory Committee (2009-2012). In 2012 he was appointed a professor of physics at the Space Research Center in Warsaw, Poland. From 2004-2010 he was editor-in-chief of the Annales Geophysicae, the journal of the European Geosciences Union. In 2015 he was appointed Knight of the Legion of Honour, the highest decoration in France. He was awarded the URSI France Medal (2016) and the Prix Académie Science (2018). He is an Honorary Fellow of Royal Astronomical Society. Asteroid 13368 was named in his honor.

Thurid Mannel, MIDAS Principal Investigator

Thurid Mannel was born in Gelnhausen, Germany, in 1989. She joined the Rosetta MIDAS team in 2014 as a student searching for a PhD project, and gained her doctorate in 2018 from the Space Research Institute of the University of Graz in Austria with the best possible grade. Over time she became a key member of MIDAS operating team, focusing on analyzing new data and suggesting future measurements in order to achieve the best scientific impact. After MIDAS successfully completed its checkout following Rosetta's hibernation, she moved to Graz. In the summer of 2017 she became principal investigator for the instrument.

When the active part of the Rosetta mission ended, the operating team of MIDAS shrank in size. In particular, Mark Bentley, the principal investigator responsible for the instrument and head of the MIDAS group departed. As his successor, Mannel became the youngest PI of the mission. She says she is very thankful towards the MIDAS team and the Rosetta community for their great support and encouragement during her steep career rise.

Willi (Willibald) Riedler: MIDAS Principal Investigator

Willi Riedler was born in Vienna, Austria, in 1932. He was one of the leading space scientists in Austria and was at the forefront of that country's space endeavors for many years. He died on 24 January 2018.

After obtaining an engineering degree from the Technical University of Vienna in 1956, he was appointed assistant professor at the Technical University's Institute of High Frequency Techniques. During this period (1956-1962), he gained a doctor of technology degree with a dissertation in microwave physics. Two major changes occurred in 1962. Firstly he started to study meteorology and geophysics at Vienna University. Then he began his 'space career' by moving to Kiruna in Sweden. One of the main reasons for this move to Scandinavia was to enable his wife, Lilli, to be closer to her Finnish homeland. From 1962-1969, he was a group leader in experimental space research at the Kiruna Geophysical Observatory. This involved research into the physics of the ionosphere and the magnetosphere. Towards the end of this period, he served as a consultant to the European Space Research Organization (ESRO) with responsibility for launching sounding rockets from Kiruna.

He returned to Austria to become a professor of Telecommunications and Wave Propagation at the Technical University of Graz in 1968, and held a series of prestigious positions there in the 1970s. He also held a number of important advisory and technical positions. In 1975, he became the Austrian delegate on the Technical Committee on Telecommunication for the European Community. In 1979 he was appointed the chairman of the COSPAR (Committee on Space Research) Panel on Scientific Ballooning. From 1979-1993 he was a corresponding member of the Austrian Academy of Sciences. During this time, he was the scientific leader of the Austromir project, where an Austrian cosmonaut spent a week on the Soviet Mir space station. He then became a full member of the Austrian Academy of Sciences and chairman of several of its committees. After acting as Vice-Director of the Institute of Space Research for three years, he was appointed Director in 1984. His other responsibilities included consultant to the United Nations Committee on the Peaceful Uses of Outer Space, and Austrian delegate to the ESA Science Program Committee.

He was involved in magnetometer experiments on various ground-breaking space missions, including the first flight of Spacelab, the Soviet Venera 13 and 14 missions to Venus, the two Vega missions to Venus and Halley's Comet, and the Phobos and Mars-96 missions. He was a co-investigator on the Cassini-Huygens mission and the Cluster missions, and was principal scientific investigator for the MIDAS experiment on Rosetta.

In September 2000 he retired his directorship of the Institute of Communications and Wave Propagation at the Technical University of Graz. At the end of the year he retired directorship of the Space Research Institute of the Austrian Academy of Sciences, but he continued as a guest scientist.

S. Alan Stern: ALICE Principal Investigator

Alan Stern was born in New Orleans, Louisiana, in 1957. He earned various degrees from the University of Texas at Austin 1978-1981. On completing twin master's degrees in aerospace engineering and atmospheric sciences, he spent six years as an aerospace systems engineer, concentrating on spacecraft and payload systems at the NASA Johnson Space Center, Martin Marietta Aerospace, and the Laboratory for Atmospheric and Space Physics at the University of Colorado in Boulder. He gained his doctorate in astrophysics and planetary science from the University of Colorado in 1989. From 1983-1991 he held positions at the University of Colorado in the Center for Space and Geoscience Policy, the office of the Vice President for Research, and the Center for Astrophysics and Space Astronomy. From 1991-1994 he led the Astrophysical and Planetary Sciences group at Southwest Research Institute in San Antonio.

In 1995, he was selected to be a Space Shuttle mission specialist finalist and in 1996 he was a candidate Space Shuttle payload specialist.

He was involved in instrument development, in particular involving ultraviolet technologies, and participated in numerous space missions: e.g. he was a principal investigator in NASA's ultraviolet astronomy sounding rocket program, and was project scientist for the SPARTAN astronomical satellite that was deployed and retrieved by the Space Shuttle. He then became the principal investigator for the ALICE ultraviolet spectrometer on Rosetta. He served on a number of NASA committees, including the Lunar Exploration Science Working Group, the Discovery Program Science Working Group, the Solar System Exploration Subcommittee, the New Millennium Science Working Group, and the Sounding Rocket Working Group. He chaired the NASA Outer Planets Science Working Group 1991-1994. He became NASA's Associate Administrator for the Science Mission Directorate in April 2007, but after facing internal battles regarding policy and funding, on 26 March 2008 it was announced that he had resigned his position the previous day, effective 11 April.

He is the principal investigator and driving force behind NASA's New Horizons mission. It made a Pluto fly-by in July 2015. He had a key role the Lunar Reconnaissance Orbiter and is a participant in the Europa Clipper mission. He has served as a co-investigator on numerous NASA and ESA planetary missions.

Since 2009, he has been an Associate Vice President and Special Assistant to the President at the Southwest Research Institute. From 2008-2012 he served on the Board of Directors of the Challenger Center for Space Science Education, and as chief scientist and mission architect for Moon Express from 2010-2013. He served as the Director of the Florida Space Institute from 2011-2013. He is chief scientist of both World View (a near-space ballooning company) and of the Florida Space Institute.

Stern has run his own aerospace consulting practice since 2008. He is also CEO of two small corporations (Uwingu and The Golden Spike Company) and serves on the Board of Directors of the Commercial Spaceflight Federation. He has received many awards and written several books.

REFERENCES

Klim Churyumov: https://sci.esa.int/web/rosetta/-/54598-klim-churyumov
http://blogs.esa.int/rosetta/2016/11/11/an-encounter-with-klim-churyumov/
https://en.wikipedia.org/wiki/Klim_Churyumov
Svetlana Gerasimenko: https://sci.esa.int/web/rosetta/-/54597-svetlana-gerasimenko
https://www.dlr.de/content/en/downloads/news-archive/2014/20141111_inter-view-with-svetlana-gerasimenko-beginner-s-luck-_12063.pdf?__blob=publicatio
nFile&v=12
John   Ellwood: https://www.esa.int/Science_Exploration/Space_Science/Rosetta/I_get_a_kick_out_of_you_An_interview_with_John_Ellwood

https://www.bythedart.co.uk/living-in-dartmouth/people/john-ellwood---engineering-for-the-stars/

Gerhard Schwehm: https://www.esa.int/Science_Exploration/Space_Science/People/Accidental_space_scientist_An_interview_with_Gerhard_Schwehm

Sylvain Lodiot: https://www.youtube.com/watch?v=_3k03Jb3dgQ
http://phelma.grenoble-inp.fr/fr/l-ecole/rosetta-sylvain-lodiot-ancien-eleve-phelma-pilote-de-la-sonde-a-l-esa-2

Claudia Alexander: https://solarsystem.nasa.gov/people/1140/claudia-alexander-1959-2015/
https://news.engin.umich.edu/2015/07/in-memoriam-claudia-alexander/

Matt Taylor: https://en.wikipedia.org/wiki/Matt_Taylor_(scientist)
https://www.theguardian.com/science/2016/oct/09/rosetta-project-scientist-matt-taylor-interview-comet-european-space-agency

Hans Balsiger: https://academieairespace.com/en/utilisateurs/hans-balsiger/?cn-reloaded=1
https://www.ae-info.org/ae/Member/Balsiger_Hans

Luigi Colangeli:
https://www.dei.unipd.it/en/distinguished?field_data_e_ora_value[value][date]=2014
https://www.esa.int/esatv/Videos/2013/01/Exobiology_and_Space_Missions/Luigi_Colangeli_Head_of_Solar_System_ESA_English

Angioletta Coradini:
https://web.archive.org/web/20180308233954/http://www.anpri.it/vecchiosito/ANPRI/Convegni/convegno2012/Coradini.html
https://web.archive.org/web/20111102135808/http://www.nature.it/scienze/angioletta-coradini-una-vita-per-le-stelle/

Eberhard Grün: https://peoplepill.com/people/eberhard-gruen/
https://en.wikipedia.org/wiki/Eberhard_Gr%C3%BCn

Horst Uwe Keller: https://de.wikipedia.org/wiki/Horst_Uwe_Keller

Jochen Kissel: https://www.researchgate.net/figure/Team-member-Jochen-Kissel-shows-a-version-of-a-modeled-comet_fig1_225980481

Wlodek Kofman: https://ipag.osug.fr/~kofmanw/Cv_Kofman_2017.pdf
https://www.ofpra.gouv.fr/fr/histoire-archives/galeries-d-images/les-refugies-celebres/wlodek-kofman

Willi Riedler: https://de.wikipedia.org/wiki/Willibald_Riedler

S. Alan Stern: https://en.wikipedia.org/wiki/Alan_Stern
https://www.nasa.gov/ames/ocs/2016-summer-series/alan-stern

Appendix 3
Fast Facts

The total cost of the Rosetta mission was €1.4 billion, of which the total Philae costs were €220 million. This included launch, spacecraft, science payload (instruments and lander) and mission and science operations.

After its launch in 2004, Rosetta traveled almost 8 billion kilometers to reach Comet 67P. Its trek included three fly-bys of Earth, one of Mars, and two of asteroids. When it ended its mission on 30 September 2016, it was making its sixth orbit of the Sun.

The craft endured 31 months in deep-space hibernation on the most distant leg of its journey, out near the orbit of Jupiter, prior to waking up in January 2014 in preparation for arrival at the comet in August 2014.

After becoming the first spacecraft to orbit a comet and the first to deploy a lander (Philae) in November 2014, the Rosetta orbiter continued to monitor the evolution of the comet through perihelion passage and as it withdrew from the Sun until sunlight was no longer sufficient to operate its scientific instruments. Rosetta operated in the harsh environment of the comet for 786 days.

Launch date: 2 March 2004

Launch vehicle: Ariane 5G+

Launch mass: 3,005 kg (including 1,670 kg propellant, 165 kg science payload and 100 kg Philae lander).

Dimensions: Orbiter: 2.8 × 2.1 × 2.0 meters with a pair of 14 meter long solar panels.
    Philae lander: 1.0 × 1.0 × 1.0 meters prior to deploying its landing gear.

Instruments:
Orbiter (11 science instrument packages):

- ALICE (Ultraviolet Imaging Spectrometer).
- CONSERT (Comet Nucleus Sounding).
- COSIMA (Cometary Secondary Ion Mass Analyzer).
- GIADA (Grain Impact Analyzer and Dust Accumulator).
- MIDAS (Micro-Imaging Analysis System).

© Springer Nature Switzerland AG 2020
P. Bond, *Rosetta: The Remarkable Story of Europe's Comet Explorer*,
Springer Praxis Books, https://doi.org/10.1007/978-3-030-60720-3

- MIRO (Microwave Instrument for the Rosetta Orbiter).
- OSIRIS (Rosetta Orbiter Imaging System).
- ROSINA (Rosetta Orbiter Spectrometer for Ion and Neutral Analysis).
- RPC (Rosetta Plasma Consortium).
- RSI (Radio Science Investigation).
- VIRTIS (Visible and Infrared Mapping Spectrometer).

Philae lander (10 science instrument packages):

- APXS (Alpha Proton X-ray Spectrometer).
- ÇIVA / ROLIS (Rosetta Lander Imaging System).
- CONSERT (Comet Nucleus Sounding).
- COSAC (Cometary Sampling and Composition experiment).
- MODULUS PTOLEMY (Evolved Gas Analyzer).
- MUPUS (Multi-Purpose Sensor for Surface and Subsurface Science).
- ROMAP (Rosetta lander Magnetometer and Plasma Monitor).
- SD2 (Sample and Distribution Device).
- SESAME (Surface Electrical Sounding and Acoustic Monitoring Experiment).

Partnerships: The scientific payload of the orbiter was provided by scientific consortia from institutes across Europe and the USA. The lander was supplied by a European consortium led by the German Aerospace Research Institute (DLR). Other members of the consortium were ESA, CNES and various institutes from Austria, Finland, France, Hungary, Ireland, Italy and the UK.

Primary mission objectives:

- Undertake a lengthy exploration of a comet at close quarters to observe how it was transformed by the warmth of the Sun along its elliptical orbit.
- Land a probe on a comet's nucleus for in-situ analysis.

Rosetta mission facts:

- Rosetta gets its name from the famous Rosetta Stone that led to the deciphering of Egyptian hieroglyphics almost 200 years ago.
- Rosetta's original target was Comet 46P/Wirtanen, but a postponement of the launch required selecting a new target: Comet 67P/Churyumov-Gerasimenko.
- Rosetta was the first spacecraft to fly close to Jupiter's orbit using only solar cells as its main power source.
- Rosetta and Philae were the first spacecraft to orbit a comet and land on its surface.
- The Philae lander was named for the island in the river Nile on which an obelisk was found bearing a bilingual inscription that enabled the hieroglyphs of the Rosetta Stone to be deciphered.

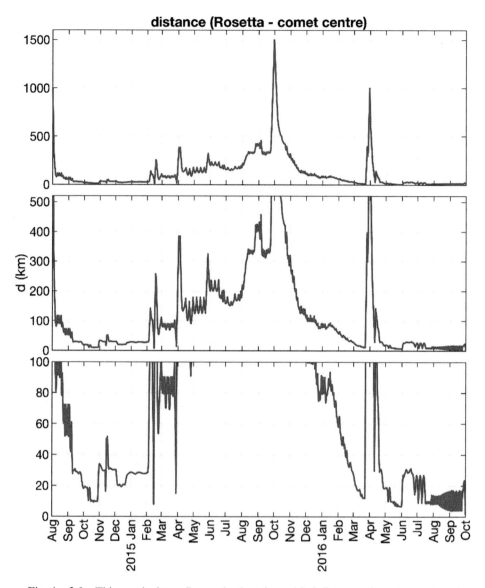

Fig. Ap.3.1: This graph shows Rosetta's changing orbital distances from the center of 67P between its arrival on 6 August 2014 and the end of mission on 30 September 2016. Notice the increase in orbital distance either side of perihelion (13 August 2015) when the comet's activity was at its highest. The sudden retreats from the nucleus were excursions to fly first 1,500 km towards the Sun (September 2015) and then 1,000 km down the tail (March-April 2016). (ESA, Charlotte Götz)

Journey milestones:

- 1st Earth gravity assist: 4 March 2005.
- Mars gravity assist: 25 February 2007.
- 2nd Earth gravity assist: 13 November 2007.
- Asteroid Steins fly-by: 5 September 2008.
- 3rd Earth gravity assist: 13 November 2009.
- Asteroid Lutetia fly-by: 10 July 2010.
- Enter deep space hibernation: 8 June 2011.
- Exit deep space hibernation: 20 January 2014.
- Comet rendezvous maneuvers: May to August 2014.
- Arrival at comet: 6 August 2014.
- Philae landing on comet: 12 November 2014.
- Closest approach to Sun: 13 August 2015.
- Mission end: 30 September 2016.

The graphics on the following pages summarize the operations and achievements of the orbiter and each instrument on the Rosetta orbiter.

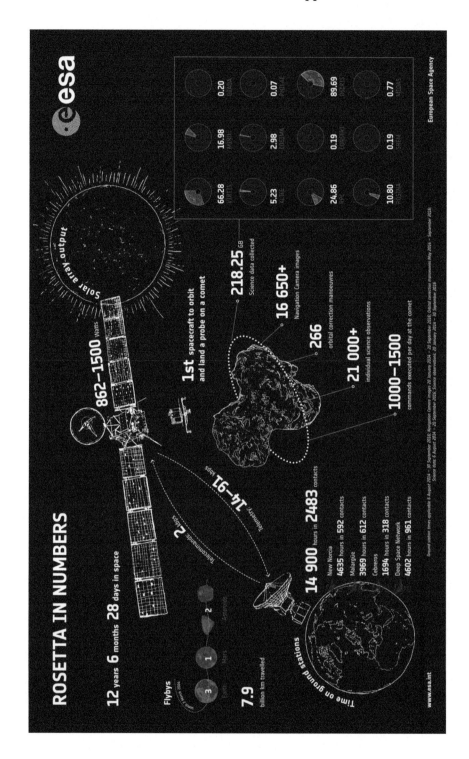

→ ROSETTA'S CONSERT
INSTRUMENT IN NUMBERS

MISSION: To study the internal structure of the comet with lander Philae

> 6 hours data during Philae's descent
> 9 hours data collected with Philae on surface
> 1 000 m distance travelled by CONSERT signals inside comet's interior
> 32 MB data collected
> 100 TB data generated by simulations needed to interpret data

CONSERT
Comet Nucleus Sounding Experiment by Radio wave Transmission

Philae

Based on numbers available during CONSERT operations 12 – 14 November 2016

ROSETTA'S MIDAS INSTRUMENT IN NUMBERS

MISSION: To make 3D profiles of dust grains collected in the comet's coma

> 8 exposed targets
> 1047 scans (863 scientific scans & 184 calibration scans)
> 100s of dust grains scanned
> 16 tips used (out of 16)
> 207 days spent exposing targets
> 382 days spent scanning targets
> 56 212 480 'touches' of the sample or target during image scans
> 9 620 392 packets of science and housekeeping data

MIDAS
Micro-Imaging Dust Analysis System

Based on numbers available 6 August 2014 – 15 September 2016

esa

→ ROSETTA'S MIRO
INSTRUMENT IN NUMBERS

MISSION: To investigate the nature of the comet's nucleus, outgassing, and development of the coma

> 1.9 billion science measurements
> 1.5 million spectra of gases in the coma
> 100 000 calibration mirror moves
> 5 700 lines of code are needed to run MIRO
> The processor in a modern cell phone runs about 750x faster than MIRO's 2.5 MHz computer

MIRO
Microwave Instrument for the Rosetta Orbiter

Based on numbers available August 2014 – end August 2016

→ ROSETTA'S OSIRIS
INSTRUMENT IN NUMBERS

MISSION: To image the comet's nucleus and its gas and dust coma

> 98 219 images taken during entire mission
> 76 308 images taken at comet
> 150 225 shutter activations
> 23 486 door operations
> 129 000 filter/band pass changes
> 115 497 telecommands sent
> 22 176 hours of operation

OSIRIS[1] WAC[2]

OSIRIS NAC[3]

[1] Optical, Spectroscopic, and Infrared Remote Imaging System
[2] Wide-Angle Camera
[3] Narrow-Angle Camera

Numbers indicate combined totals for WAC and NAC

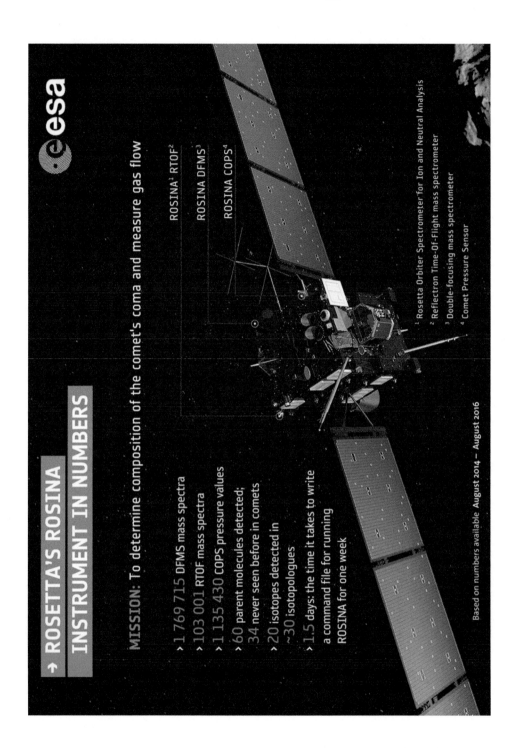

+ ROSETTA'S RPC
INSTRUMENT IN NUMBERS

MISSION: Five sensors to study the plasma environment of the comet and its interaction with the solar wind

RPC-MAG (Fluxgate Magnetometer)
> 900 000 000 magnetic field vectors collected
> 1 350 000 wave-trains of singing comet waves
  detected at 40 mHz
> 31 600 light-years: the equivalent distance all
  the magnetic field vector data have travelled during
  transmission from Rosetta to Earth

RPC-ICA (Ion Composition Analyser)
> 14 billion ions detected

RPC-MIP (Mutual Impedance Probe)
> 4 540 772 352 excitations of plasma
> 13 744 472 electric spectra
> 3 600 000 plasma density measurements

RPC-IES (Ion and Electron Sensor)
> 1 508 632 087 ions
> 88 549 757 175 electrons

RPC-LAP (Langmuir Probe)
> 88 836 404 207 433 470 000 electrons
> 3 299 041 710 586 133 000 ions
> 1 390 033 998 131 714 000 electrons
> 20 942 278 742 280 340 000 ions

RPC - LAP
RPC - MIP
RPC - IES
RPC - ICA

Rosetta Plasma Consortium

RPC - MAG
RPC - LAP

MAG (August 2014 – August 2016)   IES (May 2014 – August 2016)
ICA & MIP (March 2004 – September 2016)   LAP (Jan 2014 – August 2016)

REFERENCES

Rosetta by numbers https://sci.esa.int/s/8kKkjq8
https://blogs.esa.int/rosetta/2016/09/29/livingwithacomet-blog-post-series-summary/
ALICE instrument in numbers
https://blogs.esa.int/rosetta/2016/09/28/living-with-a-comet-an-alice-team-perspective/
CONSERT instrument in numbers
https://blogs.esa.int/rosetta/2016/09/28/living-with-a-comet-a-consert-team-perspective/
COSIMA instrument in numbers
https://blogs.esa.int/rosetta/2016/09/26/living-with-a-comet-a-cosima-team-perspective/
GIADA instrument in numbers
http://blogs.esa.int/rosetta/2016/09/26/living-with-a-comet-a-giada-team-perspective/
MIDAS instrument in numbers
https://blogs.esa.int/rosetta/2016/09/23/living-with-a-comet-a-midas-team-perspective/
MIRO instrument in numbers
https://blogs.esa.int/rosetta/2016/09/27/living-with-a-comet-a-miro-team-perpsective/
OSIRIS instrument in numbers
http://blogs.esa.int/rosetta/2016/09/29/living-with-a-comet-an-osiris-team-perspective/
ROSINA instrument in numbers
https://blogs.esa.int/rosetta/2016/09/27/living-with-a-comet-a-rosina-team-perspective/
RPC instrument in numbers
https://blogs.esa.int/rosetta/2016/09/29/living-with-a-comet-rpc-team-perspective/
VIRTIS instrument in numbers
https://blogs.esa.int/rosetta/2016/09/28/living-with-a-comet-a-virtis-team-perspective/

Acronyms

ADS	active descent system
AEKI	Hungarian Atomic Energy Research Institute
AOCS	attitude and orbit control system
AOS	acquisition of signal
AU	astronomical unit (a convenient measure of distance in the Solar System defined as the mean radius of Earth's solar orbit; approx. 150 million km)
CCLRC	Council for the Central Laboratory of the Research Councils (UK)
CME	coronal mass ejection
CNES	French Space Agency
CNR	National Research Council of Italy
CNSR	Comet Nucleus Sample Return
CRAF	Comet Rendezvous and Asteroid Flyby
DASA	Deutsche Aerospace AG
DLR	German Aerospace Center
DSM	Deep Space Maneuver
DSN	Deep Space Network
EADS	European Aeronautic Defense and Space Company
EPS	Storable Propellant Stage on Ariane rocket
ERC	Earth-Return Capsule
ESA	European Space Agency
ESAC	European Space Astronomy Center
ESO	European Southern Observatory
ESOC	European Space Operations Center
ESTEC	European Space Technology and Research Center
ESTRACK	European Space Tracking Network
GEOS	Geostationary Scientific Satellite
HGA	high-gain antenna
ICE	International Cometary Explorer
INAF	Italian National Institute for Astrophysics

© Springer Nature Switzerland AG 2020
P. Bond, *Rosetta: The Remarkable Story of Europe's Comet Explorer*,
Springer Praxis Books, https://doi.org/10.1007/978-3-030-60720-3

ISAS	Institute of Space and Astronautical Science (Japan)
JPL	Jet Propulsion Laboratory
LGA	low-gain antenna
MMH	monomethylhydrazine
MPAe	Max-Planck-Institute für Aeronomie
N	newton (SI unit of force)
NAC	narrow angle camera
NASA	National Aeronautics and Space Administration (USA)
NAVCAM	navigation camera
NEA	near-Earth asteroid
NEO	near-Earth object
OCM	orbit correction maneuver
PI	principal investigator
RCS	reaction control system
RTG	radioisotope thermoelectric generator
RWA	reaction wheel assembly
SETI	search for extraterrestrial intelligence
SOC	Science Operations Center
SPC	Science Program Committee (ESA)
TCM	trajectory correction maneuver
TWTA	Traveling Wave Tube Amplifier
UT	Universal Time
UV	ultraviolet
VLT	Very Large Telescope (Chile)
WAC	wide angle camera

About the Author

Peter Bond has written a dozen books, including the award-winning *DK Guide to Space*, and he has contributed to, or acted as consultant/editor for, many others. His earlier books include *Exploration of the Solar System* (Wiley-Blackwell) and *Distant Worlds* (Copernicus). He also edited the prestigious *IHS Jane's Space Systems & Industry* yearbook from 2008-2018.

Peter has written hundreds of articles on space and astronomy for newspapers and magazines in Britain and America, as well as material for the Sunday Times *Window On The Universe* CD-ROM, the Nature-IOP *Encyclopedia of Astronomy and Astrophysics*, and the Philips *Encyclopedia of Astronomy and Astrophysics*.

He has appeared regularly on TV and radio to comment on astronomical discoveries and events, and was Space Science Advisor/Press Officer for the Royal Astronomical Society 1995-2007. He has been a consultant and writer for the European Space Agency for many years and worked with many members of the Rosetta team.

In addition to a Group Achievement Award from NASA in 2004, he received an "Outstanding Contribution" certificate from ESA in 2005. He is a Fellow of both the Royal Astronomical Society and the British Interplanetary Society, and a member of the Planetary Society.

Website: http://www.bondspace.co.uk/

© Springer Nature Switzerland AG 2020
P. Bond, *Rosetta: The Remarkable Story of Europe's Comet Explorer*,
Springer Praxis Books, https://doi.org/10.1007/978-3-030-60720-3

Index

A
Accomazzo, Andrea, 220, 233
AEA Technology, 93
Alenia Aerospazio, 73, 75, 77
Alexander, Claudia, 362–363, 374
ALICE instrument, 105, 107, 159–161, 168, 169,
    176, 178, 179, 184, 192, 193, 208, 221,
    222, 254, 262, 376, 377, 379
Altwegg, Kathrin, 254, 322, 327
Ames Research Center, 26, 44, 362, 367
Anderson, John, 197
Antarctica, 164, 179, 194
APXS instrument, 69, 112, 116, 120–121, 380
Ariane launcher, 86, 136
Arianespace, 84, 87, 133–135, 152, 153
Ariel (spacecraft), 349
asteroid belt, 4, 17, 22, 24, 93, 142, 143, 160, 180,
    190, 197–202, 211, 323
asteroids, 1, 43, 64, 93, 131, 160, 291
Astraea (asteroid), 21
Astrium, 73, 74, 84, 86, 91, 107, 140, 146
astronomical unit (AU), 5, 6, 8, 9, 19, 20, 22, 62,
    68, 77, 89, 93, 113, 129–133, 136, 138,
    162, 175, 178, 182, 184, 191, 193, 198,
    200, 201, 203, 210, 212, 215, 216, 244,
    261, 315, 318, 319, 321, 350
ATLAS survey, 349
Auster, Hans-Ulrich, 69, 116, 119, 272, 288, 333
Australia, 72, 73, 83, 98, 100–103, 154, 155, 158,
    179, 181, 207, 246, 369

B
Balsiger, Hans, 105, 108, 363–364
Bannister, Michele, 350
Bayeux Tapestry, 8
Berner, Claude, 86

Besse, Sebastien, 189, 190
Bibring, Jean-Pierre, 70, 116, 121, 271, 275, 288
Bode, Johann, 19, 20
Bode's Law, 19, 20
Bondone, Giotto di, 36
Bonnefoy, Rene, 86
Bonnet, Roger, 49, 65
Bopp, Thomas, 7
Borisov (interstellar object), 349
Briggs, Geoffrey, 43
British Aerospace, 36
British Museum, 47, 48
Burch, James, 105, 111

C
CAESAR (Comet Atmosphere and Earth Sample
    Return), 49, 50
Canberra ground station, 103, 192, 193, 200, 221
Capaccioni, Fabrizio, 107, 222
Carr, Christopher, 331
Cassini (spacecraft), 50, 52, 53, 363
Cebreros ground station, 101, 103, 179,
    219, 246
Center Spatial Guyanais (CSG), 85, 145
Ceres (asteroid), 4, 19, 20, 22, 201, 366
Champollion, Jean François, 47, 48, 69, 148
Champollion lander, 66–71
Chelyabinsk, 28
Chernykh, Nikolai, 189
Chile, 132, 137, 179, 208, 321, 350
Churyumov, Klim, 138, 353–355
ÇIVA instrument, 114, 116, 118, 121, 164, 176,
    177, 222, 271–273, 275, 281, 284, 288, 380
Cluster (mission), 359, 364, 375
CNES, see French Space Agency
Colangeli, Luigi, 105, 111, 364–365

© Springer Nature Switzerland AG 2020
P. Bond, *Rosetta: The Remarkable Story of Europe's Comet Explorer*,
Springer Praxis Books, https://doi.org/10.1007/978-3-030-60720-3

Printed in the United States
By Bookmasters